J. Kiefer (Ed.)

Life Sciences and Radiation

Springer
Berlin
Heidelberg
New York
Hong Kong
London
Milan
Paris
Tokyo

Jürgen Kiefer (Ed.)

Life Sciences and Radiation

Accomplishments and Future Directions

With 67 Figures, 7 in Colour and 12 Tables

Springer

PROF. DR. JÜRGEN KIEFER
Strahlenzentrum
Justus-Liebig-Universität Giessen
Leihgesterner Weg 217
35392 Giessen
Germany
juergen.kiefer@strz.uni-giessen.de

ISBN 3-540-20478-4 Springer-Verlag Berlin Heidelberg New York

Cataloging-in-Publication Data applied for

Bibliographic information published by Die Deutsche Bibliothek
Die Deutsche Bibliothek lists this publication in the Deutsche Nationalbibliografie; detailed bibliographic data is available in the Internet at <http://dnb.ddb.de>.

Springer-Verlag is a part of Springer Science+Business Media
springeronline.com

Printed in Germany

Typesetting: Data conversion by author
Cover-design: design and production, Heidelberg

Printed on acid-free paper 02/3020 ra – 5 4 3 2 1 0

Preface

Scope and ideas of the workshop

The workshop which took place at the University of Giessen from Oct. 3 to Oct. 7, 2002 and whose proceedings are collected in this volume started from the idea to convene a number of scientists with the aim to outline their "visions" for the future of radiation research on the basis of their expertise. As radiation research is a very wide field restrictions were unavoidable. It was decided to concentrate this time mainly on molecular and cellular biology because it was felt that here action is par-ticularly needed. This did not exclude contributions from neighbouring fields as may be seen from the table of contents. It was clearly not planned to have a comprehensive account of the present scientific achievements but the results presented should only serve as a starting point for the discussion of future lines of research, with the emphasis on the "outreach" to other parts of life sciences.

If you are interested in the future ask the young – we attempted, therefore, to invite mainly younger colleagues (with a few exceptions) who had, however, already left their marks in the field. They were asked to describe what they felt is important in radiation research and may have significant influences on other branches of life sciences. They were given the task to demonstrate what is lost for science "if we do no longer exist". This is clearly not a well structured programme for a scientific congress but extremely useful to stimulate discussion. The views presented are hence quite diverse, and so are the contributions gathered in this volume. The following sequence of papers is not the one how they were actually given. The structure for which the editors takes responsibility is intended to guide the reader who will quickly find out that order of subtopics is only rather vague. Many problems are addressed in nearly all contributions (e.g. DNA response, repair etc.) nevertheless it was felt that some kind of structuring might be helpful.

Most of the participants came from Germany. Radiation research is, of course, international and the problems encountered world-wide. So we invited – within the limitations of the budget also some contributors from abroad to let us share their views. Their presentations and discussion contributions were highly welcome. One paper (Srivastava and Sharan) was not actually given at the workshop but only supplied in writing. As it presents ideas from a developing country we felt it should be included as an example of another view.

The following proceedings contain 20 submitted papers loosely grouped in different categories (by the editor) plus a preface and a postscript. They all give examples how radiation research could find its way into the future, some rigorously based on scientific findings, some more philosophically in a general way. We all hope that in this way new hope may arise in the people directly concerned but – more importantly – stimulate more than one second thought in those who are now convinced that radiation research is out of date and therefore dispensable.

Acknowledgements

To organise a meeting is always a joy and a lot of work. The joy is with the main organiser, the work with his coworkers. But it also requires some money. We thank the "Bundesministerium für Bildung und Forschung" (BMBF) (Drs. Komorowski and Gelford) for financial support through "Forschungszentrum Karlsruhe (Dr. R. Papp) and the "Bundesministerium für Umwelt, Naturschutz und Reaktorsicherheit" (Dr. H. Landfermann) for co-sponsorship. Drs. Gelford, Landfermann and Papp acted also as co-organisers and gave very valuable help. Karin Rinn, and Ina Allendörfer gave more than their share to the load of administrative work involved and helped to create a good working atmosphere. The production of this volume could not have been possible without the efforts and the experience of Hermann Witzenberger and the secretarial assistance of Karin Rinn. I thanks them all and hope that the workshop and its proceedings are able to supply at least a small brick to the new building of modern radiation research.

Giessen, August 2003 *Jürgen Kiefer*

Contents

Cytogenetics

Carcinogenesis, Radiation Protection

Repair

Impact of Physical Developments

General Issues

Workshop Participants

Radiation Research – Science for the future

Gießen 03. – 07.10.2002

Dr. Andreas Artmann
Gesellschaft für Anlagen-
und Reaktorsicherheit (GRS) mbh
Schwertnergasse 1
50667 Köln
art@grs.de

Prof. Dr. Dr. Richard Bauer
Klinik für Nuklearmedizin
Justus-Liebig-Universität Gießen
Friedrichstr. 25
35385 Giessen
Richard.Bauer@radiol.med.uni-giessen.de

Dr. Christa Baumstark-Khan
Institut für Luft- und Raumfahrtmedizin
(DLR)
Linder Höhe
51147 Köln
christa-baumstark-khan@dlr.de

Dr. Ulrich Bertsche
Referatsleiter
Hess. Ministerium für Umwelt, Energie,
Jugend, Familie und Gesundheit
Mainzer Str. 80
65189 Wiesbaden
u.bertsche@mulf.hessen.de

Prof. Dr. T. Butz
Universität Leipzig
Linnéstr. 5
04103 Leipzig
butz@physik.uni-leipzig.de

Dr. Christine Blattner
Forschungszentrum Karlsruhe
Institut für Genetik und Toxikologie
Postfach 36 40
76021 Karlsruhe
christine.blattner@itg.fzk.de

Prof. Dr. Jürgen Böhm
PTB Braunschweig
Bundesallee 100
38116 Braunschweig
juergen.boehm@ptb.de

Dr. Jochen Dahm-Daphi
Universität Hamburg
Abt. f. Strahlentherapie und Radioonkologie
Martinistr. 52
20246 Hamburg
dahm@uke.uni-hamburg.de

Prof. Dr. Marco Durante
Dipartimento di Scienze Fisiche
Universitá "FedericoII"
Monte S. Angelo, via Cintia
80126 Napoli
ITALY
Marco.Durante@na.infn.it

Prof. Dr. Friederike Eckardt-Schupp
GSF
Institut of Molekular Radiobiology
P.O. Box 11 29
85764 Neuherberg
Eckardt-Schupp@gsf.de

Dr. Anna Friedl
Strahlenbiologisches Institut
Universität München
Schillerstr. 42
80336 München
anna.friedl@lrz.uni-muenchen.de

Dr. Eike Gelfort
Bundesministerium für Bildung und Forschung
Referat 413
53170 Bonn
eike.gelfort@bmbf.bund.de

Dr. Martin Gotthardt
Klinik für Nuklearmedizin
Universität Marburg
Baldingerstraße
35043 Marburg
gotthardt@mailer.uni-marburg.de

Dr. Rüdiger Greinert
Dermatologisches Zentrum
Abt. Molekulare Zellbiologie
Am Krankenhaus 1
21614 Buxtehude
dermatologie@hospit.std.shuttle.de

Prof. Dr. Thomas Haaf
Institut für Humangenetik
Johannes Gutenberg-Universität Mainz
Langenbeckstr. 1, Bau 601
55131 Mainz
Haaf@humgen.klinik.uni-mainz.de

Prof. Dr. Dietrich Harder
Konrad-Adenauer-Str. 26
37075 Göttingen
dietrich.harder@ngi.de

Dr. Ludwig Hieber
GSF, Institut f. Molekulare Strahlenbiologie
Ingolstädter Landstr. 1
85764 Neuherberg
ludwig.hieber@gsf.de

Dr. Peter Hill
Forschungszentrum Jülich GmbH
52425 Jülich
p.hill@fz.juelich.de

Prof. Dr. George Iliakis
Universitätsklinikum Essen
Inst. f. Med. Strahlenbiologie
45122 Essen
Georg.Illiakis@uni-essen.de

Dr. Peter Jacob
GSF, Institut für Strahlenschutz
Ingolstädter Landstr. 1
85764 Neuherberg
jacob@gsf.de

Dr. Peter M. Kern
Franz von Prümmer Klinik
Bahnhofstr. 16
97769 Bad Brückenau
Peter.kern@franz-von-pruemmer-klinik.de

Prof. Dr. Jürgen Kiefer
Strahlenzentrum der JLU Gießen
Institut für Biophysik
Leihgesterner Weg 217
35392 Gießen
juergen.kiefer@strz.uni-giessen.de

Prof. Dr. Gerhard Kraft
Abt. Biophysik
GSI Darmstadt mbH
Planckstraße 1
64291 Darmstadt
G.Kraft@gsi.de

MR Dr. Hans-Henning Landfermann
Bundesministerium für Umwelt, Naturschutz und Reaktorsicherheit
Postfach 12 06 29
53048 Bonn
landfermann.hans@bmu.de

Prof. Dr. Charles Limoli
Dept. of Radiation Oncology
Radiation Oncology Res. Laboratory
University of California San Francisco
1855 Folsom St., MCB-200
San Francisco, CA 94103-0806, USA
Limoli@itsa.ucsf.edu

Prof. Dr. Markus Löbrich
Universität des Saarlandes
Fachrichtung Biophysik
64421 Homburg/Saar
markus.loebrich@uniklinik-saarland.de

Prof. Dr. Wolfgang-Ulrich Müller
Universitätsklinikum Essen
Institut für Medizinische Strahlenbiologie
Hufelandstr. 55
45122 Essen
wolfgang-ulrich.mueller@uni-essen.de

Prof. Dr. Roland Netz
Sektion Physik, LMU
Theresienstraße 37
80333 München
netztheorie@physik.uni-muenchen.de

Dr. Tuomo Nikula
European Commission
JRC. Inst. for Transuranium Elements
P.O. Box 23 40
76125 Karlsruhe
nukula@itu.fzk.de

Prof. Dr. Ohtsura Niwa
Radiation Biology Center, Kyoto University
Sakyo-ku
Kyoto 606-8501
JAPAN
Oniwa@house.rbc.kyoto-u.ac.jp

Dr. Reiner Papp
Forschungszentrum Karlsruhe GmbH
Projekttrager des BMBF und BMWi für
Wassertechnologie und Entsorgung
Postfach 3640
76021 Karlsruhe
reiner.papp@ptwte.fzk.de

Prof. Dr. Herwig G. Paretzke
GSF
Institut für Strahlenschutz
Ingolstädter Landstraße 1
85758 Neuherberg
paretzke@gsf.de

Dr. Sabine Pils
Forschungszentrum Jülich GmbH
52425 Jülich
s.pils@fz-juelich.de

Dr. Kevin M. Prise
Gray Cancer Institute
P.O. Box 1000
Mount Vernon Hospital
Northwood, HA6 2JR
UNITED KINGDOM
Prise@gci.ac.uk

Prof. Dr. Hans Peter Rodemann
Sektion für Strahlenbiologie und Molekulare
Umweltforschung, Radioonkologie
Universität Tübingen
Röntgenweg 11
72076 Tübingen
hans-peter.rodemann@uni-tuebingen.de

Dr. Kai Rothkamm
Universität des Saarlandes
Fachrichtung Biophysik
66421 Homburg/Saar
kai.rothkamm@uniklinik-saarland.de

Annemarie Schmitt-Hannig
Bundesamt für Strahlenschutz
Ingolstädter Landstr. 1
85764 Oberschleissheim
schmitt@bfs.de

Dr. Robert Schupfner
Universität Regensburg
Institut für Analytische Chemie
Universitätsstr. 31
93040 Regensburg
robert.schupfner@chemie.uni-regensburg.de

Prof. Dr. Dr. h.c. Christian Streffer
Institut für Wissenschaft und Ethik
Universität Essen, FB 1
Postfach
45117 Essen
streffer@t-online.de
oder christian.streffer@uni-essen.de

Dr. Werner Wagner
European Commission Joint Research Center
ITU
Postfach 23 40
76125 Karlsruhe
wagner@itu.fzk.de

Introductory Remarks: On the State of Radiation Research

Jürgen Kiefer

Strahlenzentrum der Justus-Liebig-University, Giessen, Germany
juergen.kiefer@strz.uni-giessen.de

Starting points

Some years ago I participated in a meeting of the European Society of Radiation Biology and was late for the opening session because of flight delay. Together with me was a young, very bright, very attractive (female) colleague. When we opened the auditorium door we saw the audience already assembled and my young colleague spontaneously exclaimed (fortunately only audible to me): "My God, all old age pensioners!" I do not know where she is now, definitely no longer in radiation research, apparently she opted for a younger science...

This scene always comes to my mind when we discuss the state of our field. No doubt, radiation research does not very well nowadays. This is certainly true for Germany but also – with gradual differences – for most other countries. Japan appears to be an exception what I cannot explain. According to a survey conducted by the German Radiation Protection Commission (SSK) the number of research institutions dedicated to radiation research declined from 25 in 1993 to only 9 in 2000, that is a loss of 64%. As a consequence of this alarming situation the Federal Ministry of Education asked the "Company for Reactor Safety" (GRS) to convene a meeting with experts and state representatives to discuss possible ways out which took place in spring 2002 in Köln. All state governments which in Germany run the universities were invited – not a single one came. This demonstrates that political bodies are not interested at all although the lack of radiation experts is clearly also a political issue. More importantly, however, radiation research seems to have lost its standing in the scientific community. Otherwise it cannot be explained that many faculties greed without protest to the closure of existing, sometimes quite famous institutions. During the meeting the idea was born to have a workshop where the role of radiation research in the life sciences should be freely discussed. This workshop took place from October 3 to October 7, 2002 at the "Radiation Centre" (Strahlenzentrum) (an institution which has meanwhile also ceased to exist) of the university of Giessen, Germany. This book contains the proceedings of the meeting and is aimed at provoking discussion.

Before the contents are commented on more general remarks are in place.

Attempts of an analysis

Lamentations will not help, clarification is needed. The following presents, of course, a personal view and is not endorsed by any institution, commission, society or else. If one tries to approach the problem an a more systematic way it may be burned down to four essential points:

- Historical obesity
- Revolution of the underdogs
- Emancipation of the biologists
- Loss of political correctness

The meaning of these terms appears obscure at first sight and requires explanation:

Historical obesity

Soon after the discovery of X-rays and radioactivity these new phenomena gave rise to often mythical hopes for medicine and were sometimes seen as the ultimate answer for all kinds of human ailments. Even after the disastrous events in Hiroshima and Nagasaki nuclear energy and other radiation applications evoked great hopes for the future. The impressive Geneva meetings on the "Peaceful Uses of Atomic Energy" organised by the UN and related exhibitions in many countries stimulated public interest, reaching eventually also the universities. Many faculties decorated themselves with radiation departments which was the easier as public support was generous. Researcher had a good life and generally did not suffer for lack of money. Very many good papers were published, and the quality was good, sometimes excellent. This has to be said to avoid the impression that the money given was not well spent. Particularly in the beginning promising new techniques were established, cross-fertilising quite a number of other disciplines.

But as in normal life wealth does not necessarily result in creativity and special efforts. With the great many people working in the field it was unavoidable that not all funded work lived up to high standards and radiation research tended to become a closed shop. The field put up weight which is – as we all know – rarely healthy. It was in many places too late realised that times might change and funding has to be earned. This lead to the impression by other disciplines that the performance of radiation research had become poor. At the same time many influential people who dominated scientific decisions retired: times for a change for new tune and new players. Unfortunately the wind of change missed many of the well-established institutions who went on to work as they always did following a song popular in the late 1960ies: "These were the days my friend we thought they never end…" which was purposely sung at the workshop dinner. Obviously it could not go on like this forever.

So it came the

Revolution of the underdogs

At the same time there were revolutionary developments in biochemistry, genetics, cellular biology and (non-radiation) biophysics. The people involved had more difficulties to obtain sufficient funds compared to radiation workers and – which may be more important – to gain influence. When I joined the German Biophysical Society it was essentially an association of radiation research. This picture changed drastically in the late 1970ies, and nowadays the radiation subgroup is a "quantité negligable". Membrane biophysics, biophysical chemistry and neurological physics have taken over. This development is fully justified, the results of these fields are convincing, often breathtaking. Also new people came into leading positions and the designations of former radiation chairs were changed, often openly, not seldom undercover by choosing persons whose commitment to radiation research was rather superficial to replace the outgoing thoroughbred radiation specialist. This was generally not bad for science but it accelerated the decline of radiation research. As far as I can see there is today no branch to deal with radiation applications at our main grant-giving agency (Deutsche Forschungsgemeinschaft, DFG) with the result that proper refereeing of applications is not always easy and sometimes even inappropriate. When my university planned to build an electron linear accelerator around 1960 – mainly for research in nuclear physics – funding was only given on the condition that also applications in the life sciences should be supported and the last say was with Boris Rajewski, a radiation biophysicist. Not to be misunderstood, he was an eminent scientist, knowledgeable in many areas and well equipped for the decision. But he people concerned felt someway suppressed and it is no wonder that the situation changed quite dramatically when Rajewski and others retired. The "Max-Planck-Institute for Biophysics" in Frankfurt whose director Rajewski was changed its direction towards other areas (with great success as e.g. documented by the Nobel Prize awarded to H. Michel). Radiation equipment went to the rubbish heap, parts were recovered by W. Pohlit and we ourselves were lucky to receive a large quartz prism, literally saved from the trash. Unfortunately the serious situation was not realised by many people. As a consequence radiation researchers should have sought close cooperation with other disciplines, taking up their challenges and digesting their new approaches to make use of them for the improvement of their own science. Of course, radiation methods were still widely used by life-scientistst who came to the respective departments. They were usually well-treated as customers who needed gladly given help from the enlightened – radiation as the glowing grail. But over the years they learned the techniques, established their own laboratories and more and more the radiation colleagues were only seen as radiation protection officers, a species not always deeply loved. So even in the special field if radiation application in the life sciences the next step was bound to follow:

Emancipation of the biologists

Traditionally radiation research, also when more biologically oriented, was the domain of physicists and this is largely the case even today. There are, of course, good reasons for this: Radiation is a physical agent, handling of the equipment and dosimetry require special expertise and the quantitative interpretation of the results obtained is easier for those who had a proper mathematical training. It is, therefore, not surprising that radiation biology was presumably the first biological branch where observed phenomena were fitted into a comprehensive theory ("target theory") with far reaching consequences as outlined elsewhere in this book. But the relationship between physics and biology has been rarely a love-affair and this is even more true for the cooperation between the representatives of the two areas. Physicists tend to be rather arrogant towards scientists from other fields. A classical example is Max Delbrück's talk "In what respect is biology too difficult for the biologist?" given at the 1963 convention of the German Physical Society [1]. There was, of course, a lot of irony involved. Delbrück, a Nobel Prize awardee for Medicine or Physiology knew his subject and the main purpose was to teach his fellow physicist that there is more than physics in science. The objective of the lecture was thus exactly the contrary to what the title suggested. Also it constituted some kind of late repolishing of the physicist's self-esteem as David Hilbert, the famous mathematician in Göttingen, used to say: "Physics is much too difficult for the physicists!". But nevertheless, the remark was felt as a painful sting in the sensitive skin of biologists. And it came at a time of remarkable successes very similar to the golden era of physics when Hilbert uttered his classical statement. There were not too many contact points between physics and biology (this has fortunately changed now), radiation biology being on of them. It grew thus the feeling that this area was no longer needed as a special discipline particularly because it showed some signs of senescence, not being able to incorporate the new developments. What was left could be easily handled by the proper biologists, and the few problems of measurements and equipment could be taken care of by physicists specially employed by the biology departments.

It goes without saying that I do not subscribe to these ideas (reasons will be given in the last chapter of this book) but they remain vivid until today. Biology is the science of the 21st century (a role physics was playing in the 20th century) and radiation research will only survive if it is part of it.

Unfortunately at these critical times the public experienced a change of attitude to nuclear sciences. While atomic energy was formerly seen as the great saviour it now turned to be the culprit of nearly everything. Radiation euphoria died suddenly and was reincarnated as radiophobia. These tendencies are not equally strong everywhere, certainly in Germany they dominate the discussions and entered the political field, at the latest with the new red-green government. To be for or against radiation use was no longer a decision based on scientific or economical deliberations but a matter of faith and even of personal integrity:

Loss of political correctness

The main question was, of course, the prolonged use of nuclear energy but this touches only the surface of the underlying problems. As a consequence of the 1968 movement the attitude to science underwent a deeply rooted transformation. While it was formerly seen as providing solutions to all problems of mankind it was now held responsible for all miseries. An even cursory discussion of these phenomena is far beyond this contribution but it can be stated that radiation is one symbol of the devil science. It serves this role very well: There is no doubt that it can harm human health, it cannot be directly felt and produces its results in a mysterious way decades after the exposition. Chernobyl, surprisingly more than Hiroshima, stands for this. The dramatically blown-up death numbers which have no scientific basis remain in the papers and demonstrate the irrational public attitude. It came up even in the General Assembly of the United Nations where the 2000 UNSCEAR report found it difficult to be accepted because it "played down the effects of the Chernobyl disaster".

It is difficult for a scientific discipline to survive if it is linked to powers of the dark like radiation and to receive public funding. Radiation has become a dirty word which is not be used by well-educated people. To stand up to this attitude requires a certain degree of courage to which scientists are not always properly trained. History shows that professors do not often profess and it is hence not surprising that this dangerous ability is even less found in administrative offices. At some places it was suggested to change the name of radiation department using less incriminated terms to secure survival. Even in a letter accompanying a referee's report by the DFG (the subject was biological investigations with neutrons) it was pointed out that studies of this kind do not fit into the political picture. It seems sometimes the only way out is to work with radiation but never mention the word. Of course, this is a ridiculous suggestion but it is in this atmosphere no wonder that support for our discipline rapidly fades away. It will not do to repeat again and again that radiation research is not linked to nuclear warfare and nuclear energy, people will not believe.

Conclusions

The situation is serious, partly due to our own fault, partly due to outside influences. We have to face this and have to fight with good arguments – or to give up. Of all the factors discussed above the most dangerous appears to be the loss of acceptance in the scientific community. If our colleagues are not convinced that radiation research is able to add to the progress of science the field is bound to die.

References

1 Delbrück M (1963) Inwiefern ist die Biologie zu schwierig für die Biologen?. Physikertagung Stuttgart, Physik-Verlag Mosbach/Baden

Cellular Responses, Genomic Stability

Cellular Responses to DNA Damage – a Personal Account

George Iliakis

Institute of Medical Radiation Biology, Medical School of the University Duisburg-Essen, Essen, Germany
georg.iliakis@uni-essen.de

Abstract.

The field or Radiation Biology has seen in the last fifty years dramatic developments in its focus and a continuous movement forward internationally. Research topics and discoveries once thought to represent the interests of a highly specialized group of scientists are now main stream and attract an ever increasing number of researchers and clinicians active in areas as diverse as immunology, cancer, and radiation oncology – just to name a few. Initiated in the years during and just after the Second World War when the devastating long-term effects of the weapons of mass destruction used to end the war were just beginning to become apparent, the field of radiation biology has been transformed to a central player not only in radiation protection but also in radiation oncology and basic biology. By combining the expertise of Physicists, Chemists, Biologists and Physicians, the field has generated high standards for quantitative research and performed the seminal experiments that ultimately led to the characterization of the cell cycle and the checkpoint response – topics recently honored with the Nobel price. The field has moved from its initial focus in understanding and protecting against weapons of mass destruction to the now central role it plays in basic research, the diagnosis and treatment of disease, preservation of food and other perishables and even in recent times to the protection from biological terrorism. Truly, this is a story of how to beat swords into ploughshares. It is relatively common for fast developing fields of science to go through painful transitions in the process of transformation and adaptation to the ever-increasing demands of modern science, and Radiation Biology is not an exception. With a few years lag to North America, Radiation Biology Institutions are in the middle of a crisis in Germany and see their numbers decreasing at an alarming rate. Only targeted, active and sustained support will ensure maintenance of expertise in the country and will enable the development of competitive groups that will make their mark in modern biology and medicine internationally. The required reorientation into molecular/mechanistic studies and the needed expansion into translational investigations in the field of cancer research and treatment will only be successful if the field becomes revitalized. Here, I provide a brief outline of concepts and ideas that determine the future direction of basic research at the Institute of Medical Radiation Biology, Medical School University Duisburg-Essen, together with an account of past work that led to the formulation of these ideas.

Introduction

There is long standing tradition for high-quality Radiation Biology research at the University of Duisburg-Essen. Under the direction of its founder, Prof. Dr. Christian Streffer, the Institute of Medical Radiation Biology has set high academic standards pointing the way for further developments. Future research activities attempt to maintain these high standards, rest on a series of continuously developing research activities from my part during the last two decades and are dominated by three longer employment opportunities. The first, and formative for the development of interest in the field of Radiation Biology, was in Frankfurt/Main at a satellite of the GSF, and the Institute of Biophysics of the University, both under the Direction of Prof. Dr. Wolfgang Pohlit. Seminal for the development and expansion of the research horizons was a longer stay in the United States, briefly at the Cleveland Clinic Foundation, Cleveland, OH, and subsequently for many years in the Department of Radiation Oncology at Thomas Jefferson University in Philadelphia.

The development of these activities parallels in many ways developments in the field of Radiation Biology itself. One of the early goals of Radiation Biology was to understand the physical aspects of energy deposition, the chemical consequences of this energy deposition in the cellular environment, and the characterization of animal responses with respect to killing and cancer induction. The field received a significant boost when methods were developed for culturing mammalian cells *in vitro*. This new methodology allowed for the first time quantitative studies in individual cells and culminated with the first dose-response curve using colony formation as endpoint. Subsequent milestones included the first observations on the variation of radiosensitivity throughout the cell cycle [1], as well as the realization that cells respond to radiation damage by initiating repair processes [2]. With the realization that DNA was one of the primary targets of radiation, the field of DNA repair was born and shortly thereafter the DNA double strand break was declared as a severe lesion likely to cause cell death [3]. At the same time important effects of radiation were discovered in various phases of the cell cycle and proved seminal for the later definition and characterization of checkpoint responses [4].

The recent development of powerful techniques allowing the characterization of biological processes at the molecular level has brought a revolution in all fields of biology including Radiation Biology. With the application of these techniques, the field is now aiming at the characterization at the molecular level of the cellular responses to DNA damage and the application of this knowledge in the field of medicine and in radiation protection. Main goals of the Institute of Medical Radiation Biology at the University of Essen include the further characterization of the mechanisms underlying repair of DNA DSBs, and their coordination with the checkpoint response. In the following sections, I provide a brief outline of the research that led to this focus and present fundamental questions, which our research will attempt to answer in the following 5–7 years.

Repair reactions at the cellular level

Experiments at the cellular level provided first evidence that living organisms repair radiation damage. Two main repair processes have been defined in cells of higher eukaryotes and are termed repair of sublethal damage [5] and repair of potentially lethal damage [6]. Whereas repair of potentially lethal damage (PLD) is defined by the increase in survival observed in cells incubated after irradiation under conditions suboptimal for cell growth, repair of sublethal damage (SLD), or split-dose recovery, is defined by the increase in survival observed when a dose of radiation is administered in two equal fractions separated by a few hours [7]. SLD is considered responsible for the formation of the shoulder frequently observed in the survival curves of mammalian cells, but as outlined below PLD repair can also contribute to shoulder formation. Both reactions are defined phenomenologically and are understood as expressions of the ability of cells to repair radiation induced damage.

Cellular repair reactions were studied intensively at the GSF in Frankfurt using yeast as a model system, and I was given the task to extend these investigations into mammalian cell systems. A focus of these initial studies was to enhance the mechanistic understanding of these repair processes and to establish possible correlations between them. By altering the growth conditions it was possible to demonstrate that shoulder formation can also be achieved using protocols allowing PLD repair, thus providing evidence for a relationship between the two processes [8–10]. First hints on the underlying mechanisms were obtained by carrying out PLD repair in the presence of substances known to act by inhibiting the DNA metabolism [11–14]. These investigations suggested that PLD repair is the cellular expression of DNA repair and provided further evidence that the shoulder in the survival curve can also be attributed to PLD repair.

Because PLD repair was known to be cell cycle dependent, the role of cell cycle progression in its fixation (i.e. its transformation to irreparable damage) was studied. Experiments using synchronized populations of cells could demonstrate that cells crossing the G1/S border and mitosis were unable to repair PLD even when given the required time [15, 16]. This led to the definition of points in the cell cycle where fixation of radiation-induced damage takes place. Subsequent studies focused, therefore, on the characterization of cell radiosensitivity through the cell cycle. Through a series of experiments carried out with caffeine and synchronized populations of cells, it was possible to demonstrate that the variation in cell radiosensitivity to killing throughout the cell cycle can be attributed to variations in the ability of cells to repair PLD [17]. Conditions supporting PLD repair were also found to reduce mutation induction [18, 19]. Using radiations of different LET it was possible to implicate proximity of lesions as an important parameter in the induction of PLD [19–22], and parallel studies identified PLD fixation as an important contributor of cell radiosensitization by incorporation into DNA of halogenated pyrimidines [23–28].

In an attempt to clarify controversies in the literature regarding correlations between repair of PLD and SLD, experiments were carried out evaluating the conditions used to measure PLD. These experiments suggested that ionizing radiation induces two forms of PLD with different sensitivity to fixation by postirradiation treatments. We termed alpha-PLD the slow form that repairs within 4–6 h and causes the formation of the shoulder in the survival curve. The term beta-PLD was given to the fast form of PLD that led to the restoration of the slope in the survival curve and was sensitive to the abrupt changes in chromatin conformation induced by high concentrations of salt [7, 29–33].

Overall these experiments allowed the phenomenological identification of a form of damage whose repair and fixation was responsible for the variations in radiosensitivity throughout the cell cycle, and the two commonly studied cellular repair reactions (PLD and SLD) [7]. Fixation of this damage leads not only to cell killing, but also to the induction of mutation and transformation. Two forms of this damage could be clearly identified with different sensitivities to postirradiation treatments. Of these forms, alpha-PLD appeared the most radiobiologically relevant as its fixation caused the cell cycle dependent variations in radiosensitivity to killing, the formation of the shoulder in the survival curve and the radiosensitization of cells treated with agents affecting DNA metabolism. The underlying molecular lesion was sensitive to agents inhibiting DNA metabolism and was influenced by proximity in the distribution of ionizations. These phenomena were compatible with the hypothesis that DNA damage underlies PLD. The next challenge was to work towards the identification and characterization of the DNA lesion underlying PLD.

Studies on the induction and repair of DNA double-strand breaks (DSB)

Studies during this period centered on the hypothesis that the DNA DSB is a potentially lethal lesion and aimed at characterizing the chain of events that leads from the induction of damage in the DNA to chromosome damage and ultimately to cell death. For this purpose existing technologies were adapted to evaluate induction and repair of DNA DSBs, and the technique of premature chromosome condensation [34] was introduced as a tool to bridge the gap between processes in the DNA and processes in the chromosomes.

Considerable effort during this period of time was invested in experiments designed to test the hypothesis that the variation in radiosensitivity throughout the cell cycle derives from differences in the induction of DNA DSBs as measured by neutral filter elution [35–40]. This was an alternative interpretation for the variations in radiosensitivity throughout the cell cycle, without invoking the fixation of PLD proposed by others and us. It was therefore important to test the validity of this model before proceeding with further experiments testing the hypothesis of PLD fixation. We were able to demonstrate that differences in elution rate often reflect differences in the ability of the detergents used to free up DNA rather than differences in the number of DNA DSB present. This interpretation could be confirmed using neutral sucrose density gradient centrifugation, as well as the newly developed technique of pulsed-field gel electrophoresis [41–43], to quantify DSBs in the DNA. We decided to adopt pulsed-field gel electrophoresis as a standard in the measurement of induction and repair of DNA DSBs. This method provides a well-defined theoretical background and is less prone to artifacts than the neutral filter elution. After calibration to provide absolute numbers of DNA DSBs per cell, pulsed-field gel electrophoresis was introduced to measure induction and rejoining of DNA DSBs in cells irradiated in various stages of the cell cycle. The results indicate strong potential to repair radiation induced DNA DSBs in all phases of the cell cycle including mitosis [41–43].

Subsequent experiments demonstrated the induction of two forms of DNA DSBs that could be linked, based on their kinetics of rejoining, to the two forms of PLD identified in the cellular studies. In an attempt to bridge the gap between effects in the DNA and cellular effects, we carried out experiments at the chromosome level using the technique of premature chromosome condensation. The results indicated two forms of chromosome breaks repaired with kinetics similar to those of the two forms of DNA DSBs suggesting that they are linked to them [44–50]. In aggregate, this set of experiments provided strong evidence for a chain of events initiated by damage in the DNA (DSBs), propagated as chromosome damage and ultimately leading to cell death. Observations along similar lines could also be made in irradiated cells exposed to hyperthermic temperatures [51, 52], suggesting the general importance of DNA repair in cell radiosensitivity to killing under quite diverse experimental conditions.

As the above outlined research was under way the field of cancer underwent an explosion in knowledge with the identification and characterization of genes acting by either promoting (oncogenes) or suppressing (tumor suppressor genes) tumor development. Naturally, studies on the role of these genes in cell radiosensitivity to killing started soon thereafter and developed to a hot topic of investigation. Together with our collaborators we evaluated the role of oncogenes of the *ras* and *myc* families in cell radiosensitivity to killing [33, 53, 54]. We observed a significant increase in radioresistance to killing in cells overexpressing *myc* and an activated form of *ras*. This increase in radioresistance could not be attributed to changes in either the induction or the repair of DNA DSBs, as cells overexpressing these genes sustained the same number of DNA DSBs per Gy as their normal counterparts and removed these lesions from the DNA with indistinguishable kinetics. On the other hand, irradiated cells overexpressing *myc* and activated *ras* showed an enhanced arrest in G2 and a prolonged inhibition of DNA replication as compared to their normal counterparts [53–59]. These observations identified checkpoint responses as an important determinant of cell radiosensitivity to killing and prompted us to increase the emphasis of our activities on the topic.

The above studies provided important insights into the mechanisms by which ionizing radiation inactivates cells of higher eukaryotes and paved the way for follow up studies. The suggested role of DNA DSB repair and checkpoint response in the endpoints investigated, pointed to the need for the molecular characterization, on the one hand, of the processes involved in the repair of DNA DSBs, and on the other hand of those involved in checkpoint activation. These studies were dramatically accelerated by parallel discoveries that brought DNA DSB repair and checkpoint response in the center of activities in the field cancer.

Molecular aspects of DNA DSB repair and checkpoint response

Three enzymatically distinct processes, homologous recombination repair (HRR), single strand annealing (SSA), and non-homologous end joining (NHEJ) can, in principle, repair DNA DSBs in eukaryotic cells [60–63]. Of these processes HRR and NHEJ have been directly implicated in the repair of DNA DSB in cells of higher eukaryotes. The characterization of the enzymatic machinery supporting these processes, as well as the coordination between them in the repair of DNA DSBs is one of the most active areas of investigation at present.

In recent years significant progress has been made towards the enzymatic characterization of NHEJ and the identification of key components including DNA-PKcs, Ku, DNA-ligase IV and Xrcc4. It has also become evident, that contrary to yeast, which uses HRR almost exclusively for the repair of DNA DSBs, cells of higher eukaryotes utilize extensively NHEJ to remove DNA DSBs from their genome.

Our work during this period aimed at the further characterization of NHEJ and the identification of components of the checkpoint response. It followed a dual approach combining genetic studies and biochemical investigations. For this purpose we introduced *in vitro* assays allowing the study of DNA DSB repair and S-phase checkpoint activation, and utilized a wide variety of cells with mutations in genes implicated in DNA DSB repair.

To characterize the function of known components of NHEJ and to identify new factors involved we introduced an assay based on the end joining of restriction endonuclease digested plasmid DNA. In parallel, we developed an *in vitro* assay in which rejoining of DNA DSBs could be measured using "naked" genomic DNA as a substrate prepared from agarose embedded cells using an extract of HeLa cells as a source of enzymes [64–68]. These assays provided important information on the functions of NHEJ and implicated replication protein (RPA) in the process. At the same time they provided evidence that end joining can take place in the absence of DNA-PKcs or Ku.

To elucidate the mechanism by which DNA damage-inducing agents inhibit ongoing DNA replication we employed a simian virus 40 (SV40) based *in vitro* DNA replication assay to specifically study biochemical aspects of this inhibition. We observed that the reduced DNA replication activity in extracts of irradiated cells is partly caused by a reduction in the amount of (RPA), and that a dominant inhibitory effect is caused by DNA-PK, which inactivates SV40 TAg by phosphorylation. The results demonstrated that RPA and DNA-PK are involved in the regulation of viral DNA replication after DNA damage and suggested that analogous processes regulate cellular DNA replication with the DNA-PK targeting the functional homologues of TAg [69–75]. Recent genetic studies by others and us extend and complement these investigations and implicate in addition to ATM also ATR in the regulation of DNA replication in irradiated cells [76–80].

Parallel genetic studies on the repair of DNA DSBs in cells of higher eukaryotes indicated that it is dominated by a fast component operating with half times of the order of a few minutes. This form of rejoining is severely compromised by defects in any of the constituents of DNA-PK, DNA-PKcs and Ku, or DNA ligase IV. Genetic, as well as a wealth of biochemical studies, assign these proteins to the same end joining pathway, which is characterized by an ability to remove DNA DSBs from the genome with extremely fast kinetics. We proposed the term D-NHEJ, for this pathway of DNA DSB rejoining [81–84].

Despite the prevalence of D-NHEJ, cells with defects in components of either DNA-PK, or DNA ligase IV complex are able to rejoin the majority of IR-induced DNA DSBs utilizing an alternative pathway operating with 20–30 fold slower kinetics [84–86]. A slow pathway of DNA DSBs rejoining is frequently discernible in wild-type cells as well, but it becomes well defined and assumes a dominant role when D-NHEJ is compromised through defects in participating factors, or when inhibitors against DNA-PK, such as wortmannin, are administered.

Because HRR is normally operating with kinetics of the order of hours, it is in principle possible that the slow component seen after inhibition of D-NHEJ reflects this process. We tested this hypothesis using the hyper-recombinogenic DT40 chicken cell line and a set of mutants defective in homologous recombination (HR). DT40 cells rejoin IR-induced DNA DSBs with kinetics similar to those of other vertebrate cells displaying 1000-fold lower levels of HR. In addition, knock-outs of *RAD51B*, *RAD52* and *RAD54* rejoin DNA DSB with kinetics similar to the wild type, as does also a conditional knock out mutant of *RAD51*. While a significant reduction in the fast component of rejoining is observed in $Ku70^{-/-}$ DT40 cells, a double mutant $Ku70^{-/-}/RAD54^{-/-}$ shows similar half times to $Ku70^{-/-}$ cells [83].

Thus, increases by several orders of magnitude in the capacity of cells to carry out HR, or defects in the proteins involved, fail to alter the rejoining kinetics in a way compatible with an involvement of HRR in the slow component of DNA DSBs rejoining, even when D-NHEJ is severely compromised [83]. Furthermore, rejoining of DNA DSBs with slow kinetics is associated with the rejoining of incorrect ends, an effect not compatible with the operation of homology directed repair of DNA DSBs [87]. To accommodate these observations, we hypothesized that the slow component of DNA DSBs rejoining reflects an alternative pathway of NHEJ, which we termed backup, B-NHEJ – to differentiate from D-NHEJ. According to this model, at least two distinct NHEJ pathways cooperate to remove IR-induced DNA DSBs from the genome of higher eukaryotes [83, 86, 88]. On the other hand, genetic studies clearly demonstrate the importance of HRR in cell radiosensitivity to killing and we were the first group to demonstrate that a component of HRR (Xrcc2) is required for the radioresistance observed in cells irradiated during S-phase [89]. Thus, addressing the coordination between HRR and NHEJ in the repair of DNA DSBs will be of paramount importance in our understanding of the actions of radiation in living cells.

Results compatible with the operation of alternative pathways of DNA end joining such as B-NHEJ have also been obtained at the biochemical level using Ku-depleted extracts of HeLa cells, or extracts of M059-J cells that lack DNA-PKcs. Extensive literature addressing DNA end joining from a different angle also points to the operation of different pathways in NHEJ and emphasizes the relevance of work along these lines to our understanding of cellular responses to DNA damage. There is increasing interest in the field at present for an alternative pathway of NHEJ utilizing short sequence homologies during end joining [90]. It will be important to characterize possible correlations between homology-directed end joining and B-NHEJ.

D-NHEJ dominates NHEJ *in-vivo* as long as its key factors are not compromised by either genetic mutation or chemical inhibition. This implies that DNA ends generated either endogenously, or by exogenous agents such as ionizing radiation (IR) and other DNA damaging agents, are quickly bound and processed by factors involved in this pathway. These observations suggest a priority binding of the factors involved to DNA ends, in line with the documented *in-vivo* dominance of D-NHEJ and the known properties of DNA-PKcs and Ku.

However, *in vitro* results cannot reproduce the slow kinetics of B-NHEJ *in vivo* suggesting limitations of the system. Also, the plasmid based *in-vitro* assay fails to reproduce the reduced fidelity of B-NHEJ observed *in vivo* after exposure to IR. An error prone character of alternative pathways of DNA DSB rejoining, perhaps equivalent to B-NHEJ, is also indicated by their involvement in genomic instability and tumor development in the absence of p53 or other components of the checkpoint response [91–97]. We attribute this difference to the kinetic advantage of B-NHEJ under *in-vitro* conditions and work towards the development in the future of *in vitro* assays better describing the *in-vivo* conditions.

The above studies suggested interactions and possible coordination between NHEJ and HRR that are not compatible with current models. Furthermore, the same studies also touched upon important aspects of checkpoint activation and its coordination with DNA repair. The follow up of these models will be the focus of our work in the future.

Future directions

Coordination between HRR and NHEJ in DNA DSB repair:

The above-discussed dichotomy in the pathway utilized to repair DNA DSBs in lower eukaryotes such as yeast and in cells of higher eukaryotes, suggest genetic and evolutionary contributions to the decision making process that require further experimental confirmation, and genetic as well as biochemical analysis. A central goal of our future work will therefore be to solidify existing evidence on preferential use of NHEJ over HRR in the removal of DNA DSBs in cells of higher eukaryotes, and subsequently begin with the characterization of the genetic components that determine this preference.

Although it has been demonstrated beyond any reasonable doubt that NHEJ has in cells of higher eukaryotes a much larger contribution than in yeast to the removal of IR-induced DNA DSBs, existing results cannot exclude HRR as a significant contributor to the overall repair process. Indeed work carried out using model systems assaying repair of a restriction endonuclease-induced DNA DSBs in repeat DNA sequences suggests that HRR is engaged in nearly 50% of the cases [98, 99]. Such a high contribution of HRR is not compatible with existing data of rejoining of IR-induced DNA DSBs. However, a 5–10% contribution cannot be ruled out. Furthermore, it remains possible, and is actually widely assumed, that the proportion of DNA DSBs repaired by HRR increases as the cell moves from G1 to S and to G2 where the formation of sister chromatids is expected to facilitate the process.

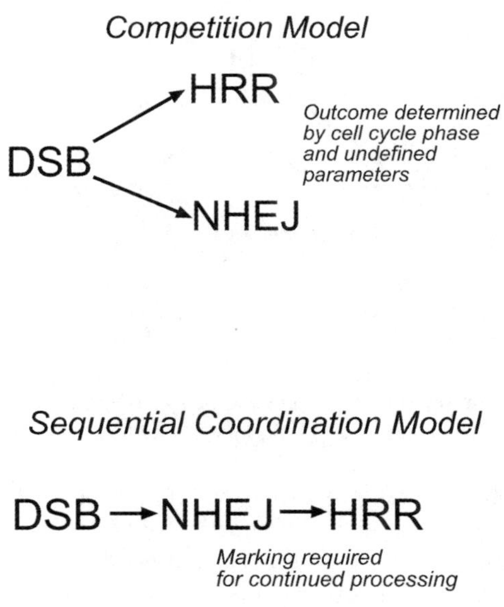

Fig. 1. Two models for the coordination between HRR and NHEJ. See text for details.

To contribute to this central question of modern radiation biology, our future work will focus on the development of methods allowing the reliable evaluation of DNA DSB rejoining in cells irradiated in various phases of the cell cycle. While initial experiments will focus on wild type cells, subsequent studies will include mutants deficient in factors involved in either NHEJ or HRR. The experiments will test two models for the utilization of HRR and NHEJ in the repair of DNA DSBs in cells of higher eukaryotes. These models are summarized in Figure 1. The first model assumes that HRR and NHEJ compete for DNA ends generated after exposure of cells to ionizing radiation. Depending on which phase of the cycle the cell was irradiated, a variable proportion of DSBs is processed by the HRR machinery with the remaining being removed by NHEJ. The second model assumes that initially all IR-induced DNA DSBs are removed by NHEJ in a reaction that does not necessarily ensure preservation of the original sequence around the DSB, and that HRR is engaged subsequently to restore the sequence around the break.

An additional issue that requires investigation is whether DNA DSBs induced by IR in different chromatin regions are repaired by different mechanisms. One could hypothesize, for example, that DNA DSBs induced in highly compacted heterochromatin are repaired by NHEJ, whereas DNA DSBs induced in euchromatin are preferentially handled by HRR (see Figure 2). While methods to address this important question reliably are still under development, the topic is highly relevant in Radiation Biology and will be part of our future research activities.

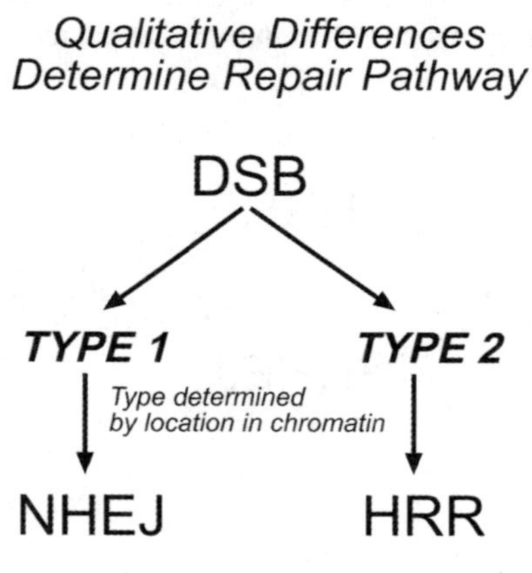

Fig. 2. It is possible that properties of the DNA DSBs associated with their location in chromatin determine the pathways utilized to repair them. See text for details.

The above-described studies will follow the general approaches adopted in recent years and will include a combination of genetics, of biochemistry and cell biology. When necessary, new mutants will be generated using knockout technologies. In addition, the newly developed powerful method of RNA interference (RNAi) will also be extensively used to knock down selected genes.

Alternative pathways of non-homologous end-joining:

The evidence for the operation of pathways of NHEJ that do not utilize DNA-PK or DNA ligase IV is strong. Since these pathways are error prone and involved in the development of tumors, it will be very important to characterize their key components. Our goal along these lines is to purify and characterize factors involved in the rejoining of DNA DSBs by such alternative pathways using biochemical approaches and then apply genetics to characterize their function in cells. Particularly important along these lines will be to elucidate the mechanisms determining which pathway of NHEJ is directed toward a given DNA DSB, and examine what role, if any, chromatin conformation plays in the process.

Although at present our studies are based on the assumption that only two pathways of NHEJ contribute to DNA DSB rejoining, it is in principle possible that additional pathways are involved. This possibility will also be carefully examined. We have also evidence suggesting that presently used biochemical assays have limitations that hinder the elucidation of the pathways utilized by the cell to repair DNA DSBs. Therefore, particular emphasis will also be placed on the development of alternative chromatin-based *in-vitro* assays.

Cross talk between checkpoints and DNA repair pathways:

Although great strides have been made towards understanding signal transduction pathways involved in the activation of checkpoint response, how the delays induced feed into and facilitate repair processes remain largely unknown. There is also only limited information regarding the signals that initiate checkpoint response, and although the DNA DSB is widely considered a strong inducer of checkpoint response, it is not known whether processed DNA DSBs also contribute to the overall effect. Our efforts along these lines will be targeted towards characterizing the signal that initiates the checkpoint response and analyzing the means by which this activation feeds back into the repair process to facilitate its successful performance. These experiments are expected to provide valuable information regarding the necessity of checkpoints for the effective coordination of the biochemical pathways required for the successful response of a eukaryotic cell to DNA damage.

Closing the gap between modern and classical Radiation Biology:

The above-described development in our research activities demonstrates that Radiation Biology has made significant progress in the last few years, but also indicates that this progress has distanced somewhat current research from classical but still very relevant questions of Radiation Biology. It is our estimation that this gap is only temporary and that it will close as the mechanisms involved in DNA DSB repair and checkpoint activation become better characterized at the molecular level. Understanding the molecular basis of cellular repair reactions such as repair of PLD and SLD, as well of the variations in radiosensitivity throughout the cell cycle remain important goals of our future research activities.

Potential contributions and utility

From its inception as a field of science, Radiation Biology has covered topics of research generated by the application of ionizing radiation in medicine for diagnosis and therapy, as well as by the necessity of defining standards for radiation protection. The recent recognition that defects in the repair of DNA DSBs are intimately associated with tumor development, and that exposure to radiation can cause genomic instability, a hallmark of cancer, has led to the definition of areas of investigation with direct connections to the field of cancer. As a result of these developments future research at the Institute of Medical Radiation Biology in Essen has the potential of contributing to key areas that are briefly outlined below:

Cancer Research:

As indicated by the clear focus of our research on basic mechanisms associated with the response of cells of higher eukaryotes to ionizing radiation, the first beneficiary of our activities will be basic research. However, the expected results should provide information useful for our understanding of cancer development. It is now clear that genes associated with DNA repair and checkpoint response, termed "caretakers" and "gatekeepers" [100], respectively, are extremely important for genomic stability and the prevention of cancer. Therefore, any information regarding their function will bring forward our understanding as to how this disease develops. This information will then be useful in devising strategies to combat it.

Cancer Treatment:

One of the earliest goals of Radiation Biology has been the generation of the theoretical background for Radiation Therapy, and the definition of ways to improve the efficacy of radiation against human tumors. Our research has the potential of contributing to this important goal. For this purpose, we will foster interactions with the Department of Radiation Oncology in Essen, as well as with Radiation Oncology clinics nationally and internationally. In addition, as pathways determining cellular responses to DNA damage are defined, targets will be identified that will allow the development of novel agents potentiating the radiation effect specifically in tumor cells. The recent characterization of key components of DNA DSB repair and checkpoint response has led to the development of a great number of such compounds that are presently tested for efficacy *in-vivo* and *in-vitro*. These novel agents promise to improve cancer therapy when combined with existing treatment modalities.

Radiation Protection:

Radiation Biology research has made seminal contributions to the generation of relevant data required for the definition of standards in radiation protection. The topic remains equally relevant today as it was fifty years ago particularly because solid information on the effects of radiation at very low doses and dose rates is lacking. The ongoing characterization of the mechanisms of DNA damage response generate the justified hope that the field of radiation protection may also be able to generate a solid theoretical background based on mechanistic information that will allow the rational definition of standards and maximally allowable exposures to radiation. This is particularly relevant because individuals are continuously exposed to cosmic rays and to an ever increasing array of man-made sources of radiation for diagnostic, therapeutic or even security purposes. Finally, it is anticipated that better radiation protection will allow the development of counter measures that will remove a major obstacle to long term space travel and space exploration.

Acknowledgements

Work in author's laboratories has been supported by grants from the NIH and NASA, USA, as well as by grants from the EU, Volkswagenstiftung, and the IFORES program of the University of Duisburg-Essen.

References

1 Terasima T, Tolmach LJ (1963) X-ray sensitivity and DNA synthesis in synchronous populations of HeLa cells. Science 140:490–492
2 Elkind MM, Sutton H (1960) Radiation response of mammalian cells grown in culture. 1. Repair of X-ray damage in surviving Chinese hamster cells. Radiat Res 13:556–593
3 Friedberg EC, Walker GC, Siede W (1995) DNA Repair and Mutagenesis.: ASM Press, Washington, D.C.
4 Tobey RA (1975) Different drugs arrest cells at a number of distinct stages in G2. Nature 254:245–247
5 Elkind MM, Sutton-Gilbert H, Moses WB, Kamper C (1967) Sub-lethal and lethal radiation damage. Nature 214:1088–1092
6 Phillips RA, Tolmach LJ (1966) Repair of potentially lethal damage in x-irradiated hela cells. Radiat Res 29:413–432
7 Iliakis G (1988) Radiation induced potentially lethal damage: DNA lesions susceptible to fixation. (Review Article). Int J Radiat Biol 53:541–584
8 Iliakis G (1980) Repair of potentially lethal damage in unfed plateau phase cultures of

Ehrlich Ascites Tumour Cells. II. Monolayer Cultures. Int J Radiat Biol 37:591–600
9 Iliakis G, Pohlit W (1979) Quantitative Aspects on Repair of Potentially Lethal Damage in Mammalian Cells. Int J Radiat Biol 36:469–658
10 Iliakis G (1980) Repair of potentially lethally damage in unfed plateau phase cultures of Ehrlich Ascites Tumor Cells. Int J Radiat Biol 37:365–372
11 Iliakis G (1980) Effects of b-arabinofuranosyladenine on the growth and repair of potentially lethal damage in ehrlich ascites tumor cells. Rad Res 83:537–552
12 Iliakis G, Nusse M (1982) Aphidicolin promotes repair of potentially lethal damage in irradiated mammalian cells synchronized in S-phase. Biochem Biophys Res Comm 4:1209–1214
13 Iliakis G, Bryant PE (1983) Effects of the nucleoside analogues a-araA, b-araA and b-araC on cell growth and repair of both potentially lethal damage and DNA double strand breaks in mammalian cells in culture. Anticancer Res 3:143–150
14 Iliakis G (1983) Effects of ß-arabinofuranosyladenine on potentially lethal damage induced in plateau phase mammalian cells exposed U.V.-light. Int J Radiat Biol 43:529–540
15 Iliakis G, Nusse M (1983) The importance of G1/S-border and mitosis in the fixation of potentially lethal damage. Radiat Environm Biophys 22:201–207
16 Iliakis G, Nusse M (1984) Arrest of irradiated G1, S, or G2 cells at mitosis using nocodazole promotes repair of potentially lethal damage. Radiat Res 99:346–351
17 Iliakis G, Nusse M (1983) Evidence that repair and expression of potentially lethal damage cause the variations in cell survival after x-irradiation observed through the cell cycle in Ehrlich Ascites tumor cells. Radiat Res 95:87–107
18 Iliakis G (1984) The influence of conditions affecting repair and expression of potentially lethal damage on the induction of 6-thioguanine resistance after exposure of mammalian cells to x-rays. Mutat Res 126:215–225
19 Iliakis G (1984) The mutagenicity of alpha particles in Ehrlich Ascites tumor cells. Radiat Res 99:52–58
20 Bertsche, U., Iliakis, G., and Kraft, G. Inactivation of Ehrlich Ascites tumor cells by heavy ions. Radiat Res 95:57–67, 1983.
21 Bertsche U. Iliakis G (1987) Modification in repair and expression of potentially lethal damage (alpha-PLD) as measured by delayed-plating or treatment with beta-araA in plateau-phase Ehrlich Ascites tumor cells after exposure to charged particles of various specific energies. Radiat Res 111:26–46
22 Iliakis G, Ngo FQ H, Roberts WK, Youngman K (1985) Evidence for similarities between radiation damage expressed by b-araA and damage involved in the interaction effect observed after exposure of V79 cells to mixed neutron and gamma radiation. Radiat Res 104:303–316
23 Iliakis G, Wright E, Ngo FQH (1987) Possible importance of PLD repair in the modulation of BrdUrd and IdUrd-mediated radiosensitization in plateau-phase 10 T1/2 mouse embryo cells. Int J Radiat Biol 51:541–548
24 Iliakis G, Kurtzman S, Pantelias GE, Okayasu R (1989) Mechanism of radiosensitization by halogenated pyrimidines: Effect of BrdUrd on radiation induction of DNA and chromosome damage and its correlation with cell killing. Radiat Res 119:286–304
25 Wang Y, Okayasu R, Iliakis G (1991) Effects of IdU on the repair of radiation induced PLD, interphase chromatin breaks and DNA double strand breaks in CGO cells. Int J Radiat Oncol Bio Phys 23:353–360
26 Iliakis G, Pantelias G, Kurtzman S (1991) Mechanism of radiosensitization by haloge-

nated pyrimidines: Effect of BrdU on cell killing and interphase chromosome breakage in radiation-sensitive cells. Radiat Res 125:56–64
27 Iliakis G, Kurtzman S (1991) Mechanism of radiosensitization by halogenated pyrimidines: BrdU and beta-arabinofuranosyladenine affect similar subsets of radiation induced potentially lethal lesions in plateau-phase CHO cells. Radiat Res 127:45–51
28 Iliakis G, Wang Y, Pantelias G, Metzger L (1992) Mechanism of radiosensitization by halogenated pyrimidines: Effect of BrdU on repair of DNA breaks, interphase chromatin breaks and potentially lethal damage in plateau-phase CHO cells. Radiat Res 129:202–211
29 Iliakis G, Bryant PE, Ngo FQH (1985) Independent forms of potentially lethal damage fixed in plateau-phase Chinese hamster cells by post-irradiation treatment in hypertonic salt solution or araA. Radiat Res 104:329–345
30 Iliakis G, Ngo FQH (1985) Evidence for the induction of two types of potentially lethal damage after exposure of plateau phase Chinese hamster V79 cells to g-rays. Radiat Environ Biophys 24:185–202
31 Iliakis, G. Evidence for the induction of two types of potentially lethal damage after exposure of plateau phase Chinese hamster V79 cells to g-rays. Radiat Environ Biophys 24:185–202, 1985.
32 Iliakis G, Wright E, Roberts WK., Ngo FQH (1986) Evidence for differences among the sectors of potentially lethal damage expressed by hypertonic treatment in Plateau Phase V79-cells after exposure to neutrons and Gamma-rays. The importance of distinction between alpha and beta-PLD forms. Radiat Res 108:23–33
33 Iliakis G, Metzger L, Muschel RJ,. McKenna WG (1990) Induction and repair of DNA double strand breaks in radiation resistant cells obtained by transformation of primary rat embryo cells with the oncogenes H-ras and v-myc. Cancer Res 50:6575–6579
34 Pantelias GE, Maillie HD(1983) A simple method for premature chromosome condensation induction in primary human and rodent cells using polyethylene glycol. Somatic Cell Genetics 9:533–547
35 Iliakis G Pantelias GE, Okayasu R, Seaner R (1987) 125IdUrd-induced chromosome fragments, assayed by premature chromosome condensation, and DNA double-strand breaks have similar repair kinetics in G1-phase CHO-cells. Int J Radiat Biol 52:705–722
36 Iliakis G, Okayasu R, Seaner R (1988) Radiosensitive xrs-5 and parental CHO cells show identical DNA neutral filter elution dose-response:Implications for a relationship between cell radiosensitivity and induction of DNA double-strand breaks. Int J Radiat Biol 54:55–62
37 Iliakis G, Okayasu R (1988) The level of induced DNA double strand breaks does not correlate with cell killing in X-irradiated mitotic and G1-phase CHO-cells. Int J Radiat Biol 53:395–404
38 Okayasu, R., Bloecher, D., and Iliakis, G. Variation through the cell cycle of DNA neutral filter elution dose response in X-irradiated synchronous CHO-cells. Int J Radiat Biol 53:729–747, 1988.
39 Okayasu R, Iliakis G (1989) Linear DNA elution dose response curves obtained in CHO cells with non-unwinding filter elution after appropriate selection of the lysis conditions. Int J Radiat Biol 55:569–581
40 Bloecher D, Iliakis G (1991) Size distribution of DNA molecules recovered from non-denaturing filter elution. Int J Radiat Biol 59:919–926
41 Iliakis GE, Cicilioni O, Metzger L (1991) Measurement of DNA double strand breaks in CHO cells at various stages of the cell cycle using pulsed field gel electrophoresis: Cal-

ibrations by means of 125I decay. Int J Radiat Biol 59:343–357
42 Metzger L, Iliakis G (1991) Kinetics of DNA double strand breaks throughout the cell cycle as assayed by pulsed field gel electrophoresis in CHO cells. Int J Radiat Biol 59:1325–1339
43 Iliakis G, Metzger L, Denko N, Stamato TD (1991) Detection of DNA double strand breaks in synchronous cultures of CHO cells by means of asymmetric field inversion gel electrophoresis. Int J Radiat Biol 59:321–341
44 Cheng X., Pantelias GE, Okayasu R, Cheong N, Iliakis G (1993) MPF activity of inducer mitotic cells may affect the radiation yield of interphase chromosome breaks in the premature chromosome condensation assay. Cancer Res 53:5592–5596
45 Iliakis G, Okayasu R, Varlotto J, Shernoff C, Wang Y (1993) Hypertonic treatment during premature chromosome condensation allows visualization of interphase chromosome breaks repaired with fast kinetics in irradiated CHO cells. Radiation Res 135:160–170
46 Okayasu R, Iliakis G (1993) Evidence that the product of the xrs gene is predominantly involved in the repair of a subset of radiation-induced interphase chromosome breaks rejoining with fast kinetics. Radiation Res 138:34–43
47 Okayasu, R., Cheong, N., and Iliakis (1993) G. Comparison of yields and repair kinetics of interphase chromosome breaks visualized by Sendai-Virus or polyethylene glycol mediated cell fusion in irradiated CHO cells. Int J Radiat Biol 64:689–694
48 Okayasu R, Iliakis G (1993) Ionizing radiation induces two forms of interphase chromosome breaks in CHO cells that rejoin with different kinetics and show different sensitivity to treatment in hypertonic medium or b-araA. Radiat Res 136:262–270
49 Okayasu R, Pantelias GE, Iliakis G (1993) Increased frequency of formation of interphase ring-chromosomes in radiosensitive irs-1 cells exposed to x-rays. Mutat Res 294:199–206
50 Okayasu, R., Varlotto, J., and Iliakis, G. Hypertonic treatment does not affect the radiation yield of interphase chromosome breaks in DNA double-strand break repair deficient xrs-5 cells. Radiat Res 135:171–177, 1993.
51 Iliakis, G. and Pantelias, G. E. Effects of hyperthermia on chromatin condensation and nucleoli disintegration as visualized by induction of premature chromosome condensation in interphase mammalian cells. Cancer Res 49:1254–1260, 1989.
52 Iliakis G., Seaner R, Okayasu R (1990) Effects of hyperthermia on the repair of radiation induced DNA single and double strand breaks in DNA double strand break repair deficient and repair proficient cell lines. Int J Hypertherm 6:813–833
53 Muschel RJ., Zhang, HB., Iliakis G, McKenna WG (1992) Effects of ionizing radiation on cyclin expression in HeLa cells. Radiat Res 132:153–157
54 McKenna WG, Iliakis G, Weiss MC, Bernhard EJ, Muschel RJ (1991) Increased G2 delay in radiation-resistance cells obtained by transformation of primary rate embryo cells with the oncogenes H-ras and v-myc. Radiat Res 125:283–287
55 Muschel RJ., Zhang HB, Iliakis G (1991) Cyclin B expression in HeLa cells during the G2 block induced by ionizing radiation. Cancer Res 51:5113–5117
56 Wang Y, Iliakis G (1992) Prolonged inhibition by x-rays of DNA synthesis in cells obtained by transformation of primary rat embryo fibroblasts with oncogenes H-ras and v-myc. Cancer Res 52:508–514
57 Wang X, Iliakis G (1993) Persistent inhibition of DNA synthesis after radiation exposure in four clones obtained from rat embryo fibroblasts by transfection with the oncogenes H-ras plus v-myc. Int J Radiat Biol 64:165–168

58 Wang Y, Cheong N, Iliakis G (1993) Persistent inhibition of DNA synthesis in irradiated rat embryo fibroblasts expressing the oncogenes H-ras plus v-myc derives from inhibition of replicon initiation and is mitigated by staurosporine. Cancer Res 53:1213–1217
59 Cheong N, Wang Y, Iliakis G (1993) Radioresistance induced in rat embryo cells by transfection with the oncogenes H-ras and v-myc is cell cycle dependent and maximal during S and G2. Int J Radiat Biol 64:623–629
60 Jackson SP (2002) Sensing and repairing DNA double-strand breaks. Carcinogenesis 23:687–696
61 Thacker JA (1999) surfeit of RAD51-like genes? Trends Genet 15:166–168
62 Thompson LH, Schild D (2001) Homologous recombinational repair of DNA ensures mammalian chromosome stability. Mutat Res 477:131–153
63 Pfeiffer P (1998) The mutagenic potential of DNA double-strand break repair. Toxicol Lett 96,97:119–129
64 Ganguly, T. and Iliakis, G. A cell-free assay using cytoplasmic cell extracts to study rejoining of radiation-induced DNA double-strand breaks in human cell nuclei. Int J Radiat Biol 68:447–457, 1995.
65 Cheong, N., Okayasu, R., Shah, S., Ganguly, T., Mammen, P., and Iliakis, G. In vitro rejoining of double-strand breaks in cellular DNA by factors present in extracts of HeLa cells. Int J Radiat Biol 69:665–677, 1996.
66 Cheong, N., Perrault, R., and Iliakis, G. In vitro rejoining of DNA double strand breaks: A comparison of genomic-DNA with plasmid-DNA-based assays. Int J Radiat Biol 73:481–493, 1998.
67 Cheong N, Perrault AR, Wang H, Wachsberger P, Mammen P, Jackson I, Iliakis G (1999) DNA-PK-independent rejoining of DNA double-strand breaks in human cell extracts in vitro. Int J Radiat Biol 75:67–81
68 Perrault R, Cheong N, Wang H, Wang H, Iliakis G (2001) RPA facilitates rejoining of DNA double-strand breaks in an in vitro assay utilizing genomic DNA as substrate. Int J Radiat Biol 77:593-607
69 Wang Y, Huq MS, Cheng X, Iliakis G (1995) Regulation of DNA replication in irradiated cells by transacting factors. Radiat Res 142:169–175
70 Cheng X, Cheong N, Wang Y, Iliakis G (1996) Ionizing radiation-induced phosphorylation of RPA p34 is deficient in fibroblasts from ataxia telangiectasia and reduced in aged normal fibroblasts. Radiother Oncol 39:43–52
71 Wang Y, Huq MS. Iliakis G (1996) Evidence for activities inhibiting in trans initiation of DNA replication in extract prepared from irradiated cells. Radiat Res 145:408–418
72 Wang Y, Perrault AR, Iliakis G (1997) Down-regulation of DNA replication in extracts of camptothecin-treated cells: Activation of an S-phase checkpoint? Cancer Res 57:1654–1659
73 Wang Y, Perrault AR, Iliakis G (1998) Replication protein A as a potential regulator of DNA replication in cells exposed to hyperthermia. Radiat Res 149:284–293
74 Wang Y, Zhou XY, Wang H-Y, Iliakis G (1999) Roles of replication protein A and DNA-dependent protein kinase in the regulation of DNA replication following DNA damage. J Biol Chem 274:22060–22064
75 Iliakis G, Wang Y, Wang H-Y (1999) Analysis of inhibition of DNA replication in irradiated cells using the SV40 based in vitro assay of DNA replication. In: Henderson DS (ed) Methods in Molecular Biology, DNA Repair Protocols: Eukaryotic Systems, Vol. 113, pp. 543–553. Totowa, NJ: Humana Press Inc.
76 Wang J-L, Wang X, Wang H, Iliakis G, Wang Y (2002) CHK1-Regulated S-phase

Checkpoint Response Reduces Camptothecin Cytotoxity. Cell Cycle 1:267–272
77 Wang X, Khadpe J, Hu B, Iliakis G, Wang Y (2003) An Over-activated ATR/CHK1 Pathway is Responsible for the Prolonged G2 Accumulation in Irradiated AT Cells. Journal of Biological Chemistry, in press
78 Wang X, Li GC, Iliakis G, Wang Y (2002) Ku Affects the CHK1-dependent G2 Checkpoint after Ionizing Radiation. Cancer Res 62:6031–6034
79 Wang H, Wang X, Zhou X-Y, Chen DJ, Li GC, Iliakis, G, Wang Y (2002) Ku Affects the Ataxia and Rad 3-related/CHK1-dependent S Phase Checkpoint Response after Camptothecin Treatment. Cancer Res 62:2483–2487
80 Zhou X-Y, Wang X, Hu B, Guan J, Iliakis G, Wang Y (2002) An ATM-independent S-Phase Checkpoint Response Involves CHK1 Pathway. Cancer Res 62:1598–1603
81 DiBiase S, Zeng Z-C, Chen R, Hyslop PA, Curran Jr. W, Iliakis G (2000) DNA-dependent Protein Kinase Stimulates an Independently Active, Nonhomologous, End-Joining Apparatus. Cancer Res 60:1245–1253
82 Wang H, Zeng Z-C, Bui T-A, DiBiase SJ, Qin W, Xia F, Powell SN, Iliakis G (2001) Nonhomologous end-joining of ionizing radiation-induced DNA double stranded breaks in human tumor cells deficient in BRCA1 or BRCA2. Cancer Res 61:270–277
83 Wang H, Zeng Z-C, Bui T-A, Sonoda E, Takata M, Takeda S, Iliakis G (2001) Efficient rejoining of radiation-induced DNA double-strand breaks in vertebrate cells deficient in genes of the RAD52 epistasis group. Oncogene 20:2212–2224
84 Wang H, Zhao-Chong Z, Perrault AR, Cheng X, Qin W, Iliakis G (2001) Genetic evidence for the involvement of DNA ligase IV in the DNA-PK-dependent pathway of non-homologous end joining in mammalian cells. Nucleic Acids Res 29:1653–1660
85 Nevaldine B, Longo JA, Hahn PJ (1997) The scid defect results in much slower repair of DNA double-strand breaks but not high levels of residual breaks. Radiat Res 147:535–540
86 DiBiase SJ, Zeng Z-C, Chen R, Hyslop T, Curran WJ Jr., Iliakis G (2000) DNA-dependent protein kinase stimulates an independently active, nonhomologous, end-joining apparatus. Cancer Res 60:1245–1253
87 Loebrich M, Rydberg B, Cooper PK (1995) Repair of x-ray-induced DNA double-strand breaks in specific Not I restriction fragments in human fibroblasts: Joining of correct and incorrect ends. Proc Natl Acad Sci US 92:12050–12054
88 Asaad NA, Zeng Z-C, Guan J, Thacker J, Iliakis G (2000) Homologous recombination as a potential target for caffeine radiosensitization in mammalian cells: Reduced caffeine radiosensitization in XRCC2 and XRCC3 mutants. Oncogene 19:5788–5800
89 Cheong N, Wang Y, Jackson, M. Iliakis G (1992) Radiation-sensitive irs mutants rejoin DNA double strand breaks with efficiency similar to that of parental V79 Cells but show altered response to radiation induced G2-delay. Mutat Research 274:111–122
90 Roth DB, Porter TM, Wilson JH (1985) Mechanisms of nonhomologous recombination in mammalian cells. Mol Cell Biol 5:2599–2607
91 Petiniot LK, Weaver Z, Vacchio M, Shen R, Wangsa D, Barlow C, Eckhaus M, Steinberg SM, Wynshaw-Boris A, Ried T, Hodes RJ (2002) RAG-Mediated V(D)J Recombination Is Not Essential for Tumorgenesis in Atm-Deficient Mice. Mol Cell Biol 22:3174–3177
92 Zhu C, Mills KD, Ferguson DO, Lee C. Manis J, Fleming J, Gao, Y, Morton CC, Alt FW (2002) Unrepaired DNA Breaks in p53-Deficient Cells Lead to Oncogenic Gene Amplification Subsequent to Translocations. Cell 109:811–821
93 Sharpless NE, Ferguson DO, O'Hagan RC, Castrillon DH, Lee C, Farazi PA, Alson S,

Fleming J, Morton CC, Frank K, Chin L, Alt FW, DePinho RA (2001) Impaired Non-homologous End-Joining Provokes Soft Tissue Sarcomas Harboring Chromosomal Translocations, Amplifications, and Deletions. Mol Cell 8:1187–1196
94 Liao M-J, Van Dyke T (1999) Critical role for Atm in suppressing V(D)J recombination-driven thymic lymphoma. Genes Developm 13:1246–1250
95 Vanasse, G. J., Halbrook, J., Thomas, S., Burgess, A., Hoekstra, M. F., Disteche, C. M., and Willerford (1999) D. M. Genetic pathway to recurrent chromosome translocations in murine lymphoma involves V(D)J recombinase. J Clin Invest 103:1669–1675
96 Gao Y, Ferguson D O, Xie W, Manis JP, Sekiguchi JA, Frank KM, Chaudhuri J, Horner J, DePinho RA, Alt FW (2000) Interplay of p53 and DNA-repair protein XRCC4 in tumorigenesis, genomic stability and development. Nature 404:897–900
97 Difilippantonio MJ, Zhu J, Chen HT, Meffre E, Nussenzweig NC, Max EE, Ried T, Nussenzweig A (2000) DNA repair protein Ku80 suppresses chromosomal aberrations and malignant transformation. Nature 404:510–514
98 Pierce AJ, Stark JM, Araujo FD, Moynahan, ME, Berwick M, Jasin M (2001) Double-strand breaks and tumorigenesis. Trends Cell Biol 11:S52–S59
99 Johnson RD, Jasin M (2001) Double-strand-break-induced homologous recombination in mammalian cells. Biochem Soc Trans 29:196–201
100 Kinzler KW, Vogelstein B (1997) Cancer-susceptibility genes. Gatekeepers and caretakers [news; comment]. Nature 386:761–763

Prospects for Research in Radiation Biology

Charles L. Limoli

Department of Radiation Oncology, University of California, San Francisco, CA 94103-0806, USA
limoli@itsa.ucsf.edu

Introduction

With the discovery of X-rays by Roentgen in 1895 came the birth of radiation research, a multidisciplinary field that has since seen remarkable progress. Over the past century our field has witnessed major advancements in the measurement of radiation fields, the identification of critical radiogenic lesions in cells, the use of therapeutic irradiation regimes to treat cancer patients, and the control of nuclear reactions used to generate energy. Despite our rich legacy of achievements, the natural flux of social and economic pressures pose new challenges to the radiation research community. What then does the future hold? Genetic and molecular techniques have hastened the identification of radiation responsive signaling pathways that regulate an impressive and intricate network of checkpoint and repair systems. Despite this progress, many of the incipient radiation-induced physical-chemical changes and/or covalent modifications that activate the upstream kinases are not known. Many of the initial damage sensors and factors regulating the recognition of a variety of damage types are incompletely understood. The interplay between multiprotein complexes acting upon specific lesions is a dynamic process and work is still needed to decipher how these complexes modify protein-protein interactions to recruit and/or reshuffle their protein subunits in efforts to fine tune repair responses to the ever changing extra- and intracellular environment. Understanding these interactions will provide insights into how the replication machinery and multiple damage-specific polymerases interface with DNA repair complexes to resolve impediments to replication and double-strand breaks that form during replication arrest.

Details concerning how the memory of prior DNA damage is propagated to progeny cells and how this ultimately elicits the aberrant phenotypes characteristic of genomic instability has yet to be unraveled. Recent data from our laboratory has pointed to the possible role of oxidative stress as a biochemical mechanism capable of sustaining chromosomal instability over multiple generations. Our data has also demonstrated a link between reactive oxygen species (ROS), genomic instability and dysfunctional mitochondria. These studies suggest a number of potentially interesting avenues for further pursuit. The remodeling of chromatin may be another factor instrumental in perpetuating genomic instability. Alterations in chromatin structure can modify the accessibility of damaged DNA to repair factors and anchor the relative mobility of subnuclear domains to hasten repair and provide an epigenetic footprint for tagging chromosomal regions for subsequent change. How the foregoing processes influence and regulate radiation-induced bystander effects provides another area of interesting research opportunity. In this regard, further work must also aim to elucidate the specific pathways and molecules that mediate radiation-induced bystander effects. The summation of these events will be critical in determining the ultimate predisposition of a cell to developing further deleterious changes and genomic instability. Just how these initial events trigger the biochemical and molecular mechanisms needed to perpetuate the memory of past insult are not known, but they are clearly important in many of the long-term pathologies associated with radiation exposure, and are certainly vital to our understanding of radiation carcinogenesis and risk.

Early events in the activation of the mammalian stress response

The interaction between ionizing radiation and the cellular environment leads to a series of physical, chemical and biological changes that transpire over a substantial relative timeframe (Table 1.).

Table 1. Relative timeframe of radiation events from the initial physical and chemical induced changes to the long-term biological consequences of exposure to ionizing radiation.

Radiation Events	Time [sec]	Radiation-induced Event
Physical and Chemical Processes: Energy Deposition and Free Radical Chemistry	10^{-16}–10^{-13}	Secondary electron energy loss and reduction to thermal energy
	10^{-9}	Hydroxyl radicals react in cell
	10^{-3}	Free radicals react with oxygen
	10^{0}	Free radical reactions complete
Early Biological Responses: Signaling and Repair	10^{2}–10^{4}	Repair of DNA damage
	10^{2}–10^{5}	Initiation of genomic instability
	10^{5}–10^{6}	Perpetuation of genomic instability
Late Biological Responses: Mutation, Transformation, Cancer	10^{6}–10^{7}	Cells die, mutations are scored
	10^{8}–10^{9}	Transformation and carcinogenesis

The initial deposition of energy leads to ionizations and free radical reactions that trigger the mammalian stress response. The cellular response to ionizing radiation exposure and the resultant cellular damage is complex and begins with the activation of multiple signaling pathways that are tightly regulated and geared to modulating cell cycle progression and the repair of DNA. Significant progress has been made in identifying and characterizing radiation responsive pathways and in determining their functional consequences in cells. Damage caused by radiation exposure is initially detected by sensor molecules (e.g. ATM and ATR) and propagated by a series of transducers (e.g. Brca1/2, Chk1/2) that then modulate the activity of multiple effector molecules (e.g. Rad51 and paralogs, Mre11 complex) that mediate physiologic change (Fig. 1.).

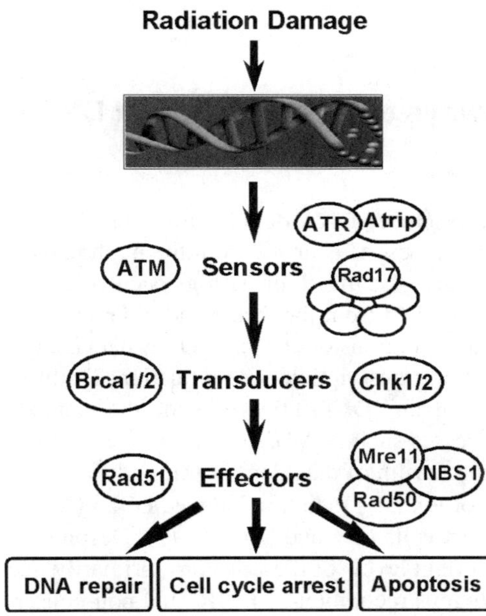

Fig. 1. Mammalian stress response. Ionizing radiation activates signaling pathways through the interplay between damage sensors, transducers and effectors. While many of the steps in these pathways have been characterized, factors acting upstream of the putative sensors remain elusive.

Despite the wealth of recent data that has improved our understanding of the molecular events occurring consequent to the induction of DNA damage, it remains unclear what activates or acts upstream of the proteins proposed to act as DNA damage sensors (see [1] for a recent review). Most models involving the recognition of DNA damage implicate ATR and ATM, members of the PI3KK family of protein kinases, as key early components in the signaling process (see [2] for review). ATM plays a more central role in responding to ionizing radiation, while ATR is more important in the UV response and in regulating replication fidelity during S-phase [2]. A recent report demonstrating that ATR binds directly to UV damaged DNA and leads to a stimulation of its kinase activity supports the idea that this kinase might function as a true DNA damage sensor [3]. These properties are likely to be attenuated by its recently identified binding partner ATRIP [4]. Less certain is how ATM is converted to the active form, but autophosphorylation occurring as a consequence of its interaction with damaged DNA is one possibility. Other radiation-induced changes that could contribute to the activation of ATM may include: 1) redox mediated conformational changes to the protein, 2) alterations in ion flux, 3) or rapid changes in chromatin structure. While impressive progress has been made toward an understanding of radiation-induced signaling pathways, many intriguing prospects still await those engaged in unraveling the details of radiation-induced signal transduction.

Alternate pathways to the formation of DNA double-strand breaks

Ionizing radiation produces DNA double strand breaks that contain a variable multiplicity of damage types. This arises from the nonhomogeneous deposition of energy that leads to highly localized ionizations occurring on the nanometer scale [5]. Reactive species derived from the direct and indirect ionizations of DNA and solvent molecules lead to regions of clustered DNA damage that challenge cellular processes of DNA repair, transcription and replication. Traditional views have held that cells contain few if any DNA DSBs outside the context of specialized and highly regulated processes such as VDJ recombination. These views are convincingly supported by the wealth of data demonstrating the toxicity of DNA DSBs to cells [6]. Low levels of ionizing radiation induced DSBs (~40 DSBs/cell) are sufficient to inactivate most cells (*i.e.* one-log kill) [7]. Despite the toxicity of DSBs, cells can withstand certain levels of these lesions, primarily due to the activities of conservative (homologous recombination, HR) and non-conservative (non-homologous end joining, NHEJ) repair processes that mediate the rejoining of broken DNA. The presence of these specialized pathways however poses a conundrum: If DSBs are toxic and cells possess relatively few of these lesions, then from where did the selective pressure arise to evolve such elegant pathways for recombinational repair? One explanation may involve the production of DSBs during the normal replication of DNA.

As cells proliferate, the DNA replication machinery can encounter a variety of endogenously and/or exogenously derived lesions. Base damage, bulky adducts, crosslinks and strand breaks represent some of the lesion types that pose problems to the fidelity of replication. Replicational stress caused by DNA damage can lead to a prolonged stalling of the fork, a potentially deleterious outcome that increases the likelihood of developing DNA DSBs. Substantial data now supports the idea that DSBs arise during the of course DNA synthesis (see [1] and references within for further reading), and suggests that S-phase derived DSBs may have provided the selective pressure to evolve the pathways for their removal.

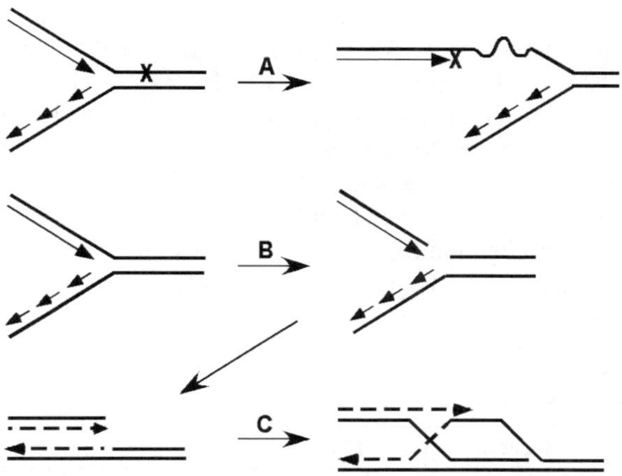

Fig. 2. Arrest of the replication fork. (a) Expansion of an assymetrical fork blocked by a lesion "X" on the leading strand; (b) Fork collapse and DSB generation upon the encounter of the replication fork with a prexisting nick or gap; and (C) Formation of a Holliday junction at the site of a broken fork.

Replication arrest can arise by the convergence of a replication fork with preexisting DNA lesions (Fig. 2. and 3). Bulky DNA lesions such as UV-induced dimers can block leading strand synthesis and lead to the generation of an asymmetrical replication fork that contains large regions (several kilobases) of unpaired single-stranded DNA (Fig. 2.a). These structures have been observed experimentally [8, 9], and predominate under certain situations where bypass of bulky lesions is impaired. Forks stalled at sites of dimer impasse have been shown to develop DSBs that undergo recombinational repair [10, 11]. Forks encountering pre-existing nicks or gaps may also collapse to generate frank DSBs (Fig. 2.b) [12]. Broken forks are then subject to strand invasion reactions to form Holliday junctions (HJs), a common intermediate produced during the repair of DSBs by HR (Fig. 2.c) [12].

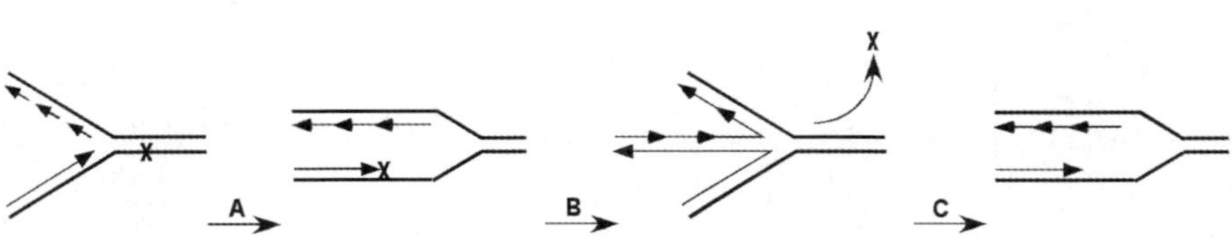

Fig. 3. Dynamics of stalled forks: (A) Replication fork encounters a leading strand lesion "X"; (B) Fork regression can generate a 4 stranded "chicken foot" structure; (C) Repair removes the lesion and branch migration resets the replication fork.

Alternatively, impediments to replication progression (Fig. 3.A) may cause stalled forks to regress generating "chicken foot" structures (Fig. 3.B) [12]. Regressed forks are structurally identical to, but topologically distinct from HJs.

These 4-stranded structures generated at sites of fork blockage can be unwound by structure-specific helicases (e.g. WRN and BLM) that facilitate branch migration [13–15] and cut by resolvases such as mus81 [16, 17]. While evidence for fork reversal in mammalian cells is not compelling [18, 19], these structures have been documented in yeast mutants [20], where they can undergo repair and branch migration to reset the fork for the resumption of replication (Fig. 3.C). While evidence supports the idea that replication arrest leads to DNA DSBs, it should be emphasized that DBSs arising at replication forks differ from those produced by ionizing radiation in that they generally involve the creation of only one double-stranded end, i.e. a broken daughter chromatid.

Table 2. Classes of DNA Polymerases.

Class	Greek	Hugo	Functions
A	Θ^*	Pol Q	Crosslink repair
	γ	Pol G	Mitochondria
B	α	Pol A	Replication/primase
	δ	Pol D	Replication/repair, 3'–5'exo
	ε	Pol E	Replication/repair? 3'–5'exo
	ζ^*	Pol Z	Mismatch extension
C			E.coli pol III
D			Archaeal polymerases
X	β	Pol B	BER (1–2 nt insertion, lyase)
	λ^*	Pol L	BER (1–2 nt insertion, lyase)
	μ^*	Pol M	DNA breaks (NHEJ)
	σ1, σ2	Pol S	Chromatids (alt. κ, *TRF4*)
	TDT	—	Antigen/antibody diversity
Y	η^*	Pol H	Bypass, low fidelity, XPV (dA)
	I^*	Pol I	Bypass, low fidelity, (dG)
	κ^*	Pol K	Low fidelity (alt. θ, *DinB1*)
	Rev1*	REV1L	dC transferase

**Damage-specific, error-prone, DNA polymerases in eukaryotic cells.*

The idea that replication arrest leads to DSBs presents some interesting opportunities for radiation biologists too. For example further work is needed to clarify the possible interactions between radiation-induced lesions and the replication machinery. Cloning of the gene (PolH) defective in the genetic disorder xeroderma pigmentosum variant (XPV) [21, 22] has led to the discovery of additional DNA damage-specific error-prone polymerases in mammalian cells (Table 2.). These specialized polymerases with lesion-bypass activities allow cells to deal with replication-interfering damages, and represent an impressive array of safeguards that help maintain the integrity of the DNA molecule [23–26].

The importance of these specialized polymerases is readily apparent in XPV cells that lack the ability to efficiently bypass UV-induced lesions. This leads to a prolonged replication arrest that increases fork breakdown, DSB formation and recombination [10, 11], factors that may underlie the marked predisposition of XPV patients to actinic skin cancer [24]. The difficulties of replicating past bulky DNA lesions may provide an explanation for the marked increase in DSBs seen after UV exposure of yeast mutants [27], and after near UV exposure hamster of cells treated with bromodeoxyuridine and a photosensitizing dye [28]. Clearly, replication poses ample opportunity for mishap within the cell, and understanding how the replication apparatus deals with regions of clustered DNA damage and what damage-specific error-prone polymerases are recruited to mediate the repair of such sites poses many interesting challenges for future work.

Genomic instability and oxidative stress

In response to radiation-induced damage cells can survive and proliferate, undergo apoptosis or accumulate a variety of genetic alterations that may predispose them to longer term deleterious change. The increased rate of acquired genetic change is a process referred to as "genomic instability", and is believed to be the driving force behind multistep carcinogenesis [29, 30]. Genomic instability encompasses a number of endpoints including gene mutation and amplification, reduced cloning efficiency, and a host of karyotypic abnormalities such as chromosomal instability [31]. Whether or not a cell eventually goes on to develop the signs characteristic of the unstable phenotype will depend on the type and level of nuclear and non-nuclear damage, on the biological response to that damage, on the genetic background and the microenvironment. Past work in hamster cells has shown that the incidence of chromosomal instability in clones surviving x-irradiation, was 3% per Gray [32], while the incidence observed after exposure of the same cells to 1 GeV/n iron ions was slightly higher at 4% Gray [33]. The linear dose responses observed in these studies could be altered by substitution of cellular DNA with BrdU. When the level of thymidine replacement by BrdU ranged from 25–66%, the incidence of genomic instability was increased at lower doses (i.e. 30% at 2 Gy) [32]. This and other work [34] clearly implicates the importance of nuclear DNA damage for the eventual manifestation of genomic instability, however, other factors including persistent oxidative stress, bystander effects and the microenvironment are likely to influence the ultimate development of genomic instability.

Significant effort has been devoted to understanding the nature of factors potentially important in the induction and maintenance of genomic instability. Much of this work has been focused on identifying the biochemical and/or molecular pathways that can perpetuate the memory of past insult over multiple cellular generations. One attractive possibility involves oxidative stress; a condition that occurs when the balance of processes that regulates the basal redox state in cells shifts in favor of a pro-oxidant state. Certain studies have found an association between genomic instability and oxidative stress [35–37]. Elevated levels of intracellular peroxy radicals and lipid peroxidation end-products have been detected in chromosomally unstable clones [36], and reactive oxygen species (ROS) have been implicated in other endpoints characteristic of unstable cells [38, 39]. Precisely how oxidative stress contributes to the development of genomic instability is uncertain, but represents an area of future work that overlaps multiple scientific disciplines.

Genomic instability and the mitochondria

One possibility linking oxidative stress to the perpetuation of other cellular abnormalities involves the mitochondria. As the site of electron transport, mitochondria constitute a logical source for the elevated ROS [40, 41]. Mitochondrial dysfunction can lead to an elevation of ROS by increasing the residence time of electrons at specific steps in the electron transport process, thereby increasing the probability of electron transfer to oxygen. Recent studies have also found that groups of chromosomally unstable clones contain elevated levels of ROS, and contain elevated numbers of dysfunctional mitochondria compared to their chromosomally stable counterparts (unpublished data).

To expand these observations we have used ethidium bromide (EtBr) to perturb mitochondrial function in efforts to elucidate the relationships between genomic instability, oxidative stress and the mitochondria. Long-term growth in the presence of EtBr has been reported to result in the generation of ρ^0 cells devoid of mitochondrial DNA (mtDNA) [42, 43]. Consequently, GM10115 cells that contain a single copy of human chromosome 4 were cultured in the presence of 50 ng/ml EtBr for 50 passages to disrupt mitochondrial function [44]. Cells were maintained in exponential growth by passing cultures twice weekly. Extra cells not re-seeded for growth in EtBr were processed periodically (every 5th or 10th passage) for the analysis of the following endpoints: 1) Mitochondrial function using the dye rhodamine 123 (R123), 2) mitochondrial content using the dye nonyl acridine orange (NAO), 3) chromosomal rearrangements involving the human chromosome, 4) loss of the human chromosome, and 5) reduced plating efficiency. Mitochondrial endpoints were measured by fluorescent automated cell sorting (FACS) analysis, chromosomal endpoints were assayed by fluorescence in situ hybridization (FISH), and plating efficiency was measured by clonogenic assay.

To assess the impact of EtBr on mitochondrial function and content cells were assayed every 10th passage using R123 and NAO respectively. FACS analysis of the R123 signal shows that mitochondrial function is inhibited by an average of 23% compared to controls (Fig. 4., light bars, page 37), and remains depressed during subsequent growth in the presence of EtBr. Interestingly, this prolonged growth in EtBr was not accompanied by a significant depletion of mtDNA (data not shown) as reported by others [45, 46]. FACS analysis of NAO fluorescence also indicated that cells grown in the presence of EtBr did not undergo significant changes in mitochondrial content compared to controls (Fig. 4., dark bars). Collectively, the data suggests that growth in the presence of EtBr impairs mitochondrial function in the absence of significant changes in the number of these organelles.

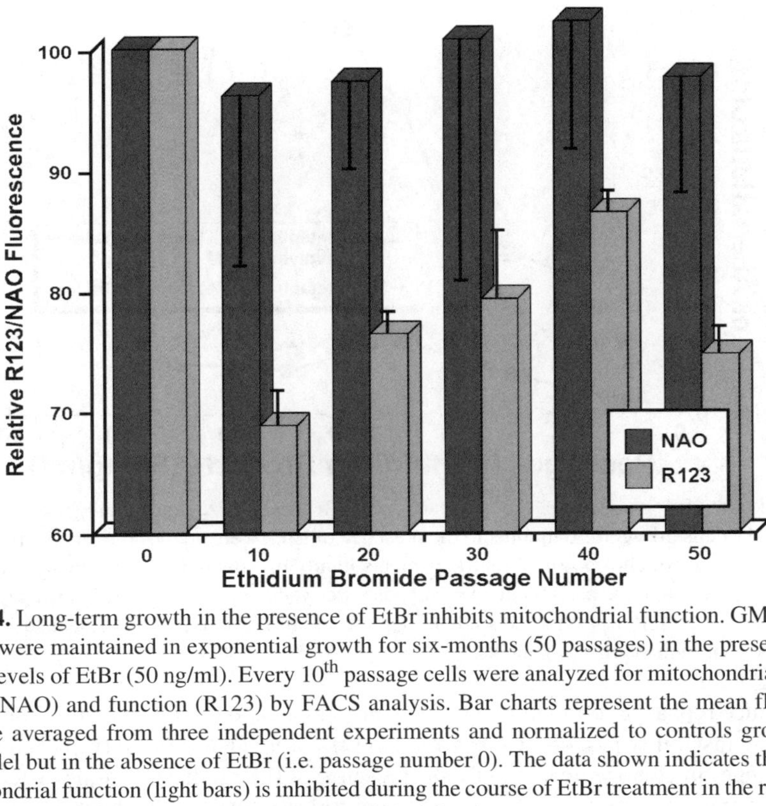

Fig. 4. Long-term growth in the presence of EtBr inhibits mitochondrial function. GM10115 cells were maintained in exponential growth for six-months (50 passages) in the presence of low levels of EtBr (50 ng/ml). Every 10th passage cells were analyzed for mitochondrial content (NAO) and function (R123) by FACS analysis. Bar charts represent the mean fluorescence averaged from three independent experiments and normalized to controls grown in parallel but in the absence of EtBr (i.e. passage number 0). The data shown indicates that mitochondrial function (light bars) is inhibited during the course of EtBr treatment in the relative absence of changes in mitochondrial content (dark bars).

To determine whether other anomalies associated with genomic instability could be associated with extended growth in the presence of EtBr, we investigated chromosomal change. Cytogenetic analysis revealed increasing levels of karyotypic abnormality in the presence of EtBr. The formation of these abnormalities was strongly dependent upon EtBr exposure time (i.e. passage number). Chromosomal aberrations increased steadily over the first 25 passages (from 0.5 to 10%) before leveling off during subsequent growth in EtBr (Fig. 5., circles). More dramatic however, was the effect that EtBr had on the retention of the human chromosome. Increased time in the presence of EtBr resulted in a steady increase in the number of cells losing human chromosome 4, from a background of 5% of metaphases without chromosome 4, to an average of 64% after 35 passages in EtBr (Fig. 5., squares). Data demonstrates that prolonged growth in the presence of EtBr can lead to an accumulation of cytogenetic abnormalities.

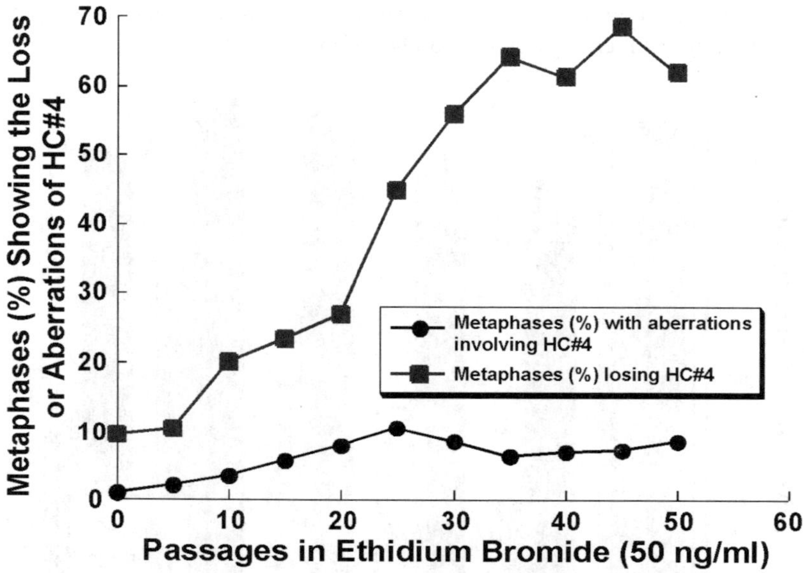

Fig. 5. Extended passage in the presence of EtBr leads to cytogenetic abnormalities. GM10115 cells grown for 6 months in the presence of EtBr were analyzed every 5th passage for the presence of chromosomal rearrangements involving human chromosome 4. The data shown indicates that the percentage of metaphases derived from cells cultured in EtBr for extended times exhibit a passage dependent increase in chromosomal aberrations and aneuploidy (loss of human chromosome 4).

Reduced plating efficiency is frequently associated with radiation-induced genomic instability [47, 48]. Cells grown in EtBr were analyzed every 10th passage for changes in clonogenic survival. The presence of EtBr reduced plating efficiency by an average of 22% (0.69) compared to controls (0.89) over the duration of treatment (see Fig. 6., page 39). These observations suggest that EtBr can induce similar changes in plating efficiency as observed previously in subsets of chromosomally unstable clones [49].

Several studies have found that tissue-specific depletion of mtDNA has been linked to a wide variety of degenerative conditions (see [50] for a review), many of which exhibit similarities to endpoints that characterize genomic instability. Our work has determined that EtBr could inhibit mitochondrial function, elicit cytogenetic change and reduce plating efficiency. While past studies have found that EtBr impairs the replication and transcription of mtDNA to a greater extent than nuclear DNA [51–53], we can not formally discount the possibility that the cytogenetic abnormalities we observed were due to the effect of EtBr on nuclear DNA. Notwithstanding the data shown in Figures 4–6 demonstrate that EtBr can elicit multiple manifestations of genomic instability, and suggests that related studies analyzing the connection between mitochondrial dysfunction, oxidative stress and genomic instability will provide useful insights.

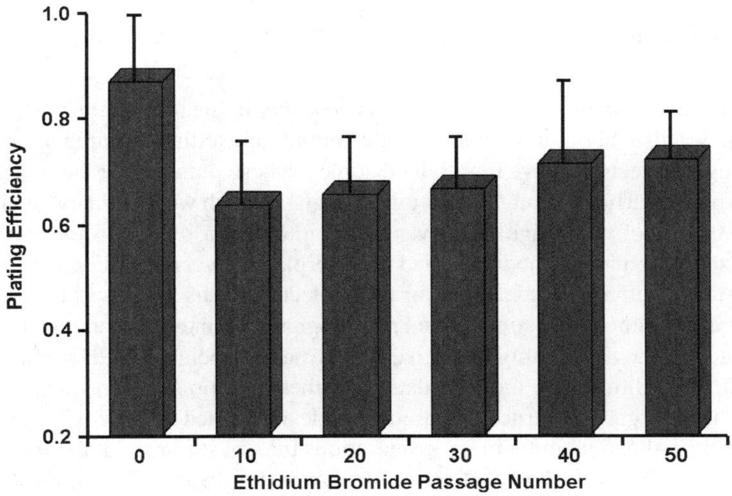

Fig. 6. Extended passage in the presence of EtBr leads to cytogenetic abnormalities. GM10115 cells grown for six-months in the presence of EtBr were analyzed every 5th passage for the presence of chromosomal rearrangements involving human chromosome 4. The data shown indicates that the percentage of metaphases derived from cells cultured in EtBr for extended times exhibit a passage dependent increase in chromosomal aberrations and aneuploidy (loss of human chromosome 4).

Radiation-induced bystander effects

Bystander or non-targeted effects induced after exposure to radiation encompass an impressive array of endpoints, have been found in a variety of cell types and vary depending on the type of radiation used. While the majority of bystander effects such as chromosomal instability [54, 55], gene induction [56] and mutation [57], and sister chromatid exchange [58] have been reported after alpha particle irradiation, X-irradiation also induces bystander effects, observed primarily as reductions in plating efficiency [59, 60]. Also worth mentioning is that bystander effects are not limited to ionizing radiation, as UVA light has been found to induce bystander effects in canine kidney cells [61], and may underlie recent observation of oxidative stress in UVA irradiated human keratinocytes [62]. Recent reviews have detailed many of the critical observations and important consideration needed to ultimately identify the initial lesions that induce bystander effects and the pathways these lesions activate that lead to the radiomimetic changes observed in unirradiated cells [63, 64]. The interaction of genetically distinct cell types with their surrounding environment and the role of bystander effects in modulating the radiation response of normal tissues and tumors are important topics that warrant further investigation.

Conclusions

While we anticipate many challenges as we usher in the next century of radiation research, we also face a future that is bright and full of exciting research opportunity. The foregoing sections have highlighted some of these potential research prospects that are particularly relevant to radiation biologists. Much work remains in elucidating the details of early signaling events, the interaction of repair and replication complexes, chromatin remodeling, and the interplay between the microenvironment and bystander effects. Understanding the molecular basis of these processes will provide insight into many complicated pathologies associated with exposure to radiation, such as the debilitating cognitive impairment associated with cranial irradiation [65, 66]. Ultimately, the elucidation of these radioresponsive pathways will identify new targets for structure-based drug design aimed at improving the therapeutic ratio and patient outcome, considerations that must remain at the forefront of radiobiology. It is the charge of those of us working in the radiation sciences to enthusiastically disseminate this potential promise to those outside our field and most importantly to the public – the future of radiation research will certainly depend upon it.

References

1 Thompson LH, Limoli CL (2003) Detection and Signaling of DNA Double-Strand Breaks in Mammalian Cells. In Press: Caldecott KW (ed.) Eukaryotic DNA Damage Surveillance and Repair. Georgetown, TX Landes Bioscience
2 Abraham RT (2001) Cell cycle checkpoint signaling through the ATM and ATR kinases. Genes Dev 15:2177–2196
3 Unsal-Kacmaz K, Makhov A.M, Griffith JD, Sancar (2001) A. Preferential binding of ATR protein to UV-damaged DNA. Proc Natl Acad Sci USA 99:6673–6678
4 Cortez D, Guntuku S, Qin J, Elledge SJ (2001) ATR and ATRIP: partners in checkpoint signaling. Science 294:1713–1716
5 Ward JF (1981) Some biochemical consequences of the spatial distribution of ionizing radiation-produced free radicals. Radiat Res 86:185–195
6 Painter, RB (1980) The role of DNA damage and repair in cell killing induced by ionizing radiation. In: Meyn RE, Withers HR (eds) Radiation Biology in Cancer Research pp. 59–68 New York Raven Press
7 Ward JF (1990). The yield of DNA double-strand breaks produced intracellularly by ion-

izing radiation: a review, Int J. Radiat Biol 57:1141–1150
8 Cordeiro-Stone M, Zaritskaya LS, Price LK, Kaufmann WK (1997) Replication fork bypass of a pyrimidine dimer blocking leading strand DNA synthesis. J Biol Chem 272:13945–13954
9 Cordeiro-stone M, Makhov AM, Zaritskaya LS, Griffith JD (1999) Analysis of DNA replication forks encountering a pyrimidine dimer in the template to the leading strand. J Mol Biol 289:1207–1218
10 Limoli CL, Giedzinski E, Morgan WF, Cleaver JE (2000) Polymerase h deficiency in the XP variant uncovers an overlap between the S phase checkpoint and double strand break repair. Proc. Natl. Acad. Sci. USA 97:7939–7946
11 Limoli C, Giedzinski E, Bonner W, Cleaver J (2002) UV-induced replication arrest in the xeroderma pigmentosum variant leads to DNA double-strand breaks, g-H2AX formation, and Mre11 relocalization. Proc Natl Acad Sci USA 99:233–238
12 Cox MM, Goodman MF, Kreuzer KN, Sherrat DJ, Sandler SJ, Marians KF (2000) The importance of repairing stalled replication forks. Nature 404:37–41
13 Karow JK, Constantinou A, Li JL, West SC, Hickson ID (2000) The Bloom's syndrome gene product promotes branch migration of holliday junctions. Proc. Natl. Acad. Sci. USA 97:6504–6508
14 Mohaghegh P, Karow JK, Brosh Jr RM, Bohr VA., Hickson ID (2001) The Bloom's and Werner's syndrome proteins are DNA structure-specific helicases. Nucleic Acids Res 29:2843–2849
15 Mohaghegh P, Hickson ID (2002) Premature aging in RecQ helicase-deficient human syndromes. Int J Biochem Cell Biol 34:1496–1501
16 Chen XB, Melchionna R, Denis CM, Gaillard PH, Blasina A, Van de Weyer I, Boddy MN, Russell P, Vialard J, McGowan CH (2001) Human Mus81-associated endonuclease cleaves Holliday junctions in vitro. Mol Cell 8:1117–1127
17 Haber JE, Heyer WD (2001) The fuss about Mus81. Cel 107:551–554
18 Tatsumi K, Strauss B (1978). Production of DNA bifilarly substituted with bromodeoxyuridine in the first round of synthesis: branch migration during isolation of cellular DNA. Nucleic Acids Res 5:331–347
19 Tatsumi K, Strauss BS (1979). Accumulation of DNA growing points in caffeine-treated human lymphoblastoid cells. J Mol Biol 135:435–449
20 Sogo JM, Lopes M, Foiani M (2002) Fork reversal and ssDNA accumulation at stalled replication forks owing to checkpoint defects. Science 297:599–602
21 Johnson RE, Prakash S, Prakash L (1999) Efficient bypass of a thymine-thymine dimer by yeast DNA polymerase eta. Science 283:1001–1004
22 Masutani C, Kusumoto R, Yamada A, Dohmae N, Yokoi M, Yuasa M, Araki M, Iwai S, Takio K, Hanaoka F (1999) The XPV (xeroderma pigmentosum variant) gene encodes human DNA polymerase eta. Nature 399:700–704
23 Woodgate RA (1999) plethora of lesion-replicating DNA polymerases. Genes Dev 13:2191–2195
24 Cleaver JE (1999) Stopping DNA replication in its tracks. Science 285:212–213
25 Patel PH, Loeb LA (2001) Getting a grip on how DNA polymerases function. Nat Struct. Biol 8:656–659
26 Friedberg EC (2001) Why do cells have multiple error-prone DNA polymerases? Environ Mol Mutagen 38:105–110
27 Kiefer J, Feige M (1993) The significance of DNA double-strand breaks in the UV inactivation of yeast cells. Mutat. Res 299:219–224

28 Limoli CL, Ward JF (1994) Response of Bromodeoxyuridine-Substituted Chinese Hamster Cells to Uva Light Exposure in the Presence of Hoechst Dye Number-33258 – Survival and Dna Repair Studies. Radiat Res 138:312–319

29 Nowell PC (1976) The clonal evolution of tumor cell populations. Science 194:23–28

30 Loeb LA (1991) Mutator phenotype may be required for multistage carcinogenesis. Cancer Res 51:3075–3079

31 Morgan WF, Day JP, Kaplan MI, McGhee EM, Limoli CL (1996) Genomic Instability Induced By Ionizing Radiation. Radiat Res 146:247–258

32 Limoli CL, Corcoran JJ, Milligan JR, Ward JF, Morgan WF (1999) Critical target and dose and dose-rate responses for the induction of chromosomal instability by ionizing radiation. Radiat Res 151:677–685

33 Limoli CL, Ponnaiya B, Corcoran JJ, Giedzinski E, Morgan WF (2000) Chromosomal instability induced by heavy ion irradiation. Int J Radiat Biol 76:1599–1606

34 Limoli CL, Kaplan MI, Phillips JW, Adair GM, Morgan WF (1997) Differential induction of chromosomal instability by DNA strand-breaking agents. Cancer Res 57:4048–4056

35 Clutton SM, Townsend KMS, Walker C, Ansell JD, Wright, EG (1996) Radiation-induced genomic instability and persisting oxidative stress in primary bone marrow cultures. Carcinogenesis 17:1633–1639

36 Limoli CL, Hartmann A, Shephard L, Yang CR, Boothman DA, Bartholomew J, Morgan WF (1998) Apoptosis, reproductive failure, and oxidative stress in Chinese hamster ovary cells with compromised genomic integrity. Cancer Res 58:3712–3718

37 Hunt CR, Sim JE, Sullivan SF, Featherstone T, Golden W. VonKappHerr C. Hock RA, Gomez RA, Parsian AJ, Spitz DR (1998) Genomic instability and catalase gene amplification induced by chronic exposure to oxidative stress. Cancer Res 58:3986–3992

38 Vafa O, Wade M, Kern S, Beeche M, Pandita TK. Hampton GM, Wahl GM (2002) c-Myc can induce DNA damage, increase reactive oxygen species, and mitigate p53 function: a mechanism for oncogene-induced genetic instability. Mol Cell 9:1031–1044

39 Karanjawala ZE, Murphy N, Hinton DR, Hsieh CL, Lieber MR (2002) Oxygen metabolism causes chromosome breaks and is associated with the neuronal apoptosis observed in DNA double-strand break repair mutants. Curr Biol 12:397–402

40 Ames, BN, Shigenaga MK, Hagen TM (1993) Oxidants, antioxidants, and the degenerative diseases of aging. Proc Natl Acad Sci USA 90:7915–7922

41 Wallace DC (1999) Mitochondrial diseases in man and mouse. Science 283:1482–1488

42 King MP, Attardi G (1989). Human cells lacking mtDNA: repopulation with exogenous mitochondria by complementation. Science 246:500–503

43 Jacobson MD, Burne JF, King MP, Miyashita T, Reed JC, Raff MC (1993) Bcl–2 blocks apoptosis in cells lacking mitochondrial DNA. Nature 361:365–369

44 King MP, Attardi G (1996) Isolation of human cell lines lacking mitochondrial DNA. Methods Enzymol 264:304–313

45 Gamen S, Anel A, Montoya J, Marzo I, Pineiro A, Naval J (1995) mtDNA-depleted U937 cells are sensitive to TNF and Fas-mediated cytotoxicity. FEBS Lett 376:15–18

46 Marchetti P, Susin SA, Decaudin D, Gamen S, Castedo M, Hirsch T, Zamzami N, Naval J, Senik A, Kroemer G (1996) Apoptosis-associated derangement of mitochondrial function in cells lacking mitochondrial DNA. Cancer Res 56:2033–2038

47 Chang WP and Little JB (1991) Delayed reproductive death in X-irradiated Chinese hamster ovary cells. Int J Radiat Biol 60:483–496

48 Mothersill C, Seymour C (1997) Lethal mutations and genomic instability. Int J Radiat

Biol 71:751–758
49 Limoli CL, Kaplan MI, Corcoran J, Meyers M, Boothman DA, Morgan WF (1997) Chromosomal instability and its relationship to other end points of genomic instability. Cancer Res 57:5557–5563
50 Wallace DC (2001) Mouse models for mitochondrial disease. Am J Med Genet 06:71–93
51 Herzberg NH, Zwart R, Wolterman RA, Ruiter JP, Wanders RJ, Bolhuis PA, van den Bogert C (1993) Differentiation and proliferation of respiration-deficient human myoblasts. Biochim Biophys Acta 1181:63–67
52 Herzberg NH, Middelkoop E, Adorf M, Dekker HL, Van Galen MJ, Van den Berg M, Bolhuis PA, Van den Bogert C (1993) Mitochondria in cultured human muscle cells depleted of mitochondrial DNA. Eur J Cell Biol 61:400–408
53 Wang H, Morais R (1997) Up-regulation of nuclear genes in response to inhibition of mitochondrial DNA expression in chicken cells. Biochim Biophys Acta 1352:325–334
54 Kadhim MA, Macdonald, DA, Goodhead DT, Lorimore SA, Marsden SJ, Wright EG (1992) Transmission of chromosomal instability after plutonium alpha-particle irradiation. Nature 355:738–740
55 Lorimore SA, Kadhim MA, Pocock DA, Papworth D, Stevens DL, Goodhead DT, Wright EG (1998) Chromosomal instability in the descendants of unirradiated surviving cells after alpha-particle irradiation. Proc Natl Acad Sci USA 95:5730–5733
56 Azzam EI, de Toledo SM, Gooding T. and Little JB (1998) Intercellular communication is involved in the bystander regulation of gene expression in human cells exposed to very low fluences of alpha particles. Radiat Res 150:497–504
57 Huo L, Nagasawa H, Little JB (2001s) HPRT mutants induced in bystander cells by very low fluences of alpha particles result primarily from point mutations. Radiat Res 156:521–5
58 Deshpande A, Goodwin EH, Bailey SM, Marrone BL, Lehnert BE (1996) Alpha-particle-induced sister chromatid exchange in normal human lung fibroblasts: evidence for an extranuclear target. Radiation Research 145:260–7
59 Mothersill C, Seymour C (1997) Medium from irradiated human epithelial cells but not human fibroblasts reduces the clonogenic survival of unirradiated cells. Int J Radiat Biol 71:421–427
60 Mothersill C, Seymour CB (1998) Cell-cell contact during gamma irradiation is not required to induce a bystander effect in normal human keratinocytes: Evidence for release during irradiation of a signal controlling survival into the medium. Radiat Res 149:256–262
61 Bagdonas S, Dahle J, Kaalhus O, Moan J (1999) Cooperative inactivation of cells in microcolonies treated with UVA radiation. Radiat Res 152:174–179
62 Phillipson RP, Tobi SE, Morris JA, McMillan TJ (2002) UV-A induces persistent genomic instability in human keratinocytes through an oxidative stress mechanism. Free Radic Biol Med 32:474–480
63 Mothersill C, Seymour C (2001) Radiation-induced bystander effects: past history and future directions. Radiat. Res 155:759–767
64 Ward JF (2002) The radiation-induced lesions which trigger the bystander effect. Mutat Res 499:151–154
65 Tofilon PF. and Fike JR (2000). The radioresponse of the central nervous system: a dynamic process. Radiat Res 153:357–370
66 Monje ML, Mizumatsu S, Fike JR, Palmer TD (2002) Irradiation induces neural precursor-cell dysfunction. Nat. Med 8:955–962

Cellular and Molecular Aspects of the Anti-inflammatory Action of Low Dose Ionizing Irradiation

Peter M. Kern

Dept. of Internal Medicine/Rheumatology,
Franz von Pruemmer Klinik, Bad Brueckenau, Germany
peter.kern@franz-von-pruemmer-klinik.de

Abstract

Low dose radio therapy (LD-RT) has been used since the early 30's of the past century for the amelioration of inflammatory processes, e.g. abscesses or arthritidies. The mechanism of action, however, remained completely unclear. We have shown in vitro that LD-RT in the dose range between 0.3 and 0.7 Gy inhibits adhesion of lymphocytes to endothelial cells by discontinuous induction of apoptosis and consecutive shedding of L-selectin, downregulates the expression of E-selectin on endothelial cells by induction of TGFβ1 and induces an antiinflammatory cytokine profile in lymphocytes with a decrease of TNFα production and an increase of IL-10 production. This is the first basic scientific approach to an explanation of the antiinflammatory action of LD-RT. With the growing knowledge of the tremendous risks of prostaglandine inhibitors for lethal gastrointestinal bleedings, it may prepare a way for LD-RT back to clinical use in inflammatory conditions that are currently treated with these drugs.

Introduction: Anti-inflammatory action of drugs and radiation

Considerable research has been undertaken during the past decades to study the effects of high dose irradiation on malignant as well as normal cells and tissues. Despite its biologically interesting inverse effect of ameliorating inflammation [20 instead of inducing it, little attention has been paid to the low dose range between 0 and 1 Gy, at least since the late 1960ies. Only in some countries, predominantly in the eastern hemisphere, low dose radiotherapy (LD-RT) was maintained as a means of anti-inflammatory therapy in selected conditions (predominantly abscesses, arthritidies, soft tissue inflammation), probably because of the low costs of that treatment modality.

In most western countries, however, these modalities were abandoned because of the risk of inducing malignancy following the studies on LD-RT in patients with ankylosing spondylitis (see e.g. [21]). Instead one preferred treatment by the new class of Non Steroidal Anti-Inflammatory Drugs (NSAIDs). For 30 years NSAIDs have been the most frequently prescribed drugs.

It was only in the last few years that we became aware of the dimension of severe gastrointestinal side effects including lethal bleeding with 1000 to 2000 deaths per year in Germany [3] due to the widespread use of these drugs. On the other hand, radiobiology research has helped to develop application and dosing techniques for therapeutic irradiation that minimize the risk of malignancies.

The time may have thus come to reevaluate the use of LD-RT for anti-inflammatory intervention. Some groups still have been working on the clinical application of LD-RT, producing encouraging evidence for its safety and efficacy [12, 13, 17]. The main obstacle for LD-RT to compete with drugs, however, has been the lack of any plausible hypothesis on the mode of action.

References:

- Ibuprofene [10]
- Benoxaprofene [22]
- Tenidap [23]
- Methylprednisolone [11]
- Gold [9]
- Piroxicam, Phenylbutazone, Diclofenac, Indometacine, ASA [5]

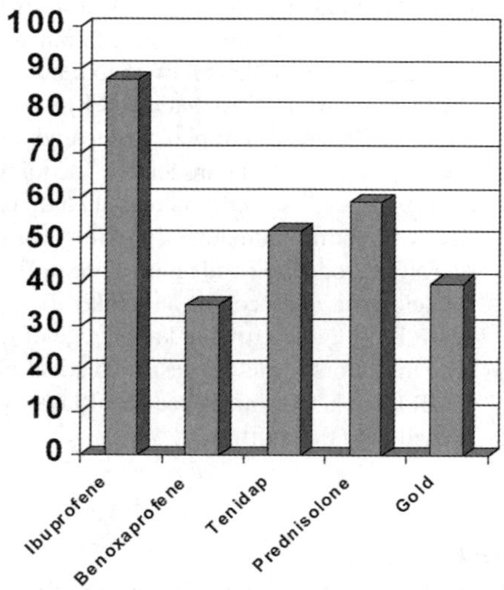

Fig. 1. Reduction of adhesion by antiinflammatory drugs.
Reduction of adhesion of leukocytes to IL1β -activated endothelial cells by antiinflammatory drugs in vitro. The bars indicate the precentage of reduction of adherent cells as compared to the non treated control.

Basic science has provided a rapidly expanding understanding of the role of adhesion for the regulation of inflammation [18]. Adhesion describes the interaction of blood leukocytes and endothelial cells that leads to the arrest of leukocytes at the vessel wall, thus allowing for the subsequent diapedesis. Selectins, a family of adhesion molecules, seem to play a key role in initiating adhesion [1]. While they are constitutively expressed on the surface of leukocytes [6], endothelial cells will express them only upon "activation" through pro-inflammatory cytokines such as IL-1β or TNF-α [2]. Adhesion of leukocytes to endothelial cells and its dependence on the cytokine network have become a focus of interest for anti-inflammatory intervention. Numerous drugs, including corticosteroids and NSAIDs, have been shown to exert their anti-inflammatory effects in part by reducing adhesion between leukocytes and endothelial cells [5, 9, 10, 11] (Fig. 1., page 46).

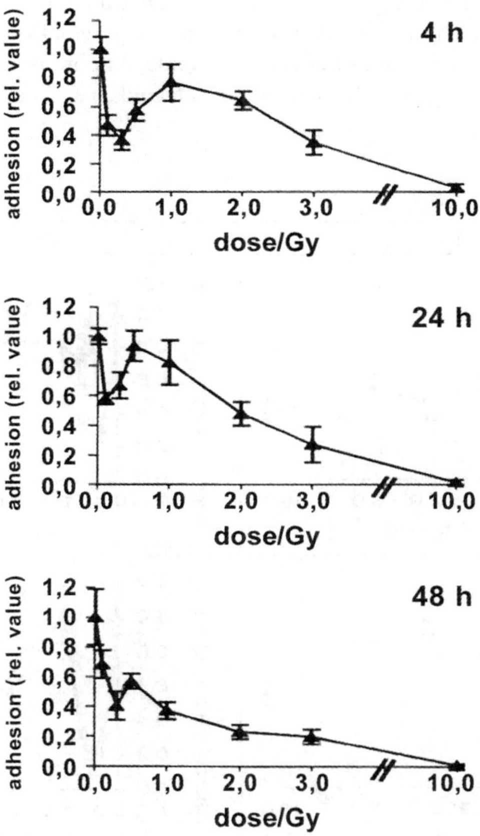

Fig. 2. Adhesion of PBMC to EC – LD-RT of PBMC.
Adhesion of peripheral blood mononuclear cells (PBMC; predominantly lymphocytes) to IL1b-activated endothelial cells after LD-RT of PBMC prior to the adhesion assay.
The curves indicate the proprtion of adherent cells as compared to the non irradiated control (0Gy, relative value 1.0).

To obtain a better insight into the underlying mechanism, we investigated the influence of LD-RT on adhesion in a modified in vitro assay [8] using lymphocytes isolated from the peripheral blood of normal healthy donors and a murine endothelial cell line that had been shown previously to behave identically to human endothelial cells [7]. We could demonstrate that LD-RT, both of lymphocytes or endothelial cells, significantly decreases adhesion especially in the dose range between 0.3 and 0.7 Gy which is exactly the dose range applied therapeutically in the clinical use (Fig. 2. and 3.).

Apoptosis of lymphocytes can be easily induced by irradiation [4], especially when they are activated [19]. LD-RT is able to induce apoptosis in lymphocytes, but very surprisingly not in a dose dependent manner as would have been expected. There is a relative maximum of irradiation induced apoptosis exactly in the dose range between 0.3 and 0.7 Gy which can be detected as early as 12 hours after irradiation [14] (Fig. 4.).

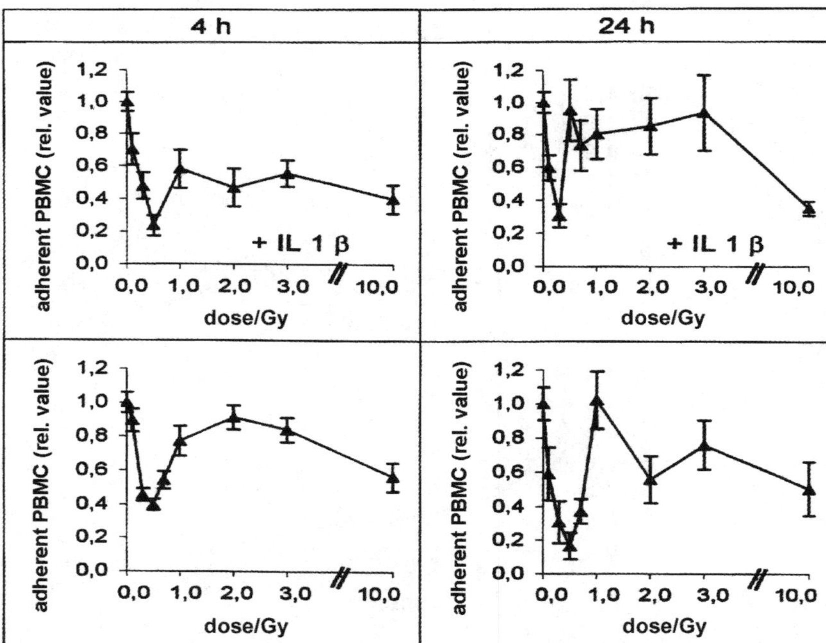

Fig. 3. Adhesion of PBMC to EC – LD-RT of EC.
Adhesion of peripheral blood mononuclear cells (PBMC; predominantly lymphocytes) to IL1β-activated endothelial cells after LD-RT of endothelial cells prior to the adhesion assay. The curves indicate the proprtion of adherent cells as compared to the non irradiated control (0 Gy, relative value 1.0).

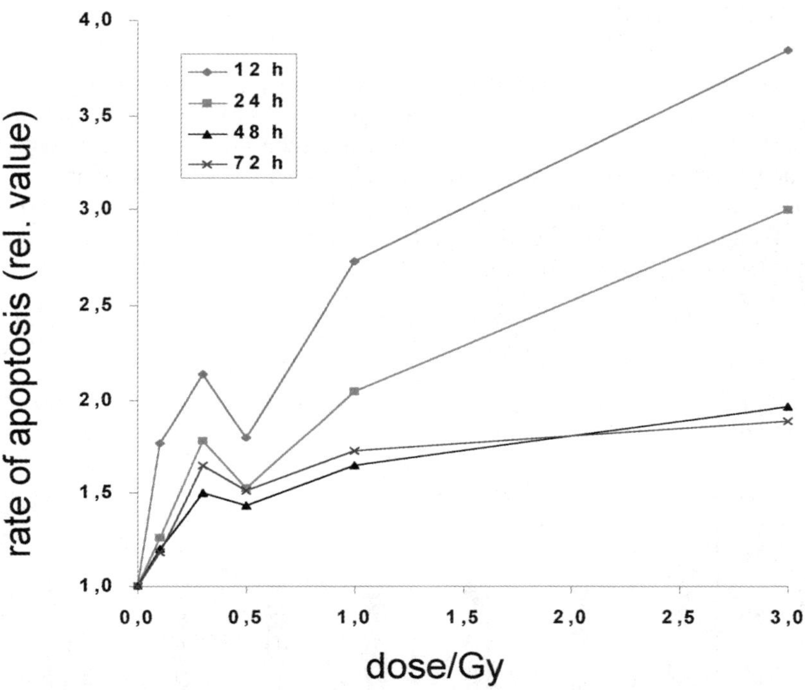

Fig. 4. Induction of apoptosis in PBMC after LD-RT (Triton-PI staining).
Induction of apoptosis in periperal blood mononuclear cells (PBMC; predominantly lymphocytes) after LD-RT.
The curves indicate the rate of apoptosis as compared to the non irradiated control (0 Gy, relative value 1.0) at the time points given. Apoptosis was determined by propidiuum iodide (PI) staining of the nuclei.

Experiments on the mechanism of anti-inflammatory radiation action

4 hours after irradiation, lymphocytes bound for apoptosis start losing L-selectin as could be shown by the loss of expression on the cell surface (Fig. 5.) and simultaneously by an increase of the soluble molecule in the supernatant of the cultured lymphocytes (data not shown). These cells also down regulate the expression of TNFα, a major pro-inflammatory cytokine, and upregulate the expression of the anti-inflammatory cytokine IL-10, as can be demonstrated both on the protein level and on the mRNA level. [15] (Fig. 6. and 7).

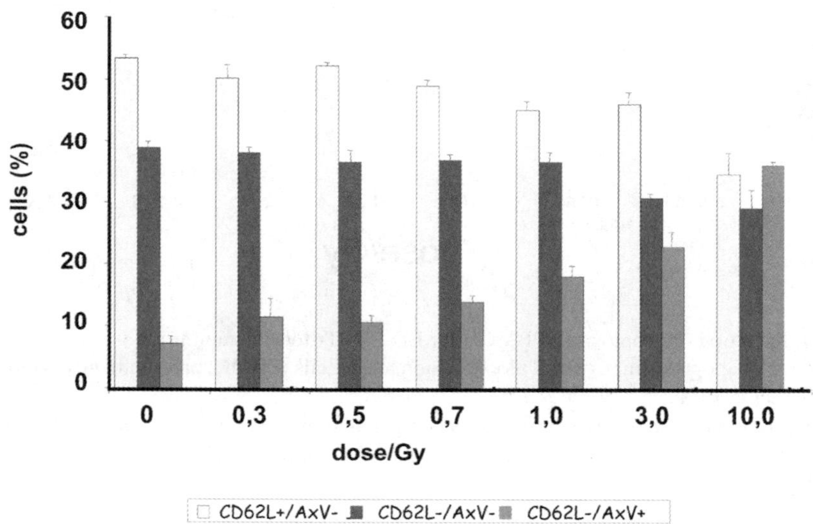

Fig. 5. Expression of L-selectin in PBMC after LD-RT.
Expression of L-selectin (CD62L) in peripheral blood mononuclear cells (PBMC; predominantly lymphocytes) 24hrs after LD-RT. The bars indicate the percentage of the the respective population according to the dose applied. Positive Annexin V (AxV) staining identifies cells undergoing apoptosis. Only vialble cells (AxV-) express L-Selectin (CD62L). There is no cell undergoing apoptosis (AxV+) and still expressing L-selectin (CD62L), indicating that induction of apoptosis is associated with the loss of L-selectin.
Again, the discontinuity in the rate of apoptosis (AxV+) with a relative maximum at 0.3 Gy can be seen.

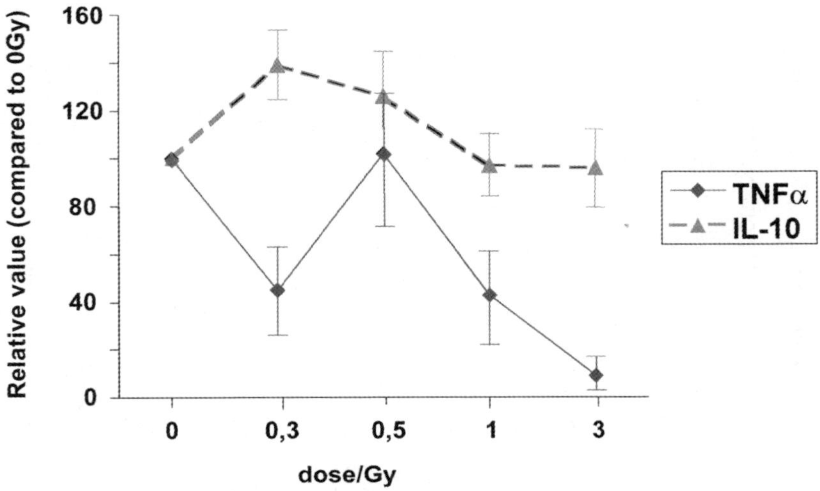

Fig. 6. Discontinuous Expression of TNFα and IL-10 in PBMC after LD-RT.
Expression of TNFα (proinflammatory cytokine) and IL-10 (antiinflammatory cytokine) in peripheral blood mononuclear cells (PBMC; predeminantly lymphocytes) after LD-RT. The curves indicate the amount of protein detected in the cell culture supernatant by ELISA as compared to the non irradiated control (0 Gy; relative value 100%). There is a relative maximum of IL-10 and a relative minimum of TNFα at 0.3 Gy.

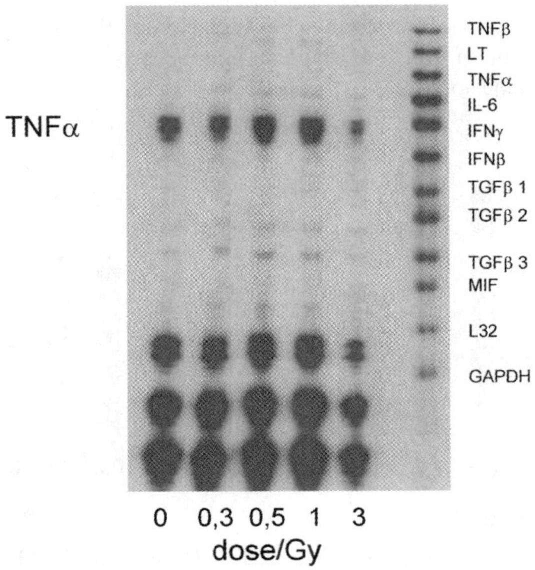

Fig. 7. Expression of TNFα mRNA in PBMC after LD-RT.
Corresponding picture to Fig. 6. on the mRNA level with a relative minimum of mRNA at 0.3 Gy.

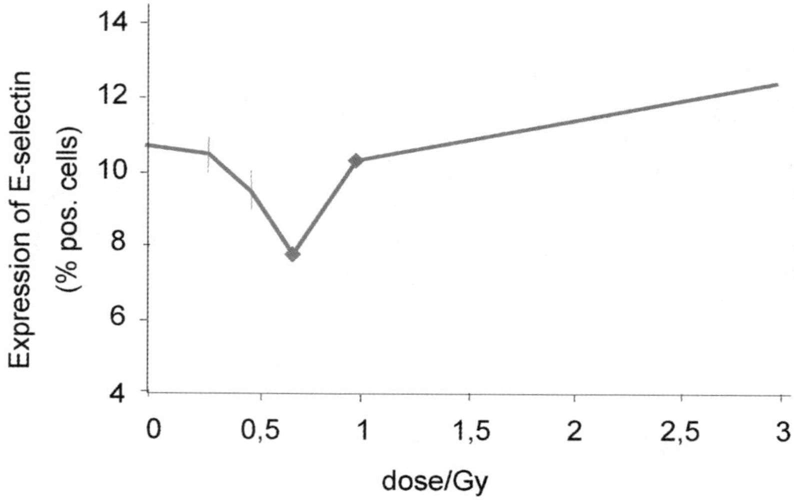

Fig. 8. Reduced Expression of E-selectin after LD-RT.
Expression of E-selectin on endothelial cells after LD-RT. The curve indicates the percentage of cells expressing E-selectin. There is a relative minimum between 0.3 and 0.7 Gy.

Irradiated endothelial cells undergo similar processes with loss of E-selectin (Fig. 8.) and upregulation of IL-6 and TGFβ_1 (Fig. 9. and 10), a cytokine that renders endothelial cells inactive and nonadhesive [16]. Again, this effect is markedly expressed in the therapeutical dose range between 0.3 and 0.7 Gy, indicating a certain regulatory function of this dose range on lymphocytes and endothelial cells in terms of inhibiting adhesion.

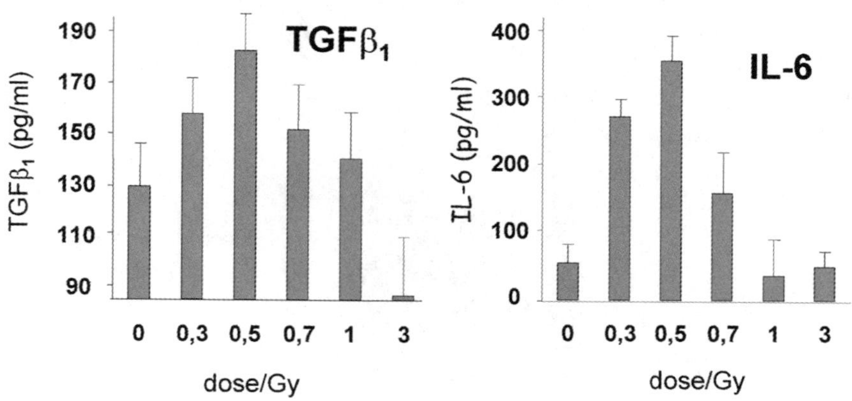

Fig. 9. IL-6 and TGFβ_1 in the supernatant of Endothelial Cells after LD-RT.
Expression of TGFβ_1 and IL-6 (both antiadhesive cytokines) in endothelial cells after LD-RT. The bars indicate the amount of protein detected in the cell culture supernatant by ELISA. There is a relative maximum of expression in the dose range between 0.3 and 0.7 Gy.

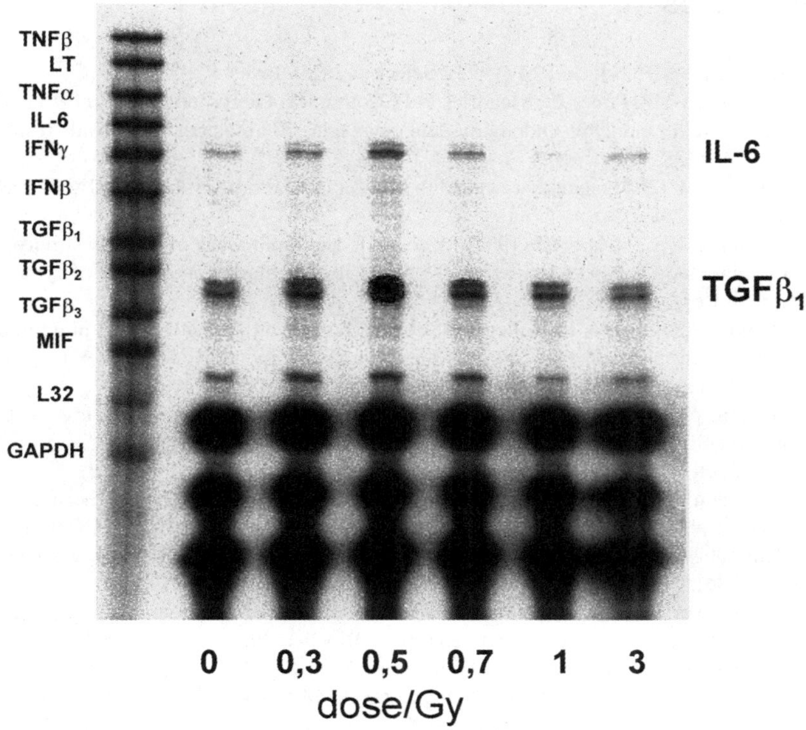

Fig. 10. Expression of IL-6 and TGFβ_1 mRNA in Endothelial Cells after LD-RT. Corresponding picture to Fig. 9. on the mRNA level with a relative maximum around 0.5 Gy.

Conclusions

These data are the first approach to a scientific explanation of the anti-inflammatory effect of LD-RT based on immunology and adhesion biology. They may help to satisfy our needs to understand the mechanisms of therapeutic interventions. Together with the up-to-date technical devices for minimizing the risks of irradiation and the growing awareness of the rate of lethal side effects of NSAIDs, it may help to give way of LD-RT back to its clinical use in anti-inflammatory therapy. This reopens a window for radiobiology as a basic and an applied science in relation to the treatment of benign diseases that had been closed 40 years ago.

References

1 Bevilacqua MP, Nelson RM (1993). Selectins J Clin Invest 91: 379.

2 Bevilacqua MP, Pober JS, Mendrick D L, Cotran RS, Gimbrone MA, Jr. (1987). Identification of an inducible endothelial-leukocyte adhesion molecule Proc Natl Acad Sci U S A 84: 9238.

3 Bolten WW (2001). Number needed to kill. Individual drug risk with NSAIDs Z Rheumatol. 60: 288.

4 Crompton NE, Ozsahin M (1997). A versatile and rapid assay of radiosensitivity of peripheral blood leukocytes based on DNA and surface-marker assessment of cytotoxicity Radiat Res 147: 55.

5 Dapino, P, Ottonello, L, Dallegri F (1994). The anti-inflammatory drug nimesulide inhibits neutrophil adherence to and migration across monolayers of cytokine-activated endothelial cells Respiration 61: 336.

6 Gallatin WM, Weissman IL, Butcher EC (1983). A cell-surface molecule involved in organ-specific homing of lymphocytes Nature 304: 30.

7 Hallmann R (1997). Habilitationsschrift Universität Erlangen-Nürnberg .

8 Hallmann R, Jutila MA, Smith CW, Anderson DC, Kishimoto TK, Butcher EC (1991). The peripheral lymph node homing receptor, LECAM-1, is involved in CD18-independent adhesion of human neutrophils to endothelium Biochem Biophys Res Commun 174: 236.

9 Heimburger M, Lerner R, Palmblad J (1998). Effects of antirheumatic drugs on adhesiveness of endothelial cells and neutrophils Biochem Pharmacol 56: 1661.

10 Hofbauer R, Hofbauer L, Moser D, Salfinger H, Speiser W, Frass M, Kapiotis S (1998). Leukocyte migration: a new triple migration chamber assay allows investigation of various cell interactions simultaneously Life Sci 63: 1725.

11 Kadioglu A, Sheldon P (1998). Steroid pulse therapy for rheumatoid arthritis: effect on lymphocyte subsets and mononuclear cell adhesion Br J Rheumatol 37: 282.

12 Keilholz L, Seegenschmiedt H, Sauer R (1998). [Radiotherapy for painful degenerative joint disorders. Indications, technique and clinical results] Strahlenther Onkol 174: 243.

13 Keilholz L, Seegenschmiedt MH, Kutzki D, Sauer R (1995). [Periarthritis humeroscapularis (PHS). Indications, technique and outcome of radiotherapy] Strahlenther Onkol 171: 379.

14 Kern P, Keilholz L, Forster C, Seegenschmiedt MH, Sauer R, Herrmann M (1999). In vitro apoptosis in peripheral blood mononuclear cells induced by low-dose radiotherapy displays a discontinuous dose-dependence Int J Radiat Biol 75: 995.

15 Kern PM, Keilholz L, Forster C, Hallmann R, Hermann M, Seegenschmiedt M (2000). Low-dose radiotherapy selectively reduces adhesion of peripheral blood mono-nuclear cells to endothelium in vitro Radiother Oncol 54: 273.

16 Roedel F, Kley N, Beuscher HU, Hildebrandt G, Keilholz L, Kern P, Voll R, Herrmann M, Sauer R (2002). Anti-inflammatory effect of low-dose X-irradiation and the involvement of a TGF-beta1-induced down-regulation of leukocyte/endothelial cell adhesion Int J Radiat Biol 78: 711

17 Seegenschmiedt MH, Keilholz L (1998). Epicondylopathia humeri (EPH) and peritendinitis humeroscapularis (PHS): evaluation of radiation therapy long-term results and literature review Radiother Oncol 47: 17.

18 Springer TA (1994). Traffic signals for lymphocyte recirculation and leukocyte emigration: the multistep paradigm Cell 76: 301.

19 Stewart CC, Stevenson AP, Habbersett RC (1988). The effect of low-dose irradiation on unstimulated and PHA-stimulated human lymphocyte subsets Int J Radiat Biol 53: 77.
20 von Pannewitz G (1933). Die Röntgentherapie der Arthritis deformans. Klinische und experimentelle Untersuchungen Erg. Med. Strahlenforschung 6: 62.
21 Weiss HA, Darby SC, Doll R (1994). Cancer mortality following X-ray treatment for ankylosing spondylitis. Int J Cancer 59: 327–338.
22 Brown KA, Ferrie J, Wilbourn B, Dumonde DC (1989). Benoxaprofen inhibits the adhesion of human monocytes to cultured vascular endothelium. Agents Actions 28: 115–120.
23 Kyan-Aung U, Lee TH, Haskard DO (1993). The inhibitory effect of tenidap on leukocyte-endothelial cell adhesion. J Rheumatol 20: 1014–1019.

The Network of Radiation Responses and Genomic Stability

Friederike Eckardt-Schupp, Simone Mörtl, and Eberhard Fritz

GSF-Institute of Molecular Radiobiology, P.B. 1129, 85758 Neuherberg, Germany
eckardt-schupp@gsf.de

Abstract.

The genetically determined balance of genomic stability and variation has been very important in evolution. The maintenance of this balance is essential for human health; disturbances give multiple causes for disease. In the past, radiobiological research contributed significantly to the elucidation of the mechanisms for genomic stability. Future research will continue to identify genes and polymorphisms of functional relevance as well as the networks of molecular and cellular responses to radiation. Thus, radiobiology will contribute to an improved understanding of the molecular and genetic basis of human disease and of individual responses to radiation and other environmental insults.

Introduction

Elucidation of the processes of genetic stability is one of the central aspects of modern radiobiological and biomedical research. Heredity is a manifestation of the stability of genes from one generation to the next. Genetic variation, on the other hand, is achieved by the instability of genes and the "fluidity" of the genomes, thus acting as the driving force of evolution and, in the worst and ultimate case, causing human disease. In times when the nature of the heriditary material was still a matter of speculation, physicists like Ernst Schrödinger and Max Delbrück were intrigued by its presumptive physico-chemical stability under physiological conditions. They postulated an unusual thermodynamic stability of the genetic substance capable of extremely accurate replication in order to guarantee inheritance [15].

Nowadays it is proven that the genetic material consists of desoxyribonucleic acid (DNA), a polymer with the "normal" thermodynamic features of organic macromolecules. Thus, DNA is amazingly unstable under physiological conditions, despite its stabilizing double-helical structure. Endogenous DNA damage occurs at a frequency of about 20.000 events per human cell per day [29]. However, the genetic programmes of pro- or eukaryotic cells encode an enormous variety of enzymatic processes which can cope with the permanent spontaneous decay of DNA and the insults of numerous endogenous and exogenous DNA-damaging agents (for comprehensive information see [9, 34, 5]).

The complexity of molecular and cellular responses to radiation

Comprehensive damage surveillance and repair processes have evolved to maintain the macromolecular organisation of pro- and eukaryotic cells. Upon specific recognition of radiation-induced macromolecular damages, complex signal transduction pathways trigger a variety of possible cellular responses (Fig.1).

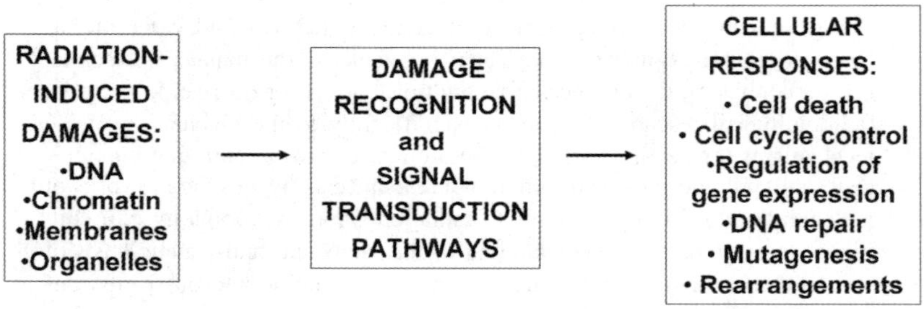

Fig. 1. Network of responses to radiation-induced damages

Most importantly, repair, which normally counterbalances the damaging effects of radiation on the molecular level, is coordinated and fine-tuned with radiation responses at the cellular level. E.g., radiation-induced cell cycle regulation provides additional time for DNA damage repair [8] and transcription-coupled repair allows preferential repair of active genes [41] Cell suicide mechanisms eliminate cells that have sustained unrepairable damage or severe deterioration of their macromolecular organisation [20, 54].

In the case of DNA damage, insufficient or error-prone repair leading to mutations and genome rearrangements, or failure to destroy damaged cells, may result in the clonal expansion of genetically aberrant somatic cells possibly leading to neoplastic transformation and cancer. In meiotic cells, such failure may allow transmission of mutations through the germ-line.

In view of genetic stability and inheritance, radiation-induced damage and repair of DNA molecules have been in the focus of radiobiological research in the last decades. However, knowledge about radiation-mediated changes in the supramolecular organization of membranes and about cellular organelles as targets of irradiation gain increasing importance [45]. Thus, for an integral understanding of the counterbalances between restorative processes and suicide mechanisms in radiation-exposed cells, all putative cellular radiation targets have to be considered.

Adding to the complexity of signal transduction pathways, changes in radiation responses can also be due to alterations of pathways dependent on cytokine receptors, which trigger pro- and anti-apoptotic signals. As exemplified for TNFR (tumour necrosis factor a receptor), radiation effects on membranes may lead to modifications of signal transduction pathways [6]. It is of high interest to analyse the relationship of signal transduction initiated by damages of membranes and organelles with those initiated by DNA/chromatin damage and to elucidate whether a cross talk exists between DNA damage-induced and membraneous signals. The understanding of the complex network of signal transduction pathways seems to be of particular importance for the analysis of intercellular communication which commonly is known as "bystander effect" [2].

There are still many open questions regarding the detailed molecular basis of the complete network of radiation responses. For instance, how is the large variety of different types of DNA damages specifically recognized? How does a cell achieve a differential response to low and high radiation doses resp. dose-rates – does it "count" the DNA damages and other molecular damages despite their large variety? How does this damage sensing – specific recognition and "counting" – influence the choice of further signal transduction pathways? How are the different types of damages channeled into "their" specific repair pathways? Specific responses may depend on the physiology of the cell, the status of nutrition, chromatin condensation, replication or transcription, etc.. How does a particular signal transduction pathway relate to the final cellular response? The responses can be numerous and, consequently, decision making will be complex.

The role of radiobiology for elucidating the mechanisms of genomic stability

Ionizing radiation and UV light, due to its direct interaction with DNA, have always been extremely useful tools for the identification and characterization of the numerous types of DNA damages which endanger the integrity of the DNA structure and function. Many of these types of DNA damage are also induced by chemical agents as well as by radicals and reactive oxygen species (ROS) [42]. The latter occur spontaneously during cellular metabolism, supporting the notion of DNA decay under physiological conditions, and they are generated as well by chemicals and radiation.

The genetic, biochemical and functional characterization of the multiplicity of enzymatic processes dealing with DNA damage in living cells has been one of the most important issues of molecular radiobiology over the last decades. Radiobiologists have accumulated tremendous knowledge on different types of DNA damage, DNA repair pathways coping with these damages and the complexity of cellular responses. DNA repair processes belong to the "basic equipment" of pro- and eukaryotic cells and constitute integral parts of the network of DNA metabolic processes, i.e. replication, transcription and recombination, and all of them have been highly conserved in evolution.

There is no doubt that DNA damage repair and tolerance mechanisms as well as the processes which guarantee the fidelity of DNA replication are extremely important for the maintenance of genomic stability. On the other hand, by their genetically controlled error-proneness, e.g. limited accuracy, they participate in those processes which are the driving forces of evolution [15]. Thus, radiobiological research has always contributed to the understanding of how mechanisms controlling mutations, chromosomal rearrangements, genomic instability and the integration of DNA fragments are kept in an optimized balance, specific for each cell type and organism, by evolution. Disturbance of this balance in man, for instance by increased exposure to radiation and other environmental genotoxic agents, is undoubtedly one cause for human disease.

Likewise, radiobiological research will profit enormously from the use of molecular biological technologies, particularly of genomics and proteomics, of modern imaging techniques and of new irradiation devices like microbeams, which allow to irradiate single cells and monitor their response. Such new approaches will help to unravel the molecular basis of radiation effects on cells and organisms, many of which had been described thoroughly by classical radiobiology. By combination of the classical with the new methods, highly integrated perspectives of the molecular, cellular and organismic responses to radiation will be achieved.

Networks of DNA double-strand break repair pathways

In order to cope with the constant threat of a great variety of DNA damages cells have evolved different strategies of DNA repair. DNA double-strand breaks (DSB) will be used to illustrate the system of mechanistically diverse repair pathways which are highly conserved in evolution. As compared to radiation-induced base damages, DSB are low in number, however, they bear a high cytotoxic, mutagenic and tumorigenic potential if they are misrepaired or remain unrepaired [35, 25]. DSB arise endogenously during replication, as intermediates during various recombination processes (meiosis, V(D)J recombination and class switching) and due to radicals produced in normal cell metabolism. Exogenously, DSB are induced by ionizing radiation, radicals and chemicals [Jackson 2002; Pfeiffer et al. 2000].

Repair processes dealing with DSB can be formally divided in homology-dependent and homology-independent processes [Kanaar et al. 1998, Karran 2000, Pastink et al. 2001]. Their genetic control and molecular mechanisms differ, though they share a number of proteins which contribute to the processing of break-induced free DNA ends and to the modeling of chromatin structure. Therefore, DSB repair pathways contribute, to different degrees, to the generation of genetic alterations, by their genetically programmed failure in restoring broken DNA molecules absolutely correctly. Furthermore, the efficiency and mode of DNA integration relies on the activities of DSB repair processes [Kiechle et al. 2000; Kiechle et al. 2002].

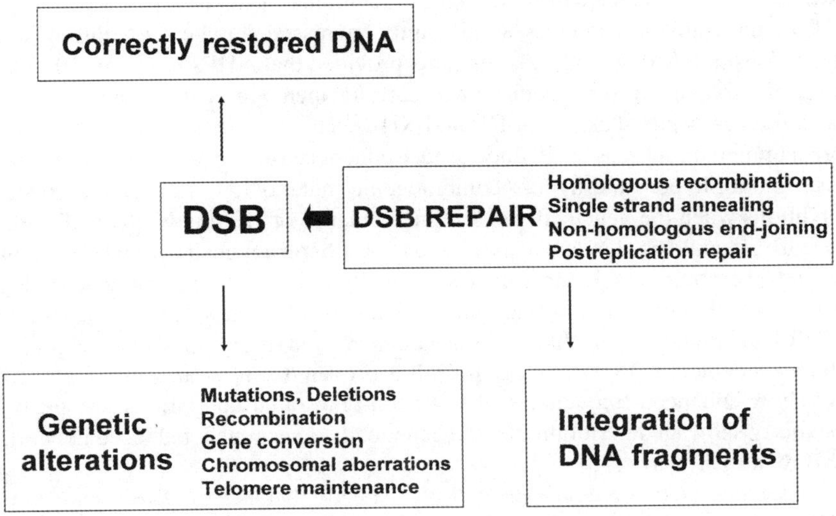

Fig. 2. Network of repair pathways for DNA double-strand breaks (DSB) and their roles for the generation of genetic alterations

Homologous recombination repair (HR), also called homology-dependent repair in mammalian cells, requires long stretches of homology of several hundreds of base pairs in a second molecule, at best in a sister chromatid with identical sequence or in a homologous chromosome. At the DSB site, repair by HR is correct according to the sequence homology of the template DNA, thus, gene conversion occurs if some bases of the template DNA differ. HR can be accompanied by the recombination of outside markers and may lead to chromosomal rearreangements if streches of homology are used on non-homologous chromosomes [14]. Single-strand annealing (SSA) depends on homology stretches (direct repeats) in the vicinity of the break, and DSB repair by SSA is always accompanied by deletion of one of the repeats and the intervening sequence [36]. The major homology-independent DSB process is so-called non-homologous end joining (NHEJ). DNA ends without any homology or with microhomologies at the ends are rejoined by NHEJ [3]. This may lead to mutations and deletions in the joining region, and due to the joining of "false" DNA ends to chromosomal aberrations. In general, NHEJ is considered to be more error-prone than HR.

How is the decision made on the choice of the specific repair mechanism for each individual DSB? This question is of high interest due to various reasons. The basic genetic repertoire of DSB repair genes is very similar in all types of cells, but the relative importance of HR and NHEJ differs considerably. For instance, in the lower eukaryote yeast Saccharomyces cerevisisae, HR is the dominant DSB repair pathway, and NHEJ serves as back-up only in situations when HR is impossible [43]. In contrast, mammalian cells efficiently make use of NHEJ, however, DSB are extremely potent inducers of homologous recombination as well [19, 22].

Gamma-irradiation induces significantly increased numbers of chromosomal aberrations in HR-deficient yeast mutants provided that NHEJ is active [10]. Obviously, the frequencies of chromosomal rearrangements in yeast are determined by the cell type-specific balance of HR and NHEJ. This hypothesis seems to hold true for mammalian cells as well. Increased frequencies of chromosomal aberrations arise in both hyper- and hypo-recombinogenic mutants [31, 12], maybe under all conditions when the genetically determined balance of HR and NHEJ is disturbed. Recently, it could be shown that extensive loss of **heterozygosity** (LOH), a common genetic alteration in tumour cells, is normally suppressed during DSB-induced allelic recombination in dividing mammalian cells. This indicates that DSB repair by HR in mammalian cells like in yeast safeguards the structural interity of genomes which encountered DNA breakage [46]. It is proven that tumour cells which generally show enhanced frequencies of LOH, chromosomal aberrations and genomic instability show an deterioration in the genetically determined, balanced network of DSB repair pathways [25].

And which factors regulate the cell-type specific balance of DSB repair by HR, NHEJ and, possibly, SSA which as a putative cause of LOH may play a major role in tumour than in normal cells? Work with the "model eukaryote" yeast indicates that ubiqitination is a crucial factor for the response to DNA damage in general [16, 49, 50]. Genes coding for ubiquitin ligases are required for maintaining the balance of HR and end joining in yeast – RAD5 [1], for protection against genomic instability of embryonic stem cells in mice – RAD18 [47], and for avoidance of cancer in man – BCRA1, just to mention a few examples [13; 11; 7]. Improved knowledge on the role of ubiquitination for the maintenance of genomic stability in human cells may help to further elucidate the basic mechanisms of radiation-induced cancerogenesis, and possibly, may be used for improved diagnosis and treatment of tumours.

Effects of disturbances of genomic stability on human health

A considerable number of human DNA repair genes are known in structure and cellular function [53; 48]. Frequently, these genes have been identified due to their high conservation during evolution, and it is proven that the organisation into pathways of their proteins is conserved from yeast to man. Additionally, the analysis of human hereditary diseases like Xeroderma pigmentosum, Ataxia telangiectasia, Nijmegen breakage syndrome, Blooms syndrome and others has added enormously to our knowledge on the role of DNA repair in human cells and individuals [51; 4].

The pleiotropic phenotypes of cells from such affected individuals confirm the interconnection of DNA damage repair and tolerance pathways within the cellular response network for the maintenance of genomic stability [17].

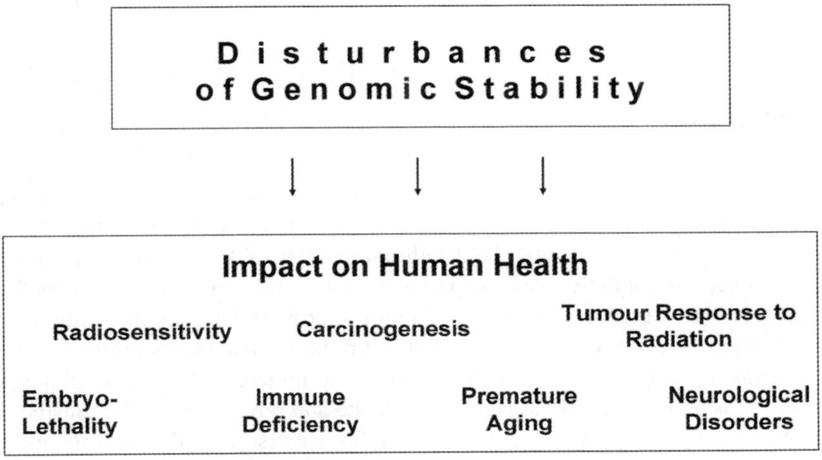

Fig. 3. Effects of disturbances of genomic stability on human health

The pleiotropic clinical phenotypes of these so-called repair deficiency and chromosomal instability syndromes have given sound evidence for the fact that efficient and accurate DNA repair processes are essential for the maintenance of human health (Fig. 3). Patients suffering from the chromosomal instability disorder Ataxia telangiectasia (A-T) are clinically characterized by increased radiosensitivity, immunodeficiency, neurological defects, premature aging and increased tumour incidence [44], thus combining the pleiotropic clinical phenotypes outlined below in one entity. Functional characterization of the gene underlying this disease, ATM, revealed that the encoded lipid kinase domain functions as a protein kinase involved in signalling DNA damage and trigger cellular responses. Since DNA repair functions are attributed to the ATM protein as well, AT may thus be an example of how both signalling and DNA repair functions encoded in one protein may control genomic stability in response to DNA damage. Furthermore, good evidences exist for links of DNA repair with immune gene diversity and immunodeficiency [21; 39], premature aging [52] and neurological disorders [40].

Polymorphisms in human genes encoding damage surveillance and repair processes could easily lead to varying capacities of radiation responses. In addition to those genes which have a severe impact if mutated and which have been identified by the analysis of repair deficiency and cancer syndromes, a large number of modifier genes can be expected. It will be a demanding goal for molecular epidemiology to identify all these genes and their functional polymorphisms, and, based on these data, molecular radiobiology and radiomedicine to provide markers and assays that will allow evaluation of an individual's ability to respond to radiation [30]. Future research aims at providing individualized strategies for radioprotection and radiotherapy. Finally, this knowledge will provide a basis for the estimation of risk to human health due to genomic instabilities occurring in either somatic or germ cells [28].

Conclusions

In the past, radiobiology has been very important for the generation of mutations in order to identify controlling genes and to elucidate the factors and pathways which determine the molecular and cellular responses to radiation damage. In future, the network of these response pathways and its regulation will be in the focus of research. With the powerful tools of genomics, proteomics, genome-wide screening techniques and the information avaible from the Human Genome Project, modern radiobiological research will gain a complete understanding of all genes and their proteins which control damage surveillance and repair. This will contribute to an improved awareness of the genetic basis of human diseases. The identification of functional polymorphisms will help to define the molecular basis of individual susceptibility to environmental and medical radiation exposure and of the responses of tumours to radiation and chemotherapy. Like in the past, the various areas of radiation research will maintain and even gain more importance by their interplay with a number of other disciplines like toxicology and pharmacology, cancer research and tumour biology, gene therapy, immunology, molecular epidemiology and others.

References

1 Ahne F, Jha B, Eckardt-Schupp F (1997) The RAD5 gene product is involved in the avoidance of non-homologous end-joining of DNA double strand breaks in the yeast Saccharomyces cerevisiae. Nucl Acid Res 25:743–749

2 Ballarini F, Biaggi M, Ottolenghi A, Sapora O (2002) Cellular communication and bystander effects: a critical review for modelling low-dose radiation action. Mutat Res 501:1–12

3 Critchlow SE, Jackson SP (1998) Endjoining from yeast to man. Trends Bio-chem Sci. 23:394–398

4 De la Torre C, Pincheira J, Lopez-Saez JF (2003) Human syndromes with genomic instability and multiprotein machines that repair DNA double-strand breaks. Histol Histopathol 18:225–243

5 Dizdaroglu M. and Karakaya A.E. (eds.) (1999) Advances in DNA Damage and Repair, Academic/Plenum Press, New York

6 Eichholtz-Wirth H, Sagan D (2002) Altered signaling of TNFa-TNFR1 and SODD/BAG-4 is responsible for radioresistance in human HT-R15 cells. Anti-cancer Res 22:235–240
7 Fang S, Lorick KL, Jensen JP, Weissman AM (2003) RING finger ubiquitin ligases: implications for tumorigenesis, matastasis and for molecular targets in cancer. Semcancer 13:5–14
8 Favaudon V (2000) Regulation of cell cycle and radiation-induced cell death. Cancer Radiother 4:355–368
9 Friedberg EC, Walker GC, Siede W (1995) DNA Repair and Mutagenesis. ASM Press, Washington D.C.
10 Friedl AA, Kiechle M, Fellerhoff B, Eckardt-Schupp F (1998) Radiation-induced chromosome aberrations in Saccharomyces cerevisiae: influence of DNA repair pathways. Genetics 148:975–988
11 Futaki M, Liu JM (2001) Chromosomal breakage syndromes and the BRCA1 genome surveillance complex. Trends Molec Med 7(12):560–565
12 Godthelp BC, Wiegant WW, van Duijn-Goedhart A, Schärer OD, van Buul PPW, Kanaar R, Zdzienicka MZ (2002) Mammalian Rad51C contributes to DNA cross-link resistance, sister chromatid cohesion and genomic stability. Nucl Acid Res 30:2172–2182
13 Gonzalez R, Silva JM, Dominguez G, Garcia JM, Martinez G, Vargas J, Provencio M, Espana P, Bonilla F (1999) Detection of loss of heterozygosity at RAD51, RAD52, RAD54 and BRCA1 and BRCA2 loci in breast cancer: pathological correlations. Br J Cancer 81:503–509
14 Haber JE (2000) Partners and pathways repairing a double-strand break. Trends Genet 16:259–264
15 Haynes RH (1991) Modes of Mutation and Repair in Evolutionary Rhythmys. In: Margulis L, Fester R (eds) Symbiosis as a Source of Evolutionary Innovation, MIT Press, pp 40–56
16 Hoege C, Pfander B, Moldovan G-L, Pyrowolakis, Jentsch S (2002) Rad6-dependent DNA repair is linked to modification of PCNA by ubiquitin and SUMO. Nature 419:135–141
17 Hoeijmakers JH (2001) Genome maintenance mechanisms for preventing cancer. Nature 411:366–374
18 Jackson SP (2002) Sensing and repairing DNA double-strand breaks. Carcinogenesis 23:687–696
19 Jasin M (2001) Double-strand break repair and homologous recombination in mammalian cells, p. 207-235. In: Nickoloff JA, Hoekstra MF (eds) DNA Damage and Repair. Vol 3. Humana Press Inc., Totowa, N.J.
20 Jaattela M (2002) Programmed cell death: many ways for cells to die decently. Ann Med 34:480–488
21 Jeggo P, Concannon P (2001) Immune diversity and genomic stability: opposite goals but similar paths. J Photochem Photobiol 65:88–96
22 Johnson RD, Jasin M (2001) Double-strand break-induced homologous recombination in mammalian cells. Biochem Soc Transact 29:196–201
23 Kanaar R., Hoeijmakers J., van Gent D. (1998) Molecular mechanisms of DNA double strand break repair. Trends Cell Biol 8:483–489
24 Karran P. (2000) DNA double strand break repair in mammalian cells. Curr Opin Genet Dev 10:144–150

25 Khanna KK, Jackson SP (2001) DNA double strand breaks: signalling, repair and the cancer connection. Nat Genet 27:247–254
26 Kiechle M, Friedl AA, Manivasakam P, Eckardt-Schupp F, Schiestl R (2000) DNA integration by Ty Integrase in yKu70 mutant Saccharomyces cerevisiae cells. Molec Cell Biol 20, 8836–8844
27 Kiechle M, Manivasakam P, Eckardt-Schupp F, Schiestl R, Friedl AA, (2002) Promoter-trapping in Saccharomyces cerevisiae by radiation-assisted fragment insertion. Nucl Acid Res 30 (24) e136
28 Knudsen LE, Loft SH, Autrup H (2001) Risk assessment: the importance of genetic polymorphisms in man. Mutat Res 482:83-88
29 Lindahl T. (2000) Suppression of spontaneous mutagenesis in human cells by DNA excision-repair. Mutat Res 462:129–135
30 Lutz WK (2001) Susceptibility differences in chemical carcinogenesis linearize the dose-response relationship: threshold doses can be defined only for individuals. Mutat Res 482:71–76
31 Meyn MS (1993) High spontaneous intrachromosomal recombination rates in ataxia telangiectasia. Science 260:1327–1330
32 Nickoloff J.A. and Hoekstra M.F. (eds.) (1998) DNA Damage and Repair. Vol 1. Humana Press Inc., Totowa, N.J.
33 Nickoloff J.A. and Hoekstra M.F. (eds.) (1998) DNA Damage and Repair. Vol 2. Humana Press Inc., Totowa, N.J.
34 Nickoloff J.A. and Hoekstra M.F. (eds.) (2001) DNA Damage and Repair. Vol 3. Humana Press Inc., Totowa, N.J.
35 Nunez MI, McMillan TJ, Valenzuela MT, Ruiz de Almodovar, JM, Pedraza V. (1996) Relationships between DNA damage, rejoining and cell killing by radiation in mammalian cells. Radiother Oncol 39:155–165
36 Paques F, Haber J.E. (1999) Multiple pathways of recombination induced by double-strand breaks in Saccharomyces cerevisiae. Microbiol Mol Biol Rev 63:349–404
37 Pastink A., Eeken JC, Lohman P (2001) Genomic integity and the repair of double strand DNA breaks. Mutat Res 480–481:37–50
38 Pfeiffer, P, Goedecke W, Obe G (2000) Mechanisms of DNA double-strand break repair and their potential to induced chromosomal aberrations. Mutagenesis 15:289–302
39 O'Driscoll M, Jeggo P (2002) Immunological disorders and DNA repair. Mutat Res 509:109–126
40 Rolig RL, McKinnon PJ (2000) Linking DNA damage and neurodegeneration. Trends Neurosci 23:417–424
41 Scicchitano DA, Mellon I, (1997) Transcription and DNA damage: a link to a kink. Environ Health Perspect 105 Suppl 1, 145–153
42 Sekiguchi M, Tsuzuki T (2002) Oxidative nucleotide damage: consequences and prevention. Oncogene 21:8895–8904
43 Siede W, Friedl AA, Dianova I, Eckardt-Schupp F, Friedberg EC (1996) The Saccharomyces Ku autoantigen homologue affects radiosensitivity only in the absence of homologous recombination. Genetics 142:91–102
44 Shiloh Y, Kastan MB (2001) ATM: genome stability, neuronal development, cancer cross paths. Adv Cancer Res 83:209–254
45 Somosy Z (2000) Radiation response of cell organelles. Micron 31, 165–181.
46 Stark JM, Jasin M (2003) Extensive loss of heterozygosity is suppressed during homologous repair of chromosomal breaks. Molec Cell Biol 23:733–743

47 Tateishi S, Niwa H, Miyazaki J-I, Fujimoto S, Inoue H, Yamaizumi M (2003) Enhanced genomic instability and defective postreplication repair in RAD18 knockout mouse embryonic stem cells. Molec Cell Biol 23:474–481
48 Thompson LH, Schild D (2002) Recombinational DNA repair and human disease. Mutat Res 509:49–78
49 Ulrich HD (2002) Degradation or maintenance, action of the ubiquitin system on eukaryotic chromatin. Eukaryotic Cell 1:1–10
50 Ulrich HD, Jentsch S (2000) Two RING finger proteins mediate acooperation between ubiquitin-conjugating enzymes in DNA repair. EMBO J 19:3388–3397
51 Vessey CJ, Norbury CJ, Hickson ID (1999) Genetic disorders associated with cancer predisposition and genomic instability. Prog Nucl Acid Res Mol Biol 63:189–221
52 Vijg J, van Orsouw N (2002) Searching for genetic determinants of human aging and longvity: opportunities and challenges. Mech Age Dev 123:195–205
53 Wood RD, Mitchell M, Sgouros J, Lindahl T (2001) Human DNA repair genes. Science 291, 1284–1289
54 Yang J, Yu Y, Duerksen-Hughes PJ (2003) Protein kinases and their involvement in the cellular responses to genotoxic stress. Mutat Res 543:31–58.

Topological Factors in Radiation Biology

Anna A. Friedl

Strahlenbiologisches Institut der Universität München, Munich, Germany
anna.friedl@lrz.uni-muenchen.de

Introduction

In recent years, we have accumulated a wealth of information concerning DNA damage induced by ionising radiation. It can be expected that the biochemistry of DNA repair pathways and the proteins involved will soon be elucidated in their entirety. In contrast, our knowledge on the impact of topology on the biological consequences of irradiation is still scarce. This is somewhat unexpected, as the importance of proximity of damaged sites for the generation of exchange-type aberrations had been acknowledged already in the earliest days of radiation biology [16]. One explanation for the lack of experimental data is that topology may affect induction and repair of damage on different levels, and that net-like interdependencies complicate the analysis of individual effects. In addition, because of lack of appropriate technologies until recently many topics could not be addressed experimentally. This article aims at reviewing our current knowledge on potential topological factors and remaining questions which are important for our radiobiological understanding.

Proximity effects

It is now generally accepted that structural chromosome aberrations result from the mis-repair of DNA double-strand breaks (DSB). As will be discussed later, all known DSB repair pathways can potentially lead to chromosomal rearrangements. To start with, we will consider only mis-rejoining of free DNA ends originating from two or more break sites. In classical terms, this corresponds to the breakage-and-reunion model [35], in molecular terms this corresponds mainly to DSB repair by non-homologous end-joining (NHEJ) mechanisms [39]. Presumably the only prerequisite for this mis-rejoining is that the free ends have to come close in time and space at some time of the joining process. The first topic of our considerations is the question how close the later interaction partners must be at the time of damage induction for exchanges to occur.

Several lines of evidence indicate that mis-rejoining probabilities are strongly influenced by initial DSB separation. This was termed proximity effects [32, 35]. For example, after irradiation the ratio of interchromosomal versus intrachromosomal exchanges is smaller than expected for purely random interactions within the nucleus, demonstrating a bias towards intrachromosomal events. Given the territorial organisation of the interphase chromosomes in mammalian cells (see below), two breaks present on the same chromosome are, on average, closer to each other than two breaks located on different chromosomes. In addition, high-LET radiation is more potent in inducing aberrations than low-LET irradiation. This has been taken to reflect smaller average distances between the correlated DSB in the case of high-LET irradiation, although, to some extent also differences in DSB induction may be responsible. While the existence of proximity factors is widely accepted, attempts aiming at deriving preferential or maximal interaction distances yield variable results, depending on the models used [32, 35]. A possible explanation is that these models mainly concentrated on physical parameters. Information on biological parameters such as chromatin structure or nuclear architecture and their roles in aberration formation has been emerging only in recent years, and there is still much to be learned.

Nuclear architecture

The biological consequences of exchange-type aberrations are determined by the sites where the exchange takes place. Balanced exchanges between heterochromatic regions without involvement of genes or regulatory sequences will probably not raise serious problems in somatic cells, as long as mitotically stable chromosomes are formed. In contrast, exchanges involving genes or regulatory sequences may result in loss of gene activity, or in deregulation events that, if relevant genes were affected, eventually lead to cancerogenesis. Nuclear architecture, i.e. the positioning of chromosomes within the nucleus, together with proximity effects affects the probability with which certain regions interact. To elucidate the underlying rules, one should first determine whether and how the distribution of aberration breakpoints in radiation-induced aberrations reflects known features of the nuclear architecture. In doing so, it is important to bear in mind that there may be cell-type specific differences in nuclear architecture.

In mammalian cells, the individual chromosomes occupy distinct territories which do not intermingle significantly [6, 26]. The radial positioning of individual chromosome territories within the nucleus appears not to be random. Rather, it has been observed that gene-rich chromosomes tend to locate in the interior of the nucleus, while gene-poor chromosomes lie at the nuclear periphery. This arrangement is evolutionary conserved [36], thus suggesting a functional significance. Chromosomes located at the nuclear periphery, which are in contact with the nuclear envelope, should have fewer possibilities to interact with other chromosomes than those located in the interior. This should affect the probability with which a specific chromosome is involved in an interchromosomal exchange-type aberration. Indeed, various authors reported that the distribution of breakpoints in radiation-induced interchromosomal exchanges is not random, i.e. that it deviates from expectations based on DNA-proportionality or chromosome surface area. A recent meta-analysis claimed, however, that there is disagreement between these studies as to which chromosomes are over- or underrepresented [14]. Unfortunately, this analysis comprised studies on different cell types and with differing degrees of post-irradiation selection, and it remains unclear to what extent aberration patterns in the first metaphase post-irradiation reflect cell-type specific chromosome arrangements.

Positioning of chromosome territories within the nucleus may also lead to specific neighbourhood relations. Two chromosome territories that share a common boundary more often than if positioning were random should have a higher frequency of interchange. Information on the relative frequencies of pairwise interactions is scarce. In a study investigating 10 chromosomes by repeated FISH (fluorescence in situ hybridisation), significant increases in exchanges between certain chromosomes were detected after neutron irradiation, and interestingly translocations of these chromosomes are often found in haematologic neoplasms [19]. In contrast, in a study performed by the same group on gamma-irradiated cells no preferential formation of exchanges in specific pairs was observed [4]. The authors concluded that preferential chromosomal associations are more relevant for exchanges induced by high-LET irradiation, where the initiating breaks supposedly are very close to each other, than in the case of low-LET irradiation. In contrast, after sparsely ionising irradiation in a recent study involving multiplex FISH (with painting of each chromosome in a different colour), a small but significant increase in interactions was observed between those chromosomes which are known to preferentially locate in the interior of the nucleus [5].

The next parameter of relevance is the positioning of specific chromosomal regions within a territory. Chromosomal territories appear to exhibit a certain polarity in that early-replicating, transcription-active regions tend to orient towards the nuclear interior while late-replicating regions tend to occupy so-called peripheric compartments [9, 33]. Together with the general chromosome positioning, this polarity leads to an enrichment of gene-density in a subvolume, namely the interior region, of the nucleus. As compared to a random situation, where the few percent of coding DNA are scattered within the nucleus, the accumulation of gene-rich regions in the nuclear interior should significantly increase the probability of gene-gene interactions (and therefore of interactions relevant for carcinogenesis). One prediction arising from this model is that breakpoints of exchange-type aberrations should accumulate in G-light bands. This has indeed been postulated by various authors [25]. Because of optical artefacts associated with assignment of breakpoints in banded chromosomes [34], however, this matter is still under debate. Using the panoply of new cytogenetic methods [38], it can be hoped that the assignment of breakpoints will become more accurate in future.

To what extent the positioning of individual chromatin regions within the territories is determined by functional constraints is not clear. It has been proposed that the interchromosomal domain, which separates adjacent territories, and a channel-like interchromatin space, which invaginates sponge-like chromosome territories, form a continuous reticular system by which protein complexes needed for transcription, RNA processing and other processes gain access to the chromatin [6]. The necessity to be close to this nuclear non-chromatin domain would thus, for certain chromatin regions, further restrict the number of potential locations and increase the probability of physical closeness between these regions. For example, transcription-active regions appear to locate preferentially at the territory surfaces [40]. Because of the sponge-like structure of the territories, a location on the surface does not necessarily mean a location on the periphery, close to other territories. Still, as compared to gene-poor or heterochromatic regions, more gene-rich regions should lie close to regions on other chromosomes, again facilitating interchromosomal gene-gene interactions.

Up to now, we have concentrated on the role of nuclear architecture on interchromosomal exchanges, because this endpoint is easy to detect and a large amount of data is available. It should be noted, however, that our considerations also apply to intrachromosomal aberrations. Once the rules governing nuclear architecture are fully deciphered, the degree of influence of architecture on the formation of exchanges between given sequences can be predicted more accurately than possible at present. In addition, if candidate sequences are known that appear to be at higher risk for interaction, one can test whether these sequences exhibit closer proximity in the interphase nucleus than expected on the basis of a random positioning. This is exemplified by the BCR and ABL loci on human chromosomes 22 and 9, respectively. Fusion of these loci by a translocation event causes chronic myeloid leukaemia, and the average geometric distance between these loci in haematopoietic cells is smaller than would be expected by chance [18, 23]. Another example is given by the RET and H4 loci which are located on chromosome 10, about 30 Mb from each other, and whose average geometric distance in thyroid cells is smaller than expected by a random-walk model [24]. Fusions of RET and H4 are frequently found in radiation-induced papillary thyroid tumours.

Dynamics and sites of repair

So far, we have regarded proximity effects and the influence of nuclear architecture only with respect to the distance between break sites at the time of break induction. Either by passive, undirected movement or by transport in the course of repair, DNA ends may come close that initially were rather distant from each other. It is not known how much both DNA ends of one break site can separate; most probably this depends on the chromatin structure but very little is known about how the 30nm chromatin fibre is arranged to form the chromosome territory [42]. If, as has been proposed [31], higher order conformation involves anchoring or fixation of loops, the ability to move may be restricted by these structural features. In general, because of a more open chromatin structure, breaks located in euchromatin regions may display a higher end motility than those located in heterochromatin regions.

A very important but so far unresolved question concerns the site of repair. Do the repair proteins move to the site of DNA damage, or does repair take place at specific sub-nuclear sites, to which the damaged DNA (in this case the DNA ends) have to move or to be transported? By drawing all break ends present in a larger catchment-area, specific repair sites ('garages' or 'factories') may serve to increase preferential interaction distances. It has been suggested that without an active process that brings different ends together, the probability of interaction of three or more break sites would be too low to be compatible with observed frequencies of complex exchanges after irradiation [35]. A possible hint at the existence of 'repair factories' is given by the observation that various repair proteins accumulate after irradiation in the form of large foci. There appear to be two main foci types, one of them containing Rad50 and the other containing Rad51 as key component [11, 37, 27, 20], but the exact compositions of these foci is not clear. The high intensity of the signals observed when individual components of these foci are made visible, e.g. by immunofluorescence, suggests that individual protein types are present in these foci in higher number than stoichiometrically necessary for the repair of one break site (e.g. Essers 2002). One possible explanation for this high number of proteins is that these foci represent repair 'factories', but other interpretations can at present not be excluded.

Based on the fact that most DNA repair pathways involve a large number of proteins and possibly pre-assembled repair complexes, it has been proposed that repair processes may take place in or close to the non-chromatin compartment in the nucleus, similar to transcription and splicing processes [6]. So far, the localisation of repair complexes or foci in relation to chromatin domains has not been investigated. If repair processes occur preferentially in the interchromosomal space, interchromosomal interactions may be facilitated. However, recent data argue against the presence of large pre-assembled repair complexes, since for several proteins involved in DSB repair by homologous recombination, on-the-spot assembly was observed [8]. Whether and how far individual chromatin domains or loops move after damage induction is not known; in situ hybridisation studies aiming at determining whether there is large-scale movement of chromatin or chromosome domains after irradiation gave conflicting results [7, 10, 13, 17].

Influence of repair pathways

As mentioned, all known DSB repair pathways have a potential to result in structural chromosome aberrations [28]. So far, we have mainly considered mis-rejoining of DNA ends from two or more break sites by homology-independent end-joining mechanisms. In a similar manner, exchanges may arise in the course of break repair via single-strand annealing (SSA), if DNA ends originating from different break sites interact. In contrast, for repair by homologous recombination, only one radiation-induced break is required, which interacts with an intact, sequence-related region. Chromosomal exchanges may occur if this intact region is located on a so-called ectopic position (i.e. not on the sister chromatid or on the allelic position on the homologous chromosome) and if the recombination intermediate is resolved with a crossing-over. Generation of exchange-type aberrations by homologous recombination is often considered as being unlikely; there are, however, many examples that suggest a role for homologous recombination in the generation of translocations [3, 15]. The relative proneness of the various repair pathways to result in mis-repair, although not yet reliably quantified, appears to differ and the decision for applying a certain pathway for the repair of a given break has, therefore, far-reaching consequences. This decision is influenced by a variety of factors, including cell type and cell status. In the present context, it is important to note that the decision may also be influenced by topological factors such as localisation with respect to chromatin structure and nuclear architecture, or the density of break sites.

Although finding a DNA region with sequence homology that can serve as a template is a crucial step in DSB repair by homologous recombination, to date little is known on how homologous sequences are found. Sister chromatids (which are close to each other) are known to be a preferential recombination substrate, if present. There is, however, accumulating evidence for recombination events involving homologous chromosomes (thus, for example giving rise to LOH events, Vrieling 2001) although, as seen in interphase chromosome painting, these are not juxtaposed in the mammalian interphase nucleus [6]. This, as well as experimental demonstration of ectopic homologous recombination [29, 12] suggest the existence of mechanisms which allow genome-wide homology searches. Homology search and/ or the subsequent physical interaction of the recombination partners demand movement of at least one of the partners during the repair process. Therefore, the spatial constraints on potential interaction partners that are evident for end-joining repair may be much lower or even absent in the case of homologous recombination. It is unclear at present whether this applies also to SSA-like repair, which is also considered as a form of homology-dependent repair, but which depends on other genes than homologous recombination.

Future prospects

Many of the questions that have remained open so far may be answered in the near future with the help of recent technological developments. Modern cytogenetic methods like M-FISH and SKY [38] will lead to more accurate determination of the relative frequency by which individual chromosomes participate in exchange-type aberrations, and also to the identification of preferential pairwise associations. Since nuclear architecture may greatly be affected by cell type and status, these studies have to be conducted in a variety of different cell types. In order to elucidate whether there are preferential interactions between certain chromosomal sub-regions, higher resolution colour banding methods are needed. A first step in this direction was made by developing cross-species colour banding (Rx-FISH; Müller at al. 1998). Candidates for preferential vicinity may also be found by analysis of recurrent tumour-associated aberrations. Once pairs of candidate regions are identified, the geometric distance between these regions can be measured by interphase FISH and compared with theoretical expectations [24]. Identification of pairs of loci that are closer to each other than expected on the basis of random localisation will help to learn more about functional nuclear architecture.

Most of the investigations on the influence of topological factors reported so far were based on FISH and/or immunodetection, thus requiring fixation of cells and precluding the observation of an individual cell over time. In future this can be circumvented by introducing live-cell observation techniques. For example, insertion of fluorescence-labelled nucleotides into genomic DNA during replication and subsequent segregation of labelled chromosomes by cell propagation results in cells that carry one or few labelled chromosomes or chromosome domains [43, 33]. Observation of these labelled regions over time allows to analyse chromosomal dynamics. In combination with targeted delivery of damage, e.g. by micro-irradiation, this method will be useful to determine the motility of damaged chromatin.

The outcome of replication-based labelling is random in that it is not known a priori which labelled chromosomal region is retained in which progeny cell. If in vivo labelling of specific regions is required, another technique can be used which relies on insertion of large arrays of binding sites for a DNA-binding protein. If in the same cells fluorescent-tagged derivatives of the DNA-binding protein are expressed, the site of the inserted array can be visualised by the fluorescence signal [30, 2]. Localisation and dynamics of proteins can be analysed in vivo by fluorescence tagging with GFP (green fluorescent protein) or derivatives with different emission spectra [21]. Thus, the recruitment of repair complexes can be watched and, together with FRAP (fluorescence redistribution after photobleaching) or FLIP (fluorescence loss in photobleaching) also the turn-over of repair proteins can be determined [8]. In combination with labelled chromatin domains and targeted damage delivery, tagging of repair proteins could be used to investigate the localisation of repair processes with respect to the damaged chromatin.

As stated in a recent excellent overview [1], radiation has been used as an effective tool to probe fundamental biological processes, such as DNA repair, mutagenesis, genome instability, cell cycle checkpoints, transposition of genetic elements and many others. I expect that with all these new techniques, we can in the near future also put functional nuclear architecture onto this list.

References

1 Bedford JS, Dewey WC (2002) Historical and current highlights in Radiation Biology: Has anything important been learned by irradiating cells? Radiat Res 158:251–291

2 Belmontt AS (2001) Visualizing chromosome dynamics with GFP. Trends Cell Biol 11:250–257

3 Bishop AJR, Schiestl RH (2002) Homologous recombination and its role in carcinogenesis. J Biomedicine Biotechnology 2:75–85

4 Cafourkova A, Lukasova E, Kozubek S, Kozubek M, Govorun RD, et al. (2001) Exchange aberrations among 11 chromosomes of human lymphocytes induced by g-rays. Int J Radiat Biol 77:419–429

5 Cornforth MN, Greulich-Bode KM, Loucas BD, Arsuaga J, Vazquez M, et al. (2002) Chromosomes are predominantly located randomly with respect to each other in interphase human cells. J Cell Biol 159:237–244
6 Cremer T, Cremer C (2001) Chromosome territories, nuclear architecture and gene regulation in mammalian cells. Nature Rev Genet 2:292–301
7 Dolling JA, Boreham DR, Brown DL, Raaphorst GP, Mitchel REJ (1997) Rearrangement of human cell homologous chromosome domains in response to ionizing radiation. Int J Radiat Biol 72:303–311
8 Essers J, Houtsmuller AB, van Veelen L, Paulusma C, Nigg AL et al. (2002) Nuclear dynamics of RAD52 group homologous recombination proteins in response to DNA damage. EMBO J 21:2030–2037
9 Ferreira J, Paolella G, Ramos C, Lamond AI (1997) Spatial organization of large-scale chromatin domains in the nucleus: a magnified view of single chromosome territories. J Cell Biol 139:1597–1610
10 Figgitt M, Savage JRK (1999) Interphase chromosome domain re-organization following irradiation. Int J Radiat Biol 75:811–817
11 Haaf T, Golub EI, Reddy G, Radding CM, Ward DC (1995) Nuclear foci of mammalian Rad51 recombination protein in somatic cells after DNA damage and its localization in synaptonemal complexes. Proc Natl Acad Sci USA 92:2298–2302
12 Inbar O, Kupiec M (1999) Homology search and choice of homologous partner during mitotic recombination. Mol Cell Biol 19:4134–42
13 Jirsova P, Kozubek S, Bartova E, Kozibek M, Lukasova E et al. (2001) Spatial distribution of selected genetic loci in nuclei of human leukemia cells after irradiation. Radiat Res 155:311–319
14 Johnson KL, Brenner DJ, Nath J, Tucker JD, Geard CR (1999) Review: Radiation-induced breakpoint misrejoining in human chromosomes: random or non-random? Int J Radiat. Biol 75:131–141
15 Kolomietz E, Meyn MS, Pandita A, Squire JA (2002) The role of Alu repeat clusters as mediators of recurrent chromosomal alterations in tumours. Genes Chrom Cancer 35:97–112
16 Lea DE (1946) Actions of radiations on living cells. Cambridge University Press
17 Lucas JN, Cervantes E (2002) Significant large-scale chromosome territory movement occurs as a result of mitosis, but not during interphase. Int J Radiat Biol 78:449–455
18 Lukasova E, Kozubek S, Kozubek M, Kjeronska J, Ryznar L et al. (1997) Localisation and distance between ABL and BCR genes in interphase nuclei of bone marrow cells of control donors and patients with chronic myeloid leukaemia. Hum Genet 100:525–535
19 Lukasova E, Kozubek S, Kozubek M, Kroha V, Mareckova A, et al. (1999) Chromosomes participating in translocations typical of malignant hemoblastoses are also involved in exchange aberrations induced by fast neutrons. Radiat Res 151:375–384
20 Mirzoeva OK, Petrini JH (2001) DNA damage-dependent nuclear dynamics of the mre11 complex. Mol Cell Biol 21:281–288
21 Misteli T (2001) Protein dynamics: implications for nuclear architecture and gene expression. Science 291:843–847
22 Müller S, O'Brien PC, Ferguson-Smith MA, Wienberg J (1998) Cross-species colour segmenting: a novel tool in human karyotype analysis. Cytometry 33:445–452
23 Neves H, Ramos C, da Silva MG, Parreira A, Parreira L (1999) The nuclear topography of ABL, BCR, PML, and RARa genes: evidence for gene proximity in specific phases of the cell cycle and stages of haematopoietic differentiation. Blood 93:1197–1207

24 Nikiforova MN, Stringer JR, Blough R, Medvedovic M, Fagin JA et al.. (2000) Proximity of chromosomal loci that participate in radiation-induced rearrangements in human cells. Science 290:138–141
25 Obe G, Pfeiffer P, Savage JRK, Johannes C, Goedecke W, et al. (2002) Chromosomal aberrations: formation, identification and distribution. Mutat Res 504:17–36
26 Parada LA, Misteli T (2002) Chromosome positioning in the interphase nucleus. Trends Cell Biol 12:425–432
27 Paull TT, Rogakou EP, Yamazaki V, Kichgessner CU, Gellert M, Bonner WM (2000) A critical role for histone H2AX in recruitment of repair factors to nuclear foci after DNA damage. Curr Biol 10:886–895
28 Pfeiffer P, Goedecke W, Obe G (2002) Mechanisms of DNA double-strand break repair and their potential to induce chromosomal aberrations. Mutagenesis 15:289–302
29 Richardson C, Moynahan ME, Jasin M (1998) Double-strand break repair by interchromosomal recombination: suppression of chromosomal translocations. Genes Dev 12:3831–3842
30 Robinett CC, Straight A, Li G, Willhelm C, Sudlow G et al. (1996) In vivo localization of DNA sequences and visualization of large-scale chromatin organization using lac operator/repressor recognition J Cell Biol 135:1685–1700
31 Sachs RK, vd Engh G, Trask B, Yokota H, Hearst JE (1995) A random-walk/giant-loop model for interphase chromosomes. Proc Natl Acad Sci USA 92:2710–2714
32 Sachs RK, Chen AM, Brenner DJ (1997) Review: Proximity effects in the production of chromosome aberrations by ionizing radiation. Int J Radiat Biol 71:1–19.
33 Sadoni N, Langer S, Fauth C, Bernardi G, Cremer T, et al. (1999) Nuclear organization of mammalian genomes. Polar chromosome territories build up functionally distinct higher order compartments. J Cell Biol 146:1211–26
34 Savage JRK (1977) Assignment of aberration breakpoints in banded chromosomes. Nature 270:513–514
35 Savage JRK (1996) Insight into sites. Mutat Res 366:81–95
36 Tanabe H, Habermann FA, Solovei I, Cremer M, Crmer T (2002) Non-random radial arrangements of interphase chromosome territories: evolutionary considerations and functional implications. Mutat Res 504:37–45.
37 Tashiro S, Walter J, Shinohara A, Kamada N, Cremer T (2000) Rad51 accumulation at sites of DNA damage and in post-replicative chromatin. J Cell Biol 150:283–291
38 Teixeira MR (2002) Combined classical and molecular cytogenetic analysis of cancer. Eur J Cancer 38:1580–1584
39 van Gent DC, Hoeijmakers JH, Kanaar R (2001) Chromosomal stability and the DNA double-stranded break connection. Nat Rev Genet 2:196–206
40 Verschure PJ, van der Kraan I, Manders EM, van Driel R (1999) Spatial relationship between transcriptional sites and chromosome territories. J Cell Biol 147:13–24
41 Vrieling H. Mitotic maneuvers in the light (2001) Nature Genetics 28:101–102
42 Woodcock CL, Dimitrov S (2001) Higher-order structure of chromatin and chromosomes. Curr Opin Genet Dev 11:130–135
43 Zink D, Cremer T, Saffrich R, Fischer R, Trendelenburg MF et al. (1998) Structure and dynamics of human interphase chromosome territories in vivo. Hum Genet 102:241–251

Signalling

Molecular Radiation Biology – Perspectives for the Future

H. Peter Rodemann

Section of Radiation Biology and Molecular Environmental Research,
Dept. of Radiation Oncology, University of Tübingen, Germany
hans-peter.rodemann@uni-tuebingen.de

Introduction

Historically radiation biology has been an upfront discipline of life sciences and has contributed to the understanding of many basic cell biological mechanisms of normal and tumor cell development, such as cell cycle regulation, principles of DNA-damage and repair as well as genomic stability. These contributions not only increased our general knowledge about biological principles but helped to establish and develop the disciplines of cell and molecular biology as well as molecular genetics and tumor biology. Over the years these new disciplines based on the development of far more sensitive analytical methods to investigate the different biological principles and thus laid the basis for a rapid progress in understanding the basic molecular mechanisms of life. Cell and tumor biology applying molecular approaches have identified specific differences between normal and tumor cells on the basis of which now numerous new concepts could be translated into preclinical and clinical research and treatment strategies.

With few exceptions most radiation biologists shared the opinion that the new molecular approaches would not be useful to address clinical relevant questions of radiation research and thus remained focussed on descriptive aspects of dose response relationships, radiation safety and protection. The new disciplines tumor biology, cell and molecular biology, however, not only led to an enormous basic scientific progress but were additionally able to proof their clinical relevance. Based on this success and in contrast to the "old days" radiation biology research was no longer of general interest for biology and medicine. Consequently and despite the fact that a small number of research institutions in radiation biology could keep up with mainstream developments, radiation biology moved into the off-side of life science. As a result and being no longer considered as an upfront life science discipline radiation biology was cut off from major developments in the reorganization of university faculties in biology and medicine.

Radiation biology understood itself always as the prime basic discipline of radiation therapy. Many research centres – although not – all focussed their research interests on clinical aspects of radiobiology. Cooperative efforts of radiation biologists and radiation therapists towards the progress of radiobiology-based concepts in radiation therapy have been undertaken applying classical radiobiological methods and the results have been introduced into clinical practice, e.g. fractionated treatment regimes (CHART). Nevertheless, in the last 3 decades most of the success and achievements of radiation therapy in cancer treatment originated from enormous technical progress and breakthrough in medical physics and engineering developments of irradiation equipments so far. Due to these developments and the fact that not all radiation biology institutes although embedded in medical faculties really worked on clinically oriented translational aspects of radiation biology the contribution of this discipline to the progress of radiation therapy was more and more questioned. Consequently radiation biology lost its support in medical faculties and many of the university institutions for radiation biology existing so far were closed or reorganized into another life science discipline due to retirement of department heads.

Within the last ten years in Germany mainly on request of the local radiation oncology centres some medical faculties re-established radiation biology as a basic research discipline. This development could be attributed to the generation change which took place since the late eighties and early nineties of the last century. Most of the radiation therapists appointed at that time as new department heads had gone through basic preclinical radiation biology research training either in Germany or abroad. They realized and envisaged that major developments in modern radiation oncology can no longer be expected only from technical achievements in medical physics but in addition urgently need the implementation of molecular biology strategies into treatment protocols. Consequently research institutions or units for radiation biology were funded to promote basic and translation research in radiation oncology on the basis of mechanistically oriented cell and molecular biology approaches. This concept which should be continued for all university based radiation oncology centres in Germany follows international developments where research institutions for molecular radiation biology were either integrated in centres of experimental and applied radiation oncology or embedded in research centres of molecular life science disciplines. The international acknowledged scientific success of these newly established institutions proofs that molecular radiation biology primarily focussed towards basic aspects and developments in radiation oncology is indeed a vivid life science discipline and thus a science for the future. Addressing internationally competitive research topics molecular radiation biology is and will be able to regain its position amongst other molecular oriented life sciences. Some of the most important research perspectives of molecular radiation biology for radiation research in general and radiation oncology especially will be addressed in the following.

Future perspectives

Modern radiation research presenting a broad spectrum of physical, chemical, biophysical, molecular biological and epidemiological approaches is an absolute need for the understanding of interaction of ionizing and non-ionizing radiation with biological matter and living organisms. The use of radiation in medicine and technology demands radiation research in the different specialities named. Even in Germany where the end of the use of nuclear energy has politically been decided and the remaining nuclear power plants will be closed within the next 30 years radiation research has still a burning demand. The vast majority of radiation exposure in Germany is due to the medical use of radiation and only a very minor part is due to nuclear energy technology. Thus the various disciplines of radiation research including radiation safety and protection and among these especially radiation biology is of utmost importance for the future. Likewise and nevertheless, even in a situation where nuclear power plants will be shut down expertise in radiation safety and protection is still needed to provide a safe break down of power plants and proper handling of remaining nuclear waste.

The knowledge of how radiation and especially ionizing radiation interacts with biological systems has improved over the last 30 years quite substantially. Especially introducing new strategies and the application of molecular biology to specific questions of cellular radiation response pattern helped to enlighten the underlying mechanisms of so far mainly descriptive radiobiological data and phenomena. However, although many answers could be given over the recent years, far more questions arose which need to be answered for a complete understanding of the complexity of cellular radiation response mechanisms and their detailed regulation in normal and tumor cells. Three major areas for future research perspectives can be foreseen:

- the complex processes of radiation-induced DNA-damage, recognition of damaged DNA and execution of correct DNA-repair
- the complex communication processes of radiation-induced DNA-damage dependent and DNA-damage independent inter- and intracellular signalling and gene expression
- the complex network of extra cellular matrix proteins, the cellular microenvironment and intracellular signalling in regulating radiation responses.

These major topics with relevance to the understanding of principal biological processes induced by external stress factors make clear why radiation biology is a science for the future. Radiation biology will provide a detailed understanding of how radiation affects cells at the genomic level and how cells cope with radiation damage to the DNA as well as alterations of inter- and intracellular signalling mediated by radiation. Thus radiation biology will unravel the principles of genomic stability and therefore elucidate one of the most important aspects of life. Consequently radiation biology is needed as the basis of future radiation protection and especially radiation oncology and will thus help to fight cancer were it counts.

To approach these aspects of modern radiation biology research will certainly need to be structured into a framework which will allow both academic basic research and applied research towards the development of both radiation protection and radiation oncology. Basic research aspects with special importance to the development of both disciplines should certainly include

- the mechanisms of DNA-damage, recognition and repair induced by low and high radiation doses of different qualities
- the identification of genetic factors determining individual radiation sensitivity
- the characterization of specific molecular extra- and intracellular components and pathways mediating radiation-induced inter- and intracellular signalling cascades
- the identification of cell type specific mechanisms triggering the radiation response
- the identification of marker genes for radiation-induced tumors

It has long been thought that the survival probability of irradiated cells is solely due to the level of DNA-damage induced. Over the recent years, however, more and more data are accumulating indicating that besides DNA-damage dependent pathways equally important pathways and signalling cascades exist which contribute to overall cellular radiation response. In this context primarily the growth factor receptor signalling pathways need to addressed. Growth factor receptors like the EGF-receptor which are normally activated by binding of their specific ligand can efficiently be activated without ligand binding by radiation exposure and independent of damaged DNA. This has been shown by a number of studies [1, 17] indicating that radiation-induced signalling through the EGF-receptor stimulates a cell survival mechanisms via the activation of specific anti-apoptotic and pro-proliferative pathways. These results are not only of great importance for the understanding of cell and molecular biological principles of receptor signalling but also of especial interest for radiation oncology. Approximately 40 % of all human tumors are characterized by mutations of the EGF-receptor family proteins leading to over expression of highly active receptor proteins. This abnormal receptor function could be correlated in turn not only to the grade of malignancy but especially to a high degree of resistance to chemo- or radiotherapy. Recent studies addressing the specific role of EGF-receptor in cellular radiation response in more detail indicate that besides activating survival strategies in the cell, radiation-induced EGF-receptor and activation of downstream signalling cascades like the ERK1/2 pathway will lead to the stimulation of DNA-repair genes XRCC1 and ERCC1 [21]. Consequently DNA-damage repair could efficiently be blocked after radiation exposure by treatment with ERK1/2 specific inhibitors resulting in decreased cell survival. These observations clearly illustrate that cellular radiation responses are mechanistically much more complex than so far thought and have to be analysed as an integrated complex network of inter- and intracellular signalling cascades.

With respect to the identification of specific proteins, peptides and small molecules which are able to trigger the cellular radiation response progress has been made especially in the field of cytokines and growth factors as well as matrix components and adhesion molecules. Based on indirect evidence it has been postulated that specific cytokines, like TGF-b1, can modify the radiation response of normal cells especially fibroblasts. Applying recently available TGF-b1-knockout systems the first direct evidence could be provided that TGF-b1 is a major determinator of fibroblasts´ radiation sensitivity and induced terminal differentiation [20]. It has long been postulated that the growth factor supply of cells might influence their specific radiation response. First molecular biology based evidence to this topic could be reported on the basis of a collaborative effort of the Tübingen and Brisbane radiobiology groups [10]. The results clearly indicate the modulatory function of specific growth factors like EGF on the expression profile of the ATM protein, an important DNA-damage sensing protein and cell cycle regulator. The complexity of cellular radiation responses is also reflected by the fact that the composition of the extracellular matrix and the expression and activity profile of adhesion molecules like the integrin receptor binding to ECM-components clearly defines the radiation response pattern of normal and tumor cells. As demonstrated recently binding of the integrin receptor to ECM-components such as fibronectin and lamin significantly increases radioresistance of tumor cells most likely by activation of anti-apoptotic pathways [4]. Likewise and of prime importance not only for radiation oncology it could be demonstrated that irradiation promotes the invasion potential of human glioblastoma cells through interaction with the focal adhesion kinase FAK, which is at least in part regulated via the integrin receptor [22]. Taken together, these studies clearly indicate that molecular radiation biology addressing mechanistic question underlying cellular radiation responses will provide data sets of general interest for biology and tumor biology especially.

With respect to aspects of radiation protection similar molecular biology approaches are urgently needed to unravel the essential targets and mechanisms. Interference with reactive oxygen species generated intracellularly by ionizing radiation has long been experimentally addressed to promote effective radiation protection. Numerous substances have been tested for their efficient radical scavenger capacity and some have been found to be useful for general radiation protection as well as clinical use to protect normal tissue during radiotherapy. At present however, concerning the normal tissue selectivity of protective substances, such as WR-2721 (amifostine) conflicting preclinical and clinical reports have been published [15,2]. Therefore, to promote selective radioprotection for normal tissue, substances have to be designed which protect cells through normal cell specific pathways. Recent investigations indicate that the Bowman-Birk protease inhibitor (BBI) could be such a candidate molecule since it is able to selectively protect normal cells from radiation damage. BBI is a 71 amino acid peptide molecule which has originally been described as an inhibitor of serine proteases [12]. Various reports demonstrated that beside its anti-proteolytic activity BBI is able to protect cells from radiation damage through the activation of the DNA-repair machinery [9] in a p53-dependent manner [7, 10, 9, 6, 5, 8]. Consequently cells mutated in the p53 pathway are not protected. Since nearly 70 % of all human cancers are mutated in p53-signaling BBI and its peptide and amino acid derivatives [8] will provide an excellent tool for clinical radioprotection of normal tissue during radiotherapy.

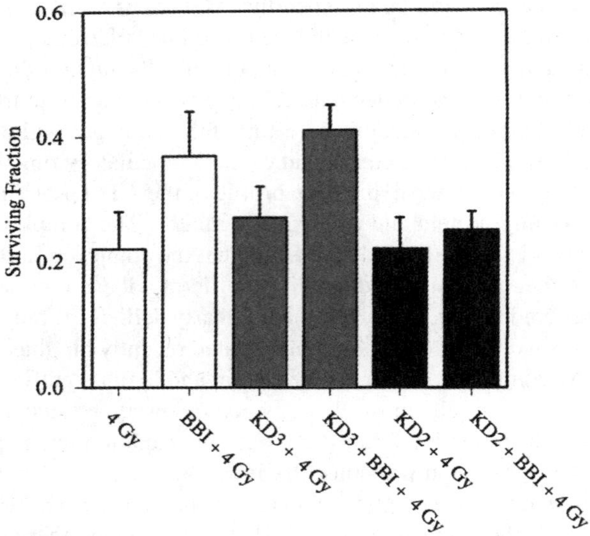

Fig. 1. p53-dependency of the radioprotective effect of BBI.

Confluent normal human fibroblasts were transfected with p53 antisense oligonucleotide (KD2) or control oligonucleotide (KD3) and irradiated with 4 Gy. Seven hours later cells were seeded for clonogenic assay. Data presented represent the mean of the surviving fraction +/- SEM of five independent experiments and statistical significance was tested by the analysis of variance (ANOVA). Statistical significance at the level of $p<0.0005$ was apparent for the conditions 4 Gy vs. 4 Gy+BBI and at the level of $p<0.0001$ for the conditions 4 Gy vs. 4 Gy+BBI+KD3. No statistical significant difference could be observed for the conditions 4 Gy vs. 4 Gy+BBI+KD2 indicating the important role of p53 in regulating the radioprotective effect of BBI. Illustration is taken from Dittmann et al. 1998b.

Although the efficacy of BBI as radioprotector in the high dose range has been documented, so far no studies have been performed to investigate its radioprotective capacity in the low dose range. Based on its biological mechanism, however, it can be expected that BBI shows a similar radioprotective potential for cells exposed to low doses of radiation. Stimulating essential intracellular pathways needed for effective radioprotection through natural or synthetic agonists will provide new perspectives for future radiation protection and safety. Of similar importance in this context will be the understanding of mechanisms of individual radiation sensitivity and solid molecular biology to find out data whether or not a threshold dose exists. Thus, achievements to be made towards the further development of radiation protection and safety as well as risk assessment based on radiobiological mechanisms will include

- the design of radioprotectors based on the understanding of the molecular mechanisms of cellular radiation responses

- the development of micro-arrays (DNA- and protein arrays) to identify markers of individual radiation sensitivity
- the provision of cellular and molecular evidence for the existence or non-existence of a threshold dose in the context of molecular biology based risk assessment.

Last but not least – and perhaps for future radiation research the most important aspect – radiation biology by applying molecular approaches will lead to further progress in radiation oncology which is the key therapy in cancer treatment. Approximately 150.000 to 160.000 cancer patients undergo radiation therapy in Germany every year. In the concert of the existing therapy options radiation therapy is an important and in combination with surgery the most successful treatment. Although in the past the success of radiation therapy was primarily based on technology progress made in the development of irradiation devices and procedures it can be foreseen that new technological achievements will only marginally develop radiation therapy further. Consequently, the understanding of molecular biological response mechanisms to radiation exposure of both tumor and normal cells will contribute to biology-based developments and treatment strategies in radiation oncology. Among these especially aspects such as

- the identification of molecular targets triggering radiation responses in tumor and normal cells
- the development of gene therapy strategies on the basis of radiation-induced therapeutic genes

will be of utmost importance. Dissecting specific molecular differences of tumor cells and normal cells and designing molecules to modulate the radiation response of tumor and normal cells selectively will allow to enhance radiation sensitivity of tumor cells and to increase radiation resistance in normal cells. Likewise the application of gene therapy approaches in radiation oncology may provide selective tools to increase radiation toxicity in tumor tissue. Per se radiation oncology is thought to be the ideal field in medicine to apply gene therapy. This assumption is based on the fact that by using radiation inducible therapeutic genes, the therapy burden can easily be restricted to the field of irradiation. Thus, several attempts are currently in progress analysing the use of specific radiation inducible promoters. Based on these studies reported by et al., especially the use of the EGR-1-promoter for radiation inducible gene therapy is under investigation [18, 14]. Although still at the slevel of basic preclinical research and provided that safe vector systems can be established gene therapy strategies can be foreseen to improve radiation toxicity specifically in tumor tissue [16]. As a consequence targeting specific signalling molecules in tumor cells selectively and applying radiation-inducible gene therapy strategies the therapeutic concept to maximize tumor radiation toxicity and to minimize normal tissue complications will widely open the therapeutic window for radiation oncology.

Molecular developments in radiation biology with respect to the identification of radiation sensitivity will furthermore provide the basis for individualizing radiation therapy. Current treatment strategies in radiation oncology are designed to allow a maximum of 5–10 percent of normal tissue complications. Thus, identification of radiation sensitive patients and determination of the radiation sensitivity profile of all potential patients prior to therapy based on simple and robust molecular predictive assays will enormously improve the treatment options and thus the success of radiation therapy.

To address these aspects of radiation biology state-of-the-art technologies of molecular biology and molecular genetics need to be applied in future research in radiation biology. Therefore, radiation biology has to implement and adopt technological developments of neighbouring discipline such as Molecular Biology, Molecular Genetics, Cell Biology and Tumor Biology. By doing this and by approaching the outlined scientific questions which are of general interest in biology, radiation biology will internationally play again a major role in academic and applied life sciences. With respect to the biomedical achievements based on molecular approaches and to be expected from research in radiation biology and its importance in terms of translational research the future academic development of this discipline will certainly be closely related to radiation oncology. With respect to future challenges in radiation research in general and especially radiation safety and protection radiation biology will still play a central role although to a lower degree when compared to the importance of the biomedical aspects of radiation biology research (Fig. 2).

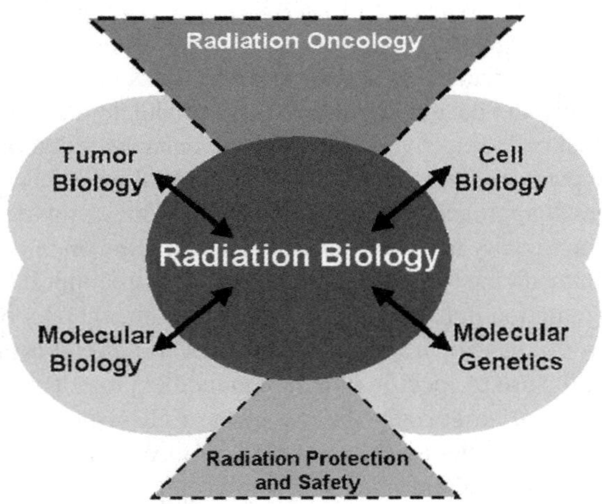

Fig. 2. Positioning of radiation biology in life science.

Radiation biology as the basic discipline of radiation oncology with respect to its future research perspectives will be an integral part of life science and interact with the disciplines molecular biology, molecular genetics, cell biology and tumor biology. Radiation Biology will also be a central discipline for general radiation research with specific molecular biology based impact especially on radiation protection and safety.

Consequences

Broken down to the number of research institutions and especially the currently given man power a comparison of German research activities in radiation biology at the international level reveals that the output and quality is almost equivalent to the United States. This indicates that in general although many research institutions have been closed in Germany (19 out of 25) during the last three decades, the remaining centres (6) and institutions newly funded (5) are able to perform internationally competitive research. This still excellent international position of German radiation biology research however is severely in danger if the current German research policy in general and with respect to radiation biology especially is further continued.

The activity to perform internationally competitive research depends not only on the specific expertise of research groups but is also a result of adequate research support and funding both from the universities and national funding agencies. Due to the financial situation of most universities support and funding of a specific discipline will only be received when it can be expected to attract research grants from external funding sources. Besides excellent ideas and project proposals another prerequisite of similar importance for successful external research grant applications, e.g. through the Deutsche Forschungsgemeinschaft or other national funding agencies, such as the Deutsche Krebshilfe, is the international rewarded standing of the applicant or applying group in the specific field of science which is judged by the scientific output, i.e. the number of peer-reviewed publications. To publish numerous highly ranked international publications requires a certain amount of man power available in the lab. Thus, attracting external research grants is more or less dependent on a threshold support of man power from the university. Based on the financial situation of the German universities at least in the field of radiation biology this threshold level is not given in the existing institutions. Consequently the current situation of radiation biology in Germany is a kind of a vicious circle.

As discussed above in Germany like all over the world a burning demand exists for basic and applied radiation research in general and in radiation biology especially. To meet this demand a political decision is necessary to establish a national radiation research program, e.g. under the roof of the Federal Ministry of Education and Research (BMBF). Such a program should allow the substantial support of man power in individual research institutions for a set period of time (e.g. 5 years). Based on such a program research institutions should be able to support outstanding young scientist and thus to build up the man power which is needed to increase the scientific output and to attract external peer-reviewed research money. If such a situation could be created the authorities at the universities, mainly the faculties of biology, medicine and physics will certainly cooperate to support and strengthen radiation research and especially radiation biology in the future.

References

1 Amorino GP, Hamilton VM, Valerie K, Dent, Lammering G, Schmidt-Ullrich RK (2002) Epidermal growth factor receptor dependence of radiation-induced transcription factor activation in human breast carcinoma cells. Mol Biol Cell 13:2233–2244

2 Bourhis J, Rosine D (2002) Radioproetctive effect of amifostine in patients with head and neck squamous cell carcinoma. Semin Oncol 29:61–62

3 Contessa JN, Hampton J, Lammering G, Mikkelsen RB, Dent, Valerie K, Schmidt-Ullrich RK (2002) Ionizing radiation activates erbB-receptor dependent Akt and p70-kinase signaling in carcinoma cells. Oncogene 21:4032–4041

4 Cordes N, Blaese MA, Meinecke V, van Beuningen D (2002) Ionizing radiation induces up-regulation of functional beta1-integrin in human lung tumour cell lines in vitro., Int J Radiat Biol 78(5):347–57

5 Dittmann K, Mayer C, Rodemann HP (2001) O-phospho-tyrosine protects TP53 wild type cells against ionizing radiation. Int J Cancer 96:1–6

6 Dittmann KH, Dikomey E, Mayer C, Rodemann HP (2000) The radioprotective Bowman-Birk protease inhibitor enhances clonogenic cell survival of ionising radiation-treated nucleotide excision repair-competent cells, but not of xeroderma pigmentosum cells. Int. J Radiat Biol 76:223–229

7 Dittmann KH., Gueven N, Mayer C, Rodemann HP (1998) The radioprotective effect of BBI is associated with the activation of DNA repair-relevant genes. Int J Radiat Biol 74:225–230

8 Dittmann KH, Güven N, Mayer C, Rodemann HP (2001) Characterisation of the amino acids essential for the photo- and radioprotective effects of a Bowman-Birk proteinase inhibitor-derived nonapeptide. Protein Eng 14(3):157–160

9 Dittmann K, Güven N, Mayer C, Ohneseit P, Zell R, Begg AC, Rodemann HP (1998) The presence of wild-type p53 is necessary for the radioprotective effect of the Bowman-Birk proteinase inhibitor in normal fibroblasts. Radiat Res 150:648–655

10 Güven N, Keating KE, Chen P, Fukao T, Khanna KK, Watters D, Rodemann HP, Lavin M (2001) Epidermal growth factor sensitizes cells to ionizing radiation by down-regulating protein mutated in Ataxia-telangiectasia. J Biol Chem 276:8884–8891

11 Keating KE., Gueven N, Watters D, Rodemann HP, Lavin M (2001) Transcriptional downregulation of ATM by EGF is defective in ataxia telangiectasia cells expressing mutant protein. Oncogene 20:4281–4290

12 Kennedy AR and Little JB (1981) Effects of protease inhibitors on radiation transformation in vitro. Cancer Res 41:2103–2108

13 Lammering G, Hewit TH, Hawkins WT, Contessa, JN, Reardon DB, Lin PS, Valerie K, Dent P, Mikkelsen RB, Schmidt-Ullrich RK (2001) Epidermal growth factor receptor as a genetic therapy target for carcinoma cell radiosensitization. J Natl Cancer Inst 93:921–929

14 Meyer R, Küpper J.-H, Kandolf R, Rodemann HP (2002) Early growth response-1 gene 1 (Egr-1) promoter induction by ionizing radiation in U87 glioma cells in vitro. Eur J Biochem 269(1):337–346

15 Pauwels B, Korst AE, de Pooter CM, Lambrechts HA, Pattyn GG, Lardon F, Vermorken

JB (2003) The radiosensitising effect of gemcitabine and the influence of the rescue agent amifostine in vitro. Eur J Cancer 39:838–846
16 Rodemann HP, Meyer R, Kandolf R (1999) Strahleninduzierte Gentherapie: Möglichkeiten und Perspektiven. Der Onkologe 5:910–914
17 Schmidt-Ullrich RK, Dent P, Grant S, Mikkelsen RB, Valerie K (2000) Signal transduction and cellular radiation responses. Radiat Res 153:245–257
18 Scott SD, Joiner MC, Marples B: Optimizing radiation-responsive gene promoters for radiogenetic cancer therapy. Gene Ther 9:1396–1402, 2002
19 Senzer N: A phase III randomized evaluation of amifostine in stage IIIA/III non-small cell lung cancer patients receiving concurrent carboplatin, paclitaxel, and radiation therapy followed by gemcitabine and cisplatin. Semin Oncol 29:38–41, 2002.
20 Von Pfeil A, Hakenjos L, Herskind C, Dittmann K, Weller M, Rodemann HP (2002) Irradiated homozygous TGF-beta1 knockout fibroblasts show enhanced clonogenic survival as compared with TGF-beta1 wild-type fibroblast. Int J Radiat Biol 78 (5): 331–9
21 Yacoub A, McKinstry R, Hinman D, Chung T, Dent P, Hagan MP (2003) Epidermal growth factor and ionizing radiation up-regulate the DNA repair genes XRCC1 and ERCC1 in DU145 and LNCaP prostate carcinoma through MAPK signaling. Radiat Res 159(4): 439–52
22 Wick W, Wick A, Schulz JB, Dichgans J, Rodemann HP, Weller M (2002) Prevention of irradiation-induced glioma cell invasion by temozolomide involves caspae 3 activity and cleavage of focal adhesion kinase. Cancer Res 62(6):1915–1919

P53 Stabilization and the Role of Radiation-Induced Signalling

Christine Blattner, Markus Winter and Roman Kulikov

Forschungszentrum Karlsruhe, Institute of Genetics and Toxicology, PO-Box 3640, 76021 Karlsruhe, Germany
christine.blattner@itg.fzk.de

Abstract.

In the past, radiation research contributed fundamentally to our understanding of biological systems. The definition of a gene and the discovery of the cell cycle were, for instance, based on observations made by radiation biologists. Today, radiation research experiences considerable cuts in funding and positions. This cutback will lead to an unbearable loss in expertise in radiation related subjects and have substantial consequences for other scientific disciplines in the near future.
With this short summary of our more recent work we will show that today radiation research contributes fundamentally to various disciplines of biomedical research and constitutes an indispensable part of modern biomedical sciences.

Introduction

Radiation of cells and organisms damages the DNA and triggers a variety of actions, which are collectively called the radiation response. These actions include alterations in the gene expression program, implementation of cell cycle arrest and repair of lesions, or elicitation of programmed cell death. The latter events are quite well understood while the immediate actions of damage recognition and signal transduction are still a mystery.

In our group we are interested in these immediate events, for example, how the damage is recognized by cellular systems and which signaling components transmit the information in the cell. Here we are particularly focusing on the signaling cascade that controls the accumulation of the p53 tumor suppressor protein.

Mechanisms of p53 stabilization in response to irradiation

DNA damage causes the accumulation of the p53 protein to high levels, since its degradation is prevented in the presence of DNA lesions.
Initially we considered two principles as being responsible for rescueing p53 from degradation after DNA damage: I) the p53 protein itself could be modified in a way that prevents its recognition by the degradation system, or II) the degradation system could be inactivated in response to DNA damage.
To test the relevance of the former option, we created a number of mutant p53 proteins. In these mutant p53 proteins sites for post-translational modifications were replaced by non-modifiable residues or parts of the protein were deleted. When we assayed whether these mutant p53 proteins could be stabilized in response to DNA damage, we found, consistent with other research groups, that all mutant p53 proteins accumulated in response to DNA damage. Both the kinetics and the extent of accumulation were similar to wild type p53 [1,4,7].
In keeping with these results we considered that DNA damage-induced modifications of the p53 protein might only have a minor impact on the stabilization of the p53 protein and we speculated that the degradation system could be inactivated upon DNA damage.

The p53 protein is destroyed by the 26S proteasome, a multifactorial protease that degrades the majority of short-lived proteins in the cell. It consists of a cylindrical catalytic core that is bound at each end by a regulatory subunit. These regulatory subunits bind the substrates, unfold them and enable their entry into the catalytic core. Binding of the substrate to the regulatory domain requires prior polyubiquitylation of the substrate, an enzymatic reaction by which a multitude of ubiquitin residues are covalently attached to the target protein [for review see: 8].
The polyubiquitylation reaction requires the concerted action of three to four enzymes of the ubiquitin system. In an initial step, the small and ubiquitously expressed protein ubiquitin is activated by the ubiquitin-activating enzyme (E1). The activated ubiquitin is then transferred to an ubiquitin-conjugating enzyme (E2), which transfers the activated ubiquitin to the ubiquitin ligase, if the ubiquitin ligase contains a HECT domain, or associates with the ubiquitin ligase, if the ligase is of the "ring-finger"-type. The proper formation of the polyubiquitin chain requires in some cases the activity of an additional protein, which is called chain elongation factor (E4).
Most components of the ubiquitin degradation system are involved in the degradation of a large number of proteins. However, only a minority of cellular proteins are stabilized in response to DNA damage. Thus, the component that is inactivated upon DNA damage should be quite specific for p53. We therefore considered the ubiquitin ligase as the most likely target of DNA damage since ubiquitin ligases are quite specific for their corresponding substrate.

The ubiquitin ligase for p53 is Mdm2, a proto-oncogene that is occassionally amplified in human tumors [see 12 for an overview]. We therefore investigated whether Mdm2 is targeted and inactivated by DNA damage.

In an initial experiment we assayed Mdm2 protein levels after irradiation with ultraviolet light (UV) by Western Blotting. When cells were irradiated with UV, we observed that Mdm2 protein abundance dropped within 3–4 hours to undetectable levels [5]. Moreover, the decline in Mdm2 expression correlated in time with the accumulation of p53, consistent with the theory of stabilization of a target after depletion of its corresponding ubiquitin ligase.

Several principles could be responsible for the decrease in Mdm2 protein levels: I) transcription could be blocked, II) the turnover of the mRNA could be increased, or III) the half-life of the protein could be reduced.

In order to investigate the mechanism that leads to Mdm2 depletion, we first measured Mdm2 protein and RNA turnover. However, no reduction in the half-life of Mdm2 RNA or protein could be observed after irradiation. The overall amount of Mdm2 RNA was, however, significantly reduced, consistent with the idea of blocked transcription.

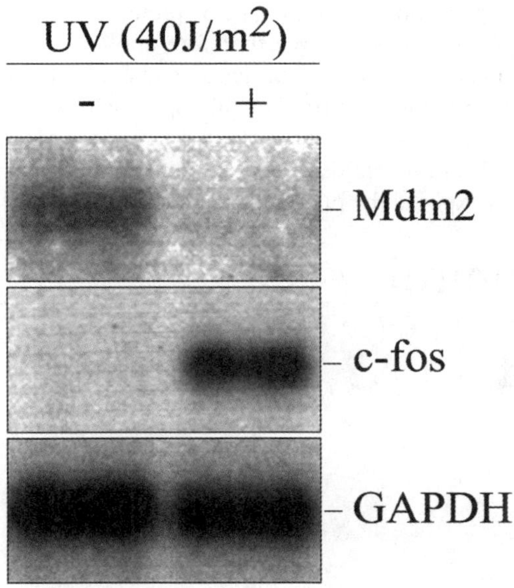

Fig. 1. Mdm2 RNA is downregulated after UV-irradiation.. NIH 3T3 cells were irradiated with 40 J/m2. 6 hours after irradiation cells were harvested and PolyA+ RNA was prepared. The RNA was separated through a formaldehyde-agarose gel, transferred to nitrocellulose membrane, consecutively probed with a Mdm2 cDNA, a c-fos cDNA and a GAPDH cDNA and exposed onto X-ray films.

UV light has been reported to be a general inhibitor of gene transcription [10]. Nevertheless, when we examined mRNA levels of the transcription factor c-fos, we found that its RNA still accumulated under conditions were Mdm2 RNA levels declined (Figure 1). Obviously, the c-fos gene can be transcribed under conditions, which interfere with transcription of the Mdm2 gene. We thus speculate that Mdm2 transcription requires a specific cellular factor whose activity or abundance is reduced upon UV-irradiation. This factor is most likely optional for transcription the c-fos gene.

After ionizing irradiation (IR), Mdm2 levels remained the same, the Mdm2 protein from irradiated cells migrated, however, faster through SDS-PAGE. Such shifts in protein migration usually indicate post-translational modifications. We therefore considered whether Mdm2 might be post-translationally modified in response to IR and thus inactivated.

Among the different kinds of modifications that are conceivable, for example, attachment or removal of methyl-, acetyl- or phosphate groups, sumo or ubiquitin or variations in glycosylation, phosphorylations seemed the most likely ones. The Mdm2 protein is a phosphoprotein with a remarkable number or putative phosphorylation sites and changes in the phosphorylation pattern of various proteins have been reported to occur in response to DNA damage [2].

A change in the phosphorylation pattern of proteins can easily be detected by two-dimensional peptide mapping. We therefore applied this method to obtain a comprehensive synopsis of all phosphorylation and dephosphorylation events that occur to the Mdm2 protein after IR. For this approach, we incubated cells with radioactive orthophosphate prior to irradiation, purified the Mdm2 protein from the cell extract, digested it into peptides, separated the peptides in two dimensions on thin layer plates and exposed the plates against an X-ray film.

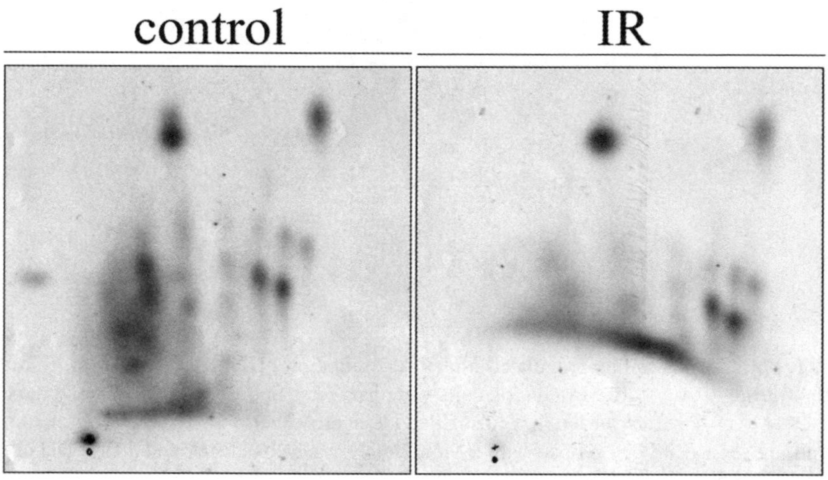

Fig. 2. Two-dimensional peptide mapping of Mdm2 from irradiated and control cells. Cells were incubated with radioactive orthophosphate. 4 hours after irradiation with 10 Gray, cells were harvested and Mdm2 was immunoprecipitated. Mdm2 immunocomplexes were separated by SDS-Page and Mdm2 was eluted from the gel. The purified Mdm2 protein was digested with chymotrypsin, the resulting peptides were separated on thin layer plates by electrophoresis and chromatography and the plates were exposed onto an X-ray film.

When we compared the phosphopeptide pattern of Mdm2 from irradiated cells with that from non-irradiated cells, we found multiple alterations (Fig. 2.). In addition to the phosphorylation of one particular peptide we also found several constitutively phosphorylated peptides of the Mdm2 protein to be hypophosphorylated in response to IR [3].

Phosphorylation of the central domain of Mdm2 is essential for p53 degradation

The hypophosphorylation of several constitutively phosphorylated residues of the Mdm2 protein after IR was very surprising and we considered whether this hypophosphorylation could contribute to the stabilization of p53 in response to DNA double strand breaks.

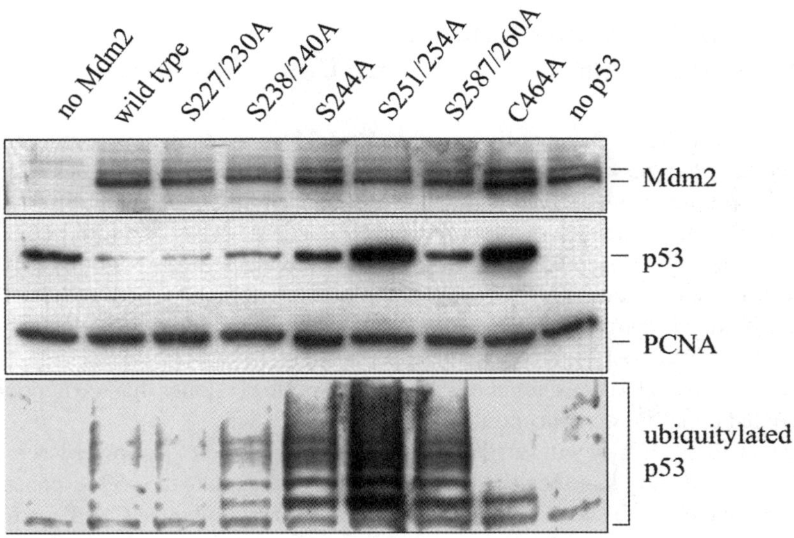

Fig. 3. Mutant Mdm2 proteins, where serines in the central domain were replaced by alanine fail to target p53 for degradation. P53, His-tagged ubiquitin and wild type Mdm2 or the indicated Mdm2 mutants were transfected into H1299 cells. 24 hours after transfection, cells were harvested and an aliquot of the cell lysat was separated by SDS page, blotted onto Immobilon-membrane and probed for the presence of Mdm2, p53 and PCNA (A). From the remaining lysate, ubiquitylated proteins were purified by adsorption to Ni2+-agarose, separated by SDS-page, transferred onto Immobilon membrane and probed for the presence of p53 (B).

With an approach involving site-directed mutagenesis and two-dimensional peptide mapping, we were able to link the hypophosphorylated residues to serines in the central domain of Mdm2. When we replaced these phosphorylation sites by alanine, which mimics a non-phosphorylated serine, we found that these Mdm2 mutants were hampered or failed completely to target p53 for degradation [3]. The severity of the phenotype varied in dependence on the residue that had been mutated with the single point mutation S244A and the double mutation S251A/S254A showing the strongest phenotype (Fig. 3.A). In contrast, mutants, where serines of the central domain were replaced by aspartic acids, which mimic phosphorylated serines were fully capable of targeting p53 for degradation indicating that the loss of the negative charge rather then the substitution per se interfered with p53 degradation.

Interestingly, all mutant Mdm2 proteins were fully capable of interacting with p53 and they all retained their capacity to act as an ubiquitin ligase for p53 (Fig. 3.B). Thus, and in contrast to previous assumptions, we have to accept, that ubiquitylation is insufficient for the degradation of p53. Our data clearly indicate that additional cellular factors are required for the degradation of ubiquitylated p53. These restrictions most likely apply also to the degradation of other cellular proteins.

The central domain of Mdm2 is phosphorylated by GSK-3 and Casein Kinase I

The importance of the central domain of Mdm2 for the regulation of p53 activity and the fact that this domain is hypophosphorylated at several contiguous residues after IR, prompted us to search for kinases, which phosphorylate Mdm2 in this domain. Given the acidity of this region, we wondered whether Mdm2 could be phosphorylated by Casein kinases. Casein kinases comprise a large group of constitutively active enzymes, which are divided in type I (CKI) and type II (CKII). Both classes of kinases phosphorylate preferentially serines and threonines that are surrounded by clusters of negative charges.

As we knew that CKII does not phosphorylate any of the residues that are hypophosphorylated upon IR, we concentrated in our study on CKI.

CKI comprises a family with several members (a, b, g, d and e) [for review see: 11]. The isoforms d and e have been reported previously to regulate the p53 response [9]. When we incubated bacterially expressed Mdm2 with recombinant CKId, we found that the central domain of Mdm2 was heavily phosphorylated by CKId. We therefore consider CKId as a possible candidate involved in the signal transduction chain that regulates p53 degradation. Whether CKId is involved in the regulation of p53 in response to IR is subject to current investigations.

In addition to the CKI phosphorylation sites, the Mdm2 protein possesses two consensus sites for the Glucagen sythase kinase 3 (GSK-3) in its central domain. GSK-3 is particularly renowned for its regulation of sugar metabolism and b-catenin degradation [reviewed in: 6].

When we incubated bacterially expressed Mdm2 with recombinant GSK3-b, the central domain of Mdm2 was heavily phosphorylated indicating that GSK3-b might be a prospective Mdm2 kinase. This idea is further supported by the observation that incubation of cells with synthetic GSK3-b inhibitors resulted in the stabilization of p53. GSK3-b activity is strongly reduced in response to IR. This inactivation could therefore be one of the principles leading to hypophosphorylation of Mdm2 in response to double strand breaks.

Conclusions

This summary of our work shows that radiation research is an innovative and modern branch of biomedical life sciences that contributes significantly to the fields of genetics, molecular biology and biochemistry. Termination of radiation research would therefore certainly lead to the disappearance of an instrumental tool for other branches of biomedical research in addition to the painful loss of an important scientific discipline.

References

1 Ashcroft M, Kubbutat MH, Vousden KH (1999) Regulation of p53 function and stability by phosphorylation. Mol Cell Biol 19:1751–1758
2 Bender K, Blattner C, Knebel A, Iordanov M, Herrlich P, Rahmsdorf HJ (1997) UV-induced signal transduction. J Photochem Photobiol B 37:1–17.
3 Blattner C, Hay T, Meek DW, Lane DP (2002) Hypophosphorylation of Mdm2 augments p53 stability. Mol Cell Biol 22:6170.6182.
4 Blattner C, Tobiasch E, Litfen M, Rahmsdorf HJ, Herrlich P (1999) DNA damage induced p53 stabilization: no indication for an involvement of p53 phosphorylation. Oncogene 18:1723–1732.
5 Blattner C, Sparks A, Lane D (1999) Transcription factor E2F-1 is upregulated in response to DNA damage in a manner analogous to that of p53. Mol Cell Biol 19:3704–3713
6 Cohen P, Frame S (2001) The renaissance of GSK3. Nat Rev Mol Cell Biol 2:769–776.
7 Hengstermann A, Whitaker NJ, Zimmer D, Zentgraf H, Scheffner M (1998) Characterization of sequence elements involved in p53 stability regulation reveals cell type dependence for p53 degradation. Oncogene 17:2933–2941
8 Hershko A, Ciechanover A (1998) The ubiquitin system. Annu Rev Biochem 67:425–479
9 Knippschild U, Milne DM, Campbell LE, DeMaggio AJ, Christenson E, Hoekstra MF, Meek D (1997) Casein Kinase 1 delta (CK1d) phosphorylates p53 in vivo and is induced by p53 in response to DNA damage. Oncogene 15:1727–1737
10 Ljungman M, Zhang F, Chen F, Rainbow AJ, McKay BC (1999) Inhibition of RNA polymerase II as a trigger for the p53 response. Oncogene 18:583–592
11 Vielhaber E, Virshup DM (2001) Casein kinase I: from obscurity to center stage. IUBMB 51:73–78
12 Vousden KH (2002) Activation of the p53 tumor suppressor protein. Biochim Biophys Acta 1602:47–59

Cytogenetics

Molecular Genetic Mechanisms of Radiocarcinogenesis and Predictive Markers for Radioresistance in Tumors – Present Work and Future Perspectives –

Ludwig Hieber and Horst Zitzelsberger

GSF-Institute of Molecular Radiobiology, Ingolstädter Landstraße 1, 85764 Neuherberg Germany
ludwig.hieber@gsf.de, zitzelsberger@gsf.de

Abstract

Mechanisms of radiation carcinogenesis are being studied in two different systems (i) radiation-transformed retina pigment epithelial cells (RPE) and (ii) primary cultures of human childhood papillary thyroid carcinomas from Belarus. Metaphase spreads are investigated with cytogenetic and molecular cytogenetic techniques, i.e. G-banding, comparative genomic hybridization (CGH) and spectral karyotyping (SKY) in order to identify chromosomal aberrations. Chromosomal breakpoints involved in chromosomal changes are characterized via positional cloning using fluorescence in situ hybridization (FISH) with breakpoint-spanning YAC and BAC clones. Candidate genes are identified via sequence database blast search and the gene expression patterns are analyzed by RT-PCR and Northern blot analyses. In order to search for predictive markers for radioresistance in tumors, CGH analysis will be performed with radiosensitive and radioresistant cohorts of a particular tumor entity (i.e. head and neck squamous cell carcinomas). Mapping of chromosomal regions distinctive for radioresistant tumors are performed using FISH analysis with YAC and BAC clones.

Introduction

The 'Cytogenetic Research Group' of the GSF-Institute of Molecular Radiobiology is interested in three major fields of cytogenetic research, **(i) quantification of radiation exposures, (ii) mechanisms of radiocarcinogenesis and (iii) tumor-specific changes and stage-specific markers of different human tumor entities**.

In the field of **quantification of radiation exposures**, dose reconstruction of accidentially or occupationally radiation-exposed individuals is performed by biological dosimetry in peripheral blood lymphocytes. This is necessary when no physical dosimetry is available. For this purpose, either the number of dicentric chromosomes, ring chromomes and acentric fragments or the frequency of translocations via FISH analysis is examined. In order to improve biological dosimetry, also issues of individual radiosensitivity, of the temporal persistence of translocations and of DNA-proportional distribution of aberrations have been addressed.

To study **mechanisms of radiocarcinogenesis** chromosomal breakpoints in cell cultures of radiation-induced thyroid tumors and in radiation-transformed cell lines have been identified. Using the technique of positional cloning chromosomal breakpoints by FISH candidate genes are identified and their involvement in the carcinogenic process is analyzed. A part of these studies are described in "Mechanisms of radiation carcinogenesis" in more detail.

A further research area are **cytogenetic studies of different human tumours**, in order to identify tumor entity-specific chromosomal aberrations which may also correlate with different stages of tumor progression. An additional aim is to find predictive markers for radiorestiance and/or radiosensitivity. A recently proposed project on head and neck cancers will be described in "Predictive markers for radioresistance in tumors", page 108.

Mechanisms of radiation carcinogenesis

Model systems

For the investigation of mechanisms underlying radiation carcinogenesis two model systems are used.

During the past 10 years we have collected about 270 tissue samples from radiation-induced childhood thyroid carcinomas from Belarus which occurred after the Chernobyl reactor accident. The fresh tumor samples were divided into three parts, (i) one was embedded in paraffin for pathological diagnosis, (ii) one was shock-frozen for molecular genetic experiments and (iii) one was used for initiation of primary cell cultures for cytogenetic analysis. Several studies have been performed, i.e. analysis of mutations of TP53 [1], analysis of rearrangements of the proto-oncogene RET, which led to the result that more than 56% of the papillary thyroid carcinomas showed ret/PTC1 and ret/PTC3 rearrangements [2]. In addition, a new rearrangement has been identified in one case, i.e. ret/PTC8 [3]. Cytogenetic analysis of metaphase spreads from the primary cultures showed that approxiamtely 25% of cases had clonal and non-clonal chromosomal aberrations in addition to RET rearrangements [4].

For in vitro radiation-induced transformation studies immortalized human retina pigment epithelial cells (RPE) are the model system of choice. RPE cells have been immortalized by transfection with a telomerase construct, containing the catalytical subunit of human telomerase (hTERT), which leads to the stabilization of the telomers [5]. Since the immortalized parental cell line exhibits a normal stabile human female karyotype 46,XX, karyotypic changes in radiation-transformed cell lines can be assigned to the radiation-induced neoplastic process. RPE cells were exposed to 10 or 15 fractionated doses of 2 Gy of Cs137 grays; anchorage-independent, transformed clones were isolated from soft agar cultures (RPE-X1.5 to RPE-X1.12) [6]. Clones were tested for tumorigenic potential in athymic nude mice [6]. Tumor cell lines were established from the isolated tumors after sacrificing the animals (for example RPE-X1.5TCl).

Identification of chromosomal changes and breakpoints in radiation-transformed RPE clones

To identify chromosomal changes the transformed RPE clones were analyzed by G-banding, comparative genomic hybridization (CGH) and/or spectral karyotyping (SKY).

For G-banding-analysis metaphase spreads were stained with Giemsa-Wright solution. Karyotyping was performed according the International System for Human Cytogenetic Nomenclature [7].

CGH analysis was performed as described earlier [8, 9]. DNA from RPE clones and human reference DNA were labeled with a green (FITC) and a red (TRITC) fluorescent dye and hybridized onto metaphases from normal human lymphocytes. Averaged profiles were generated by CGH analysis software (Isis 3, Metasystems) and interpreted according published criteria.

For SKY analysis metaphase spreads from the RPE clones were hybridized with painting probes for all human chromosomes that were differentially labeled with combinations of five fluorochromes (Spectrum Green, Spectrum Orange, Texas Red, Cy5 and Cy5.5). The slides were counterstained with DAPI. Metaphases were acquired by use of a SpectralCube system (ASI) and analyzed with the SKYView imaging software [9,10].

Whereas the parental RPE cell line exhibits a stable normal human female karyotype 46,XX, all radiation-transformed RPE clones showed chromosomal changes. RPE-X1.9 to RPE-X1.12 showed a near-tetraploid karyotype with various marker chromosomes (see Fig. 1., page 106). Chromosomal aberrations are visible as color junctions in the SKY pictures. RPE-X1.5, RPE-X1.7 and RPE-X1.8 had a identical near-diploid karyotype 45,XX. There is a common loss of chromosome 13 (see Fig. 2.A, page 106) and a marker chromosome 10 (see Fig. 2.B, page 106). CGH analysis revealed a high level amplification on chromosome 10p11.2 and a deletion of the remaining short arm segment of chromosome 10 distal to 10p11.2 resulting in a huge marker chromosome 10. This chromosome was further analyzed for breakpoint characterization.

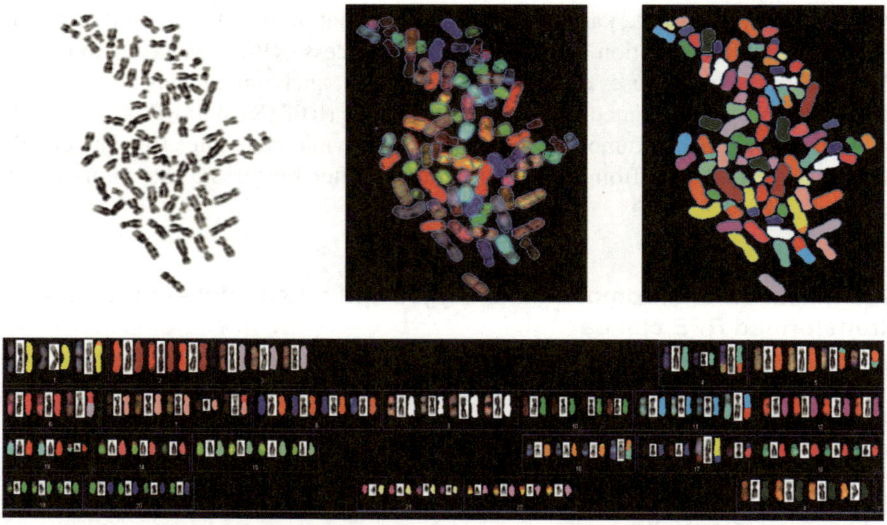

Fig. 1. Sky analysis of clone RPE-X1.10 with a near-tetraploid karyotype (75–86 chromosomes) with various marker chromosomes. A: DAPI staining corresponding to G-banding, B: RGB image, C: classified image, and D: classified spectral karyotype.

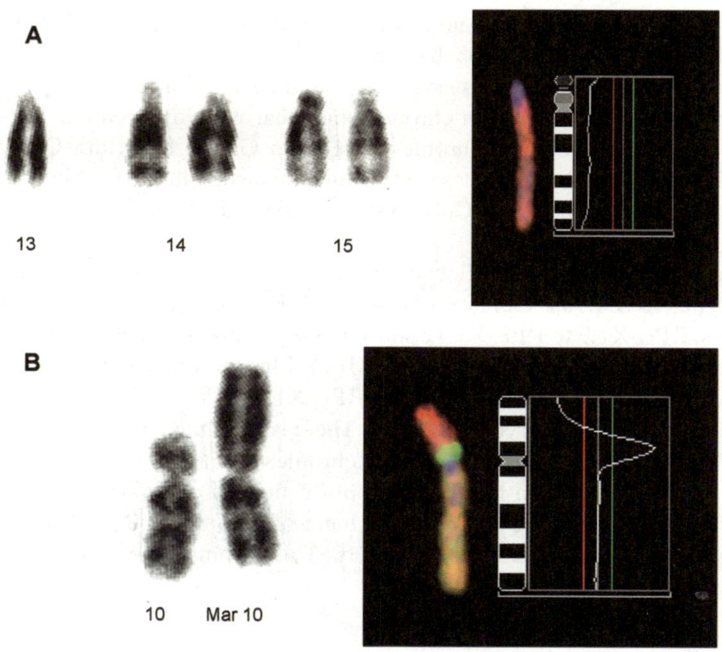

Fig. 2. G-banding and CGH analysis showing loss of chromosome 13 (A) and one marker chromosome 10 (B). CGH analysis reveals an amplification of 10p11.2 and a deletion of the distal part of 10p.

Characterization of breakpoints by positional cloning

For positional cloning of chromosome 10p11.2 several overlapping YAC clones were selected from the Whitehead Institute database (www.genome.wi.mit.edu/cgi-bin/contig/phys_map), labeled with FITC and hybridized by FISH onto metaphase spreads of RPE-X1.5, RPE-X1.7 and RPE-X1.8. Three breakpoint-spanning YAC clones exhibited highly amplified hybridization signals on chromosome 10p (see Fig. 3.). To further narrow down the chromosomal region a series of BAC clones within the sequence of the hybridizing YACs were hybridized to metaphases, two of them gave also highly amplified FISH signals similar to the YACs (not shown).

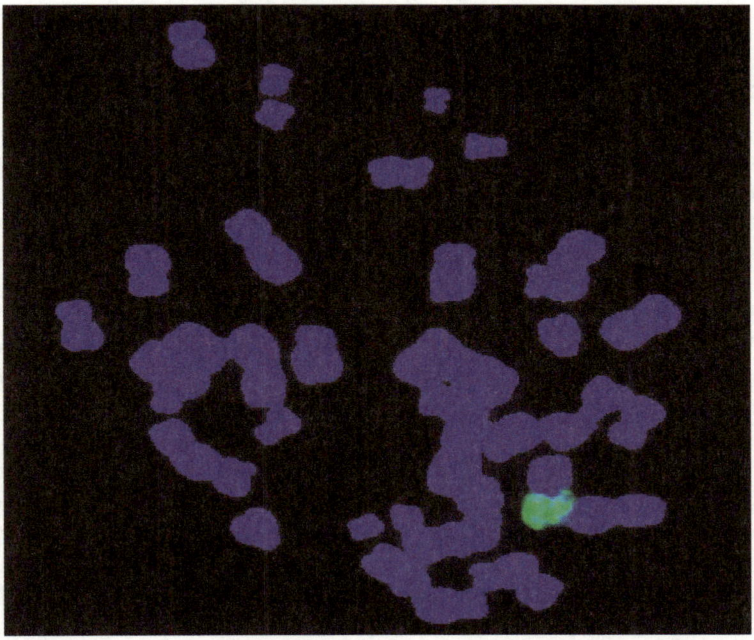

Fig. 3. FISH analysis of RPE-X1.5, RPE-X1.7, and RPE-X1.8 hybridized with a YAC clone homologous to 10p11.2 exhibiting a highly amplified signal.

Identification and verification of the candidate gene

For the identification of a candidate gene a sequence blast search of the known DNA sequence of the BAC clones revealing an amplified FISH signal has been performed with the Ensembl database (www.ensembl.org). The sequence analysis of respective BAC clones indicated that only one gene, i.e. the gene coding for the *atypical PKC isotype-specific interacting protein (Asip)* contributed to the observed amplification. Therefore, a PCR product of 2 kb from the coding region of this gene has been produced, labeled and hybridized to metaphase spreads of the RPE clones.

Also this probe exhibited a highly amplified signal on the marker chromosome 10. This gave evidence for the involvement of Asip in the transformed phenotype of the cell clones.

To improve this hypothesis gene expression is being studied by RT-PCR analysis. Total RNA has been isolated from the cell clones and transcribed into cDNA. Prelimary result indicate that an Asip-specific PCR product could be produced in the tested clones as well as in the respective tumor cell clones, but not in the parental non-transformed cell line. This is an indication that Asip might be overexpressed in the transformed RPE clones. For further support, a Northern blot as well as a quantitative comparative RT-PCR analysis are underway. Asip has been described to be involved in several cases of human hepatocellular carcinomas [11].

Predictive markers for radioresistance in tumors

Radiotherapy is one of the treatments to treat different human tumors such as head and neck squamous carcinomas or lung tumors. However, not all tumors do respond equally well to this therapy. It would be, therefore, a great advantage for therapy planning and for prognostic purposes, if predictive genetic markers for radioresistant tumors would be available.

We therefore propose a research project to identify predictive markers for radioresistance. We have access to about 180 squamous head and neck tumors, from which clinical follow up and data regarding therapy resistance is well documented. For this cohort there are pre-therapeutic tissue samples as well as samples of tumor resection after radiotherapy. As a first step, CGH analysis in radiosensitive and radioresistant tumors will be performed in order to identify chromosomal regions that exhibit changes in the tumor samples (amplified and deleted chromosomal regions). Statistical analysis of the aberration patterns will provide specific changes for the radioresistant phenotype. In the next step, these chromosomal regions will be narrowed down by positional cloning using FISH analysis with YAC and BAC clones as described in section 1 of this paper. From DNA sequence analysis candidate genes that might be responsible for a radioresistant phenotype will be identified. In addition, by the use of real-time TaqMan PCR the expression of affected genes and known repair, tumor suppressor and oncogenes will be investigated.

Future perspectives of radiation cytogenetics

Radiation Cytogenetic research in future will lead to a more detailed understanding of mechanisms underlying radiation carcinogenesis. It will be important to further identify rearranged genes either in radiation-transformed model systems as well as in human radiation-induced tumors such as thyroid carcinomas described above. Investigations will be performed to study gene interactions to understand the changes in function of rearranged gene products.

A further benefit of radiation cytogenetic research will result in the optimization of tumor therapy. If it will be possible to identify predictive markers either for radioresistant or radiosensitive tumors this will lead to great advantage for therapy planning. The development of cytogenetic diagnostic markers in biopsy samples that may correlate with tumor progression can also improve therapy planning and are of prognostic value too.

Alltogether, future radiation cytogenetic research will have great impact in several fields, such as *Cancer Genetics* and *Tumor Biology, Genomics* and *Epidemiology* as well as for *Radiation Biodosimetry* and *Radiation Protection.* Clinical impact will be brought to *Tumor Pathology, Tumor Diagnosis* and *Radiotherapy.*

References

1 Smida J, Zitzelsberger H, Kellerer AM, Lehmann L, Minkus G, Negele T,Spelsberg F, Hieber L, Demidchik EP, Lengfelder E, Bauchinger M (1997) p53 mutations in childhood thyroid tumours from Belarus and in thyroid tumours without radiation history. Int J Cancer 73:802–7

2 Smida J, Salassidis K, Hieber L, Zitzelsberger H, Kellerer AM, DemidchikEP, Negele T, Spelsberg F, Lengfelder E, Werner M, Bauchinger M.(1999) Distinct frequency of ret rearrangements in papillary thyroid carcinomas of children and adults from Belarus. Int J Cancer 80:32–8

3 Salassidis K, Bruch J, Zitzelsberger H, Lengfelder E, Kellerer AM, Bauchinger M (2000) Translocation t(10;14)(q11.2:q22.1) fusing the kinetin to the RET gene creates a novel rearranged form (PTC8) of the RET proto-oncogene in radiation-induced childhood papillary thyroid carcinoma. Cancer Res 60:2786–9

4 Zitzelsberger H, Lehmann L, Hieber L, Weier HU, Janish C, Fung J, Negele T, Spelsberg F, Lengfelder E, Demidchik EP, Salassidis K, Kellerer AM, Werner M, Bauchinger M (1999) Cytogenetic changes in radiation-induced tumors of the thyroid. Cancer Res 59:135–40

5 Bodna, A., Ouellette M., Frolkis M., Holt S., Chiu C., Morin G., Harley C., Shay J., Lichtsteiner S., and Wright W (1998) Extension of life span by introduction of telomerase into normal human cells. Science 279: 349–353

6 Riches A., Peddie C., Rendell S., Bryant P., Zitzelsberger H., Bruch J., Smida J., Hieber L., and Bauchinger M (2001) Neoplastic transformation and cytogenetic changes after gamma irradiation of human epithelial cells expressing telomerase. Radiat Res 155: 222–229

7 Mitelman, F (1995) Ed.: An International System for Human Cytogenetic Nomenclature. S. Karger, Basel

8 Walch AK, Zitzelsberger HF, Aubele MM, Mattis AE, Bauchinger M, Candidus S, Prauer HW, Werner M, Hofler H (1998) Typical and atypical carcinoid tumors of the lung are characterized by 11q deletions as detected by comparative genomic hybridization. Am J Pathol 153:1089–98

9 Zitzelsberger H, Kulka U, Lehmann L, Walch A, Smida J, Aubele M, Lorch T, Hofler H, Bauchinger M, Werner M (1998) Genetic heterogeneity in a prostatic carcinoma and

associated prostatic intraepithelial neoplasia as demonstrated by combined use of laser-microdissection, degenerate oligonucleotide primed PCR and comparative enomic hybridization. Virchows Arch 433: 297–304

10 Schrock E, du Manoir S, Veldman T, Schoell B, Wienberg J, Ferguson-Smith MA, Ning Y, Ledbetter DH, Bar-Am I, Soenksen D, Garini Y, Ried T (1996) Multicolor spectral karyotyping of human chromosomes. Science 273 (5274): 494–7

11 Fang CM, Xu YH (2001) Down-regulated expression of atypical PKC-binding domain deleted asip isoforms in human hepatocellular carcinomas. Cell Res 11: 223–229

FISH and CHIPs: Colorful Clues to Radiation-Induced Chromosomal Instability

Tanja Hardt[a], Karen Stout[a], Frank Guthermuth[b], Jürgen Kiefer[b], and Thomas Haaf[c]

[a]Charité Institut für Medizinische Genetik, Humboldt Universität, 13353 Berlin, Germany
[b]Strahlenzentrum, Justus Liebig-Universität, 35392 Giessen, Germany
[c]Institut für Humangenetik, Johannes Gutenberg-Universität, 55101 Mainz, Germany
haaf@humgen.klinik.uni-mainz.de

Abstract.

Radiation produces a variety of clonal and non-clonal chromosome aberrations that can be characterized by fluorescence in situ hybridization (FISH). Epigenetic changes affecting the expression of an essential DNA repair gene(s) may be an importantant mechanism for radiation-induced chromosomal instability. Expression profiling with specialized cDNA chips promises to identify candidate genes for the delayed effects of radiation and to provide new insights into the manifold and complex cellular responses to DNA damage. Much progress can be made by using FISH and CHIPs to study the mechanisms and biological consequences of ionizing radiation.

Introduction

Both evolutionary and pathological chromosome rearrangements are a universal result of DNA breakage and subsequent misrepair. DNA double-strand breaks arise as a direct effect of ionizing radiation, but may also occur spontaneously during DNA replication and meiotic recombination. From an evolutionary point of view, radiation-induced and other pathological chromosome rearrangements are just a price to be paid for the dynamic nature of our genome. In classical radiation biology, chromosome aberrations have been extensively used for biodosimetry and cancer risk estimation. However, more recently with the availability of powerful FISH techniques [3] it has become apparent that radiation-induced chromosome aberrations can also act as colorful clues to DNA damage processing mechanisms [6].

Radiation-induced chromosome aberrations are not only attributed to the production of double-strand breaks at the time of irradiation. Cells that have survived irradiation may produce clonal descendants in which de novo aberrations arise. These delayed effects are manifestations of a radiation-induced genomic instability, which depends on the type of radiation exposure and the genetic background of the irradiated cells. The relatively high frequency of non-clonal aberrations in the clonal descendants of irradiated cells is consistent with an epigenetic mechanism but not with direct gene mutations [9, 14]. Considering that genomic instability is one of the most important hallmarks of tumorigenesis, this phenomenon deserves to be studied in more detail. To this end, we have isolated and cytogenetically characterized stable and unstable clones from an irradiated human cell line. Expression profiling with specialized cDNA chips will be used to study whether modulation of DNA repair enzymes is a possible mechanism underlying radiation-induced genomic instability.

Materials and methods

The P3 cell line is derived from a human epithelial teratocarcinoma and extensively used in radiation biology [5]. Apart from a derivative t(15;20) chromosome, P cells do not show cytogenetically detectable gross chromosomal abnormalities (Fig. 1). In contrast to most other tumors, the P3 karyotype is relatively stable. P3 cells were grown in D-MEM medium supplemented with 10% fetal bovine serum and and antibiotics at 37°C. Subconfluent P3 cell cultures were exposed to a 250 kV X ray tube at doses of 2 Gy and 4 Gy. Following irradiation, the cells were harvested and seeded at low density in Petri dishes. After 10 days cells surviving irradiation had formed colonies. Each single clone was then transferred to a new culture flask and allowed to grow for two more weeks. Metaphase chromosomes were prepared according to standard procedures. At least 10 metaphases per clone were analyzed by spectral karyotyping (SKY). Clonal (stable) translocations which were present in all metaphases of an analyzed clone were the direct result of misrepaired DNA breaks at the time of irradiation. Non-clonal translocations that occurred in only a percentage of metaphases were formed some time after irradiation, reflecting radiation-induced chromosomal instability.

For SKY, combinatorial labeling of DNA libraries for the 24 human chromosomes was performed by directly incorporating Spectrum Green-dUTP, Spectrum Orange-dUTP, Texas Red-dUTP, biotin-16-dUTP, and digoxigenin-11-dUTP in a degenerate oligonucleotide primed PCR reaction. One microgram of each of the differentially labeled painting probes that is approximately 25 mg of DNA was coprecipitated overnight with 50 mg human cot1 DNA (as competitor) and 10 mg salmon sperm DNA (as carrier). Hybridization took place in 50% formamide, 10% dextran sulfate and 2′ SSC for three days at 37°C. The biotinylated probes were detected with Cy5-conjugated avidin, the digoxigenin-labeled probes by incubation with a mouse anti-digoxin antibody followed by Cy5.5-conjugated anti-mouse IgG. Image acquisition was performed using a custom designed optical filter set with broad emission bands and a SD200 Spectracube (Applied Spectral Imaging) attached to a CCD camera. The emission light of the illuminated sample had to pass a Sagnac interferometer in the optical head, thus generating an interferogram of all image points. This interferogram was converted by Fourier transformation into a spectrum.

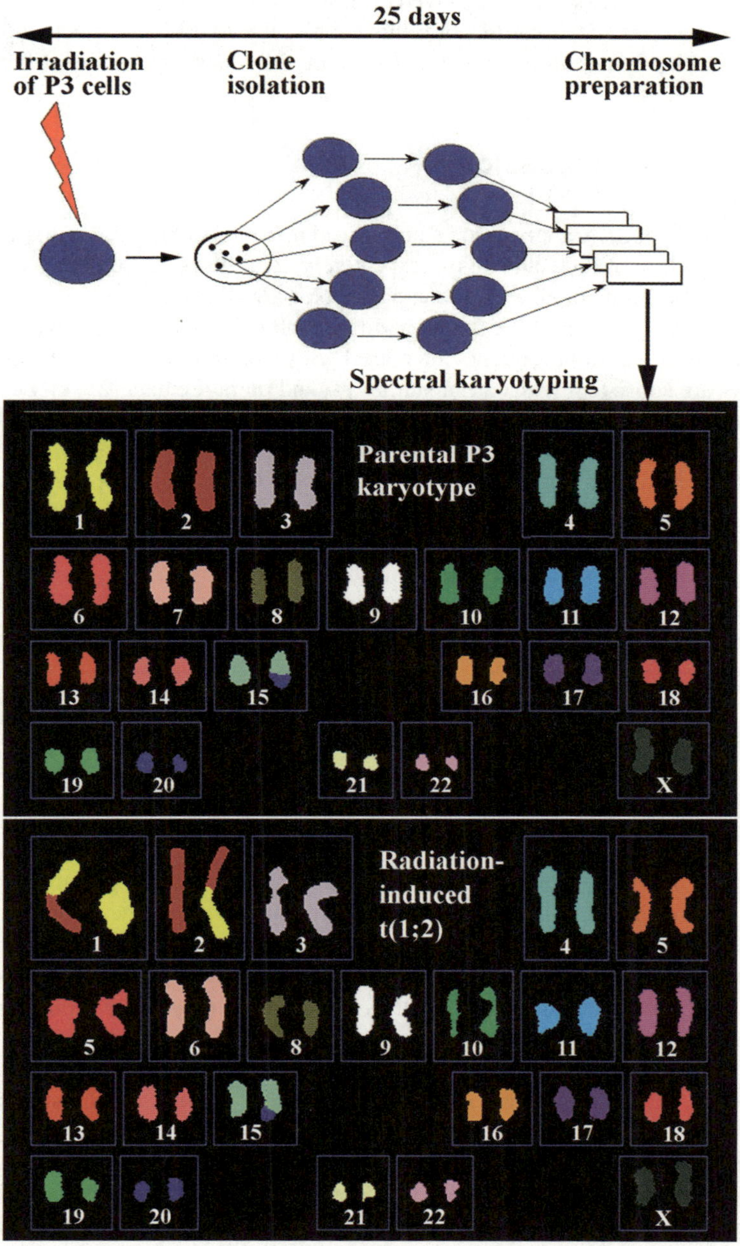

Fig. 1. Outline of the experiment. Clonal descendants of irradiated P3 cells were analyzed by spectral karyotyping. The karyotype of the parental P3 line is characterized by a derivative t(15;20) chromosome which serves as a staining control. The bottom karyotype displays an additional radiation-induced translocation between chromosomes 1 and 2.

The spectral information was displayed by assigning RGB colors to specific spectral ranges. A spectral classification algorithm enables the 24 different spectra (chromosome-specific fluorochrome combinations) in the image to be identified [12].

Results and discussion

The results of SKY analysis are summarized in Table 1. After X irradiation of P3 cells with a dose of 2 Gy and 4 Gy, respectively, 27% and 44% of clones exhibited stable translocations in all their metaphases. Non-irradiated clones showed the P3-specific t(15;20) chromosome but no additional clonal translocations. As expected, the stable translocation frequency increased with radiation dose. The observed frequencies are consistent with earlier studies [4] and literature data [2, 11]. Seventeen percent of the clones after 2 Gy exhibited non-clonal translocations in some but not all their metaphases, compared to only 13% of unstable clones in non-irradiated cultures. However, after irradiation with a dose of 4 Gy, we did not observe clones showing instability. Evidently, the radiation-induced chromosomal instability is not dose-dependent and the P3 cell line not particularly susceptible to the initiation of genomic instability by ionizing radiation.

Table 1. Clonal and non-clonal translocations in clonal descendants of irradiated P3 cells

X-ray dose	Number of clones analyzed	Number (%) of clones with clonal translocations	Number (%) of clones with non-clonal translocations
0 Gy	30	0 (0%)	4 (13%)
2 Gy	30	8 (27%)	5 (17%)
4 Gy	9	4 (44%)	0 (0%)

At present littele is understood of the mechanisms behind radiation-induced instability. Multiple DNA repair pathways are involved in the maintenance of genome stability. Both inherited mutations and epigenetic (acquired) deficiencies in these maintenance systems can lead to genome instability and carcinogenesis [8, 6]. Abnormal promoter methylation of DNA repair genes may cause their epigenetic silencing and, thus, contribute to cancer development [1]. Accumulating experimental evidence suggests that at least some variation in DNA repair gene expression is epigenetic in origin. Following irradiation of P3 cells, chromosomal instability was induced at relatively high frequencies (Table 1). This supports the idea that epigenetic changes rather than direct gene mutations are the underlying mechanism. Previously promoter methylation and gene inactivation have been demonstrated in X-ray and plutonium-induced lung tumors [7]. We therefore propose that radiation-induced chromosomal instability may be a manifestation of modulation of DNA repair. To test this hypothesis, we have constructed a high-density cDNA microarray chip for the massive parallel expression analysis of human DNA repair genes (see Table 2., page 115)).

Table 2. Selected human DNA repair genes

Base excision repair
APE1, APE2, LIG3, MBD4, OGG1, SMUG1, UNC, XRCC1
Direct repair
MGMT
Homologous recombination
BRCA1, BRCA2, DMC1, RAD51, RAD51B–RAD51D, RAD52, RAD54, XRCC2, XRCC3
Mismatch repair
MLH1, MLH3, MSH2–MSH6, PMS1, PMS2
Non-homologous end-joining
DNA-PK, Ku70, Ku80, LIG1, LIG4, MRE11, NBS, RAD50, XRCC4
Nucleotide excision repair
CSA, CSB, ERCC1, LIG1, RPA1–RPA3, TFIA, TFIIH, XPA–XPG, XPV
Repair disorders
ATM, ATR, BLM, FANCA–FANCG, RECQL4, WRN
Telomere integrity
TEL1, RAP1, RAP2, TRF1, TRF2
Others
CHK1, CHK2, GADD45, MDM2, PARP, PNKP, RAG1, RAG2, RB
A more comprehensive list of human repair genes is available at http://www.cgal.icnet.uk/DNA_Repair_Genes.html (Wood et al., 2001)

Expressed sequence tags (EST clones) of several hundred genes that are associated with DNA repair, cell cycle control and apoptosis, along with 50 control housekeeping genes were PCR amplified and then spotted onto glass chips, using a robot (Fig. 2, page 116). Polyadenylated mRNAs from a test sample, i.e. an irradiated P3 clone, and the parental control cell line were differentially fluorescent-labeled by reverse transcription and hybridized together to this chip. The mRNA levels for each gene are evaluated by calculating the ratio of spot intensity in irradiated cells to that in control cells. Overexpression of a gene in an irradiated cell clone increases the red fluorescence intensity of the tester RNA in the corresponding spot, whereas downregulation results in an elevated green fluorescence of the control RNA. Fig. 3 (page 116) shows hybridization of a tester mRNA sample to such a chip. To identify genes underlying radiation-induced genomic instability, we will perform systematic chip hybridization experiments to compare the expression patterns of a larger number of stable and unstable clones of the irradiated P3 cell line with that of the non-irradiated P3 control cell line. Genes that contribute to genomic instability should be modulated differentially in clones with an unstable phenotype.

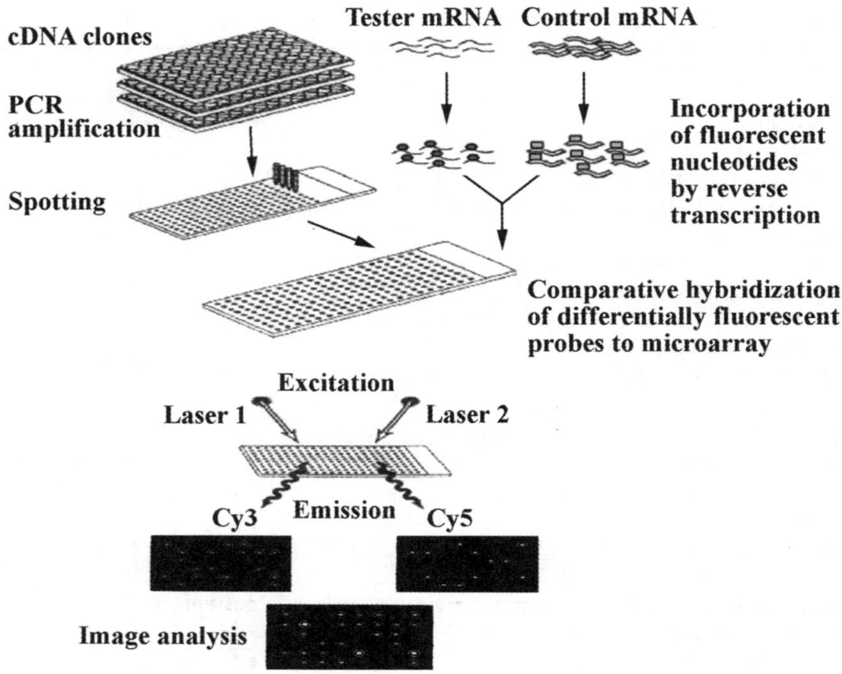

Fig. 2. Quantitative expression analysis of DNA repair genes in the clonal descendants of irradiated P3 cells using cDNA microarrays. The mRNA sample of a stable or an unstable clone is used as a tester and compared to that of the non-irradiated P3 control.

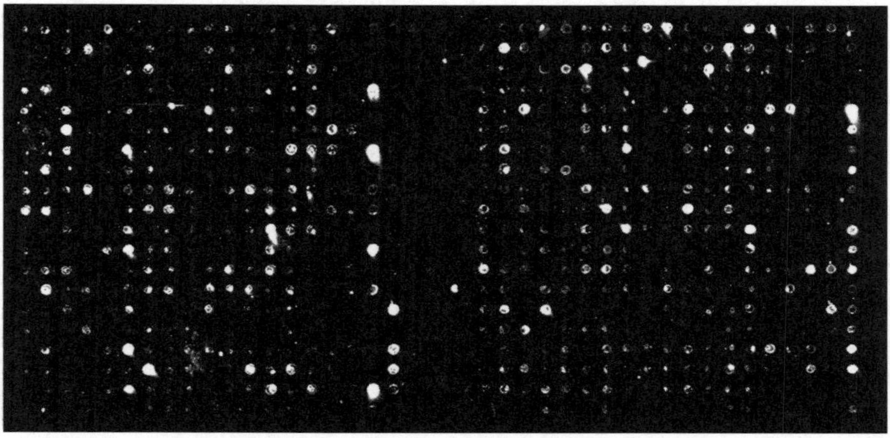

Fig. 3. Representative image of a chip hybridization experiment. Two subarrays of 20 x 20 spots (EST clones) carrying 350 DNA repair, cell cycle control and apoptosis genes and 50 housekeeping genes were hybridized with 25 mg total mRNA of a tester sample. Polyadenylated RNA was labeled with Cy3 by reverse transcription and hybridized for 12 h at 55°C in 3x SSC.

References

1 Esteller M, Gaidano G, Goodman SN, Zagonel V, Capello D, Botto B, Rossi D, Gloghini A, Vitolo U, Carbone A, Baylin SB, Herman, JG (2002) Hypermethylation of the DNA repair gene O(6)-methylguanine DNA methyltransferase and survival of patients with diffuse large B-cell lymphoma. J Natl Cancer Inst 94: 26–32
2 Finnon P, Lloyd DC, Edwards AA (1995) Fluorescence in situ hybridization detection of chromosomal aberrations in human lymphocytes: applicability to biological dosimetry. Int J Radiat Biol 68: 429–435
3 Haaf T (2000) Fluorescence in situ hybridization, in Encyclopedia of Analytical Chemistry. Volume 1. Nucleic Acids Structure and Mapping, Meyers RA, Ed., John Wiley & Sons, Chichester, pp. 4984–5006
4 Hardt T (1996) Dosis-Effekt-Beziehungen für Translokationen nach Röntgenbestrahlung unter Anwendung der Fluoreszenz-in situ-Hybridisierung. Diploma thesis, Institute for Biophysics, University of Giessen
5 Hill CK, Holland J, Chang-Liu CM, Buess EM, Peak JG, Peak MJ (1988) Human epithelial teratocarcinoma cells (P3): radiobiological characterization, DNA damage, and comparison with other rodent and human cell lines. Radiat. Res. 113, 278–288
6 Hoeijmakers JHJ (2001) Genome maintenance mechanisms for preventing cancer. Nature 411: 366–374
7 Issa JP, Baylin SB, Belinsky SA (1996) Methylation of the estrogen receptor CpG island in lung tumors is related to the specific type of carcinogen exposure. Cancer Res 56: 3655–3658
8 Jones PA, Laird PW (1999) Cancer epigenetics comes of age. Nature Genet 21: 163–167
9 Mothersill C, Seymour CB (1998) Mechanisms and implications of genomic instability and other delayed effects of ionizing radiation exposure. Mutagenesis 13: 421–426
10 Sachs RK, Hlatky LR, Trask BJ (2000) Radiation-produced chromosome aberrations: colorful clues. Trends Genet 16: 143–146
11 Schmid E, Zitzelsberger H, Braselman H, Gray JW, Bauchinger M (1992) Radiation-induced chromosome aberrations analyzed by fluorescence in situ hybridization with a triple combination of composite whole chromosome-specific DNA probes. Int J Radiat Biol 62: 673–678
12 Schröck E, du Manoir S, Veldman T, Schoell B, Wienberg J, Ferguson-Smith MA, Ning Y, Ledbetter DH, Bar-Am I, Soenksen D, Garini Y, Ried T (1996) Multicolor spectral karyotyping of human chromosomes. Science 273: 494–497
13 Wood RD, Mitchell M, Sgouros J, Lindahl T (2001) Human DNA repair genes. Science 291: 1284–1289
14 Wright EG (1998) Radiation-induced genome instability in haematopoietic cells. Int J Radiat Biol 74: 681–687

Carcinogenesis, Radiation Protection

Recombination and Radiation-Induced Cancer. Mechanisms and Genetic Testing

Jochen Dahm-Daphi , Petra Hubbe

University of Hamburg, Clinics of Radiotherapy and Radiation Oncology, Martinistr. 52, 20246 Hamburg, Germany
dahm@uke.uni-hamburg.de

Radiation-induced cancer

Radiation-induced cancer is the major problem for general radiation protection and a well recognized late effect after radiation therapy for malignant or benign diseases. A large part of the current knowledge is based on the analysis of cancer cases following the atomic bombs in Hiroshima and Nagasaki [98,]. Further understanding of carcinogenesis arose from accidental radiation exposure the most prominent of which occurred after the reactor catastrophe in Tschernobyl in 1986. During the next ten years several hundred cases of childhood thyroid cancer were observed [61, 43]. Among natural radiation sources domestic exposure to radon has the highest impact on human carcinogenesis, linked to lung cancer [55, 13]. However cancer induction is also observed as a sequel of medical application of radiation. Under rare circumstances, carcinogenic risk can even be associated with diagnostic procedures. This has been found for X-irradiation performed during pregnancy [19] and in tuberculosis patients who developed lung cancer after repeated exposure to X-rays for chest photographs, where they accumulated appreciable doses [10]. Therapeutic irradiation is better known to induce cancer due to higher doses applied leading to a higher probability of developing cancer. In several studies radiotherapy for benign diseases such as juvenile thymus enlargement [99], tonsil hypertrophy [80], tinea capitis [90], hemangioma [66]), Bechterew´s disease [109] and post-partum mastitis [100] was found to be associated with an enhanced cancer risk. Also radiotherapy for primary malignomas is, in rare cases though (1%), followed by the occurrence of a second primary tumours. This has been documented after treatment for cervix [11, 3, 41], prostate cancer [14, 83], rhabdomyosarcoma [41], Hodgkin´s lymphoma [79, 7], mamma [68], retinoblastoma [114] and CNS tumours [38].

Translocation as a prominent pathway toward carcinogenesis

The most frequent malignomas arising after radiation exposure are leukemias or lymphomas [79, 11, 3, 41, 109, 69, 7,] and a variety of solid tumours, namely thyroid cancers, [99, 80, 90, 38, 66, 7, 81], sarcomas [38, 41, 14, 7, 109, 114, 83, 37], mamma [9, 100, 7] lung [98, 109, 14] colorectal [109, 14, 50, 83] and other cancers. These radiation-induced tumours typically originate at the margins of the radiation fields or a few centimeters beyond where the absorbed dose was less than 5–6 Gy [26, 35]. Higher doses as achieved within the radiation field predominantly result in cell killing, hence, preventing the clonogenic survival of mutated cells.

Table 1. Chromosomal rearrangements in leukaemia, thyroid carcinoma and sarcoma

Leukaemia					
T-ALL	t(2;8)	(q24,q11)	**B-ALL**	t(9;22)	(q34,q11)
	t(8;22)	(q24,q11)		t(1;19)	(q23,p13)
	t(7;9)	(q35,p13)		t(17;19)	(q22,p13)
	t(8;22)	(q24,q11)		t(8;22)	(q24,q11)
	t(10;14)	(q24,q11)		t(12;21)	(p13,q22)
	t(1;14)	(p32,q11)	**CML**	t(9;22)	(q34,q11)
	t(11;14)	(p15,q11)		t(3;21)	(q26,q11)
	t(13;11)	(p13,q11)	**AML**	t(8;21)	(q22,q22)
praeB-ALL	t(19;19)	(q23,q13)		t(3;21)	(q26,q22)
	t(14;18)	(q31,q32)		t(16;21)	(p11,q22)
				t(12;22)	(q13,q11)
				t(5;17)	(q32,q12)
Sarcoma			**Papillary thyroid carcinoma**		
Osteosarcoma	t(2;13)	(q35,q14)	**RET/PTC1(H4)**	i(10;10)	(q21,q11)
	t(5;10)	(p13,q14-15)	**RET/PTC2(R1a)**	i(10;17)	(q11,q13)
	t(8;17)	(q23-24;p11-13)	**RET/PTC3(Ele1)**	i(10;10)	(q11,q11)
Ewing Sarcoma	t(11;22)	(q24,q12)	**RET/PTC4(Ele1)**	i(10;10)	(q11,q11)
	t(7;22)	(p22,q12)	**Ret/PTC5(Golga5)**	i(10;10)	
Fibrosarcoma	t(12;22)	(q13,q12)	**RET/PTC6(hTIF1)**	t(10;7)	(q11,q32)
Chondrosarcoma	t(9;22)	(q22-31,q11-12)	**RET/PTC7(KTN1)**	t(10;1)	(q11,p13)
Liposarcoma	t(12;16)	(q13,p11)	**Ter/PTC8(RFG8)**	t(10;14)	(q11,q22)
Synovial Sarcoma	t(X/18)	(p11.2,q11.2)	**Ret/ELKS (ELKS)**	t(10;12)	(q11,p13)
Rhabdomyosarcoma	t(1;13)	(p36,q14)	**TRK-T1,2,4**	i(1;1)	
	t(2;13)	(q35,q14)	**TRK-T3**	t(1;3)	
Desmoplast. round cell	t(11;22)	(q23-24;p11-13)			

The three most common radiation-induced tumours, leukaemia (mostly of the acute myleloic type, AML), papillary thyroid cancer and various kinds of sarcoma share the high probability of genomic rearrangements. This was observed for naturally as well as for radiation induced tumours. Among the chromosomal aberrations translocations of large chromosomal parts to another chromosome and inversions predominate (Tab. 1). Generally, up to 80% of leukaemia beare a translocation. The best known example is the t(9;22)(q34;q11)-translocation, known as "Philadelphia-chromosome". This typical translocation is diagnostic for CML (chronic myeloic leukaemia) or some subtypes of ALL (acute lymphoblastic leukaemia). It brings the ABL gene in direct vicinity to the BCR gene giving rise to a fusion protein. This BCR-ABL fusion constitutively activates the ABL-tyrosin kinase transmitting permanently cellular growth signals [8]. Inversions or translocations of another gene, the RET-oncogene, are the hallmark of papillary thyroid cancer (70%–90%) [49, 45] (Tab. 1). This rearrangement also leads to constitutive expression of a tyrosin kinase (RET), which is a trans-membrane receptor protein and thus usually only activated upon ligand binding. The fusion to the ELE [117] or other genes (Tab.1) likewise results in permanently enhanced growth signals [45, 8]. Those steady stimuli towards unrestricted proliferation are assumed to be one critical step for carcinogenesis. Translocations are also characteristic for various types of sarcoma (Tab.1). Many other radiation-induced tumours show frequently additional markers of genomic rearrangements such as loss of heterocygosity (LOH) with or without duplication of the second allel, and gene amplification [6]. All those genetic alterations are not only encountered in radiation-induced but also in spontaneous tumours. However, the underlying mechanisms are assumed to be commonly DNA damage and subsequent recombination processes.

Mechanisms and genetics of recombination

In somatic cells recombination processes are not part of the regular genetic program, as it is in meiosis, but rather a response to DNA damage. The low frequency of spontaneous recombination can be drastically enhanced by irradiation. In both cases a double-strand break (dsb) is the initiating event and recombination is the cellular tool to restitute chromosomal continuity. These processes are strictly controlled by a variety of cooperating repair enzymes that act upon two major repair pathways, homologous recombination (HR) and non-homologous endjoining (NHE) (Table 2.). The classical homologous recombination involves a characteristic repair structure, the Holliday-junction, enabling a complete and conservative restitution of the original sequence. Recently, two variants have been described [48, 82], termed break-induced replication (BIR) (also called replication-induced recombination) and synthesis-dependent strand annealing (SDSA). These three repair processes involve a homologous sequence normally provided by the sister chromatid, and normally allows an error-free repair. In more rare cases the second allel, a pseudogene or widespread repetitive sequences like LINE and Alu-elements can be recruited. The resulting recombination product needs not to be strictly conservative. Another pathway the single-strand annealing (SSA) also relies on homology but involves repetitive sequences, which are tandemly positioned on the same chromosome. One copy of the repeats is lost by nucleolytic cut-off or degradation, which makes SSA a non-conservative pathway.

Table 2. Genes/proteins involved in dsb repair.

Non-homologous Endjoining	**HomologousRecombination**
Ku 80	**RAD51**
Ku 70	**RAD51B, C, D**
DNA-PKcs*	**XRCC2**
Ligase IV	**XRCC3**
XRCC4	**RAD52**
Artemis	**RAD54**
RMN-complex§	**RMN-complex§**
BRCA1	**BRCA1/2**
ATM	**ATM**
p53	**p53**
Fanconi (C)	**Fanconi (A-G)**
ReQ (Wrn)#	**ReQ (Wrn, Blm)#**
Pol μ	**RPA&**

* DNA-Protein Kinase, catalytic subunit
§ RAD50/MRE11/NBS1-complex
# Werner-, Bloom-syndrome
& Replication Protein A

However, in mammalian cells the majority of repair events are considered to proceed without the use of a homologies, the pathway termed non-homologous end-joining (NHEJ). At least two variants are described. The typical way is to remove a few base pairs and to rejoin the ends with or without an overlap of one or two matching bases. The second way to repair a dsb is to make use of limited so-called micro-homologies that are located within a short distance on either site of the dsb and consist of 3–7 bp. The probability to find and to use such short homologies declines with increased distance to the dsb finally precluding the joining of proper ends (see below). Our own analysis of 58 chromosomal repair junctions formed upon NHEJ in mouse embryonic fibroblasts (MEFs) and human tumour cells (Tab. 3) showed that in most cases (26%) one or two bases were lost. Only 16% showed a deletion of more than 20 bp. 10 junctions involved a microhomology. In 7 cases times the same 3 bp triplet was used although 5 different short homologies (3–5bp micro-homologies) were positioned close to the dsb. These results generally agree with those obtained with non-chromosomal systems, such episomal plasmid restitution or repair of plasmids by nuclear extracts [91, 75, 47, 18, 17, 82, 30, 76, 116]. In some cases insertions of moderate length (8–40 bp) were observed, consisting either of a short direct duplication or a capture of a nearby chromosomal sequence stretch. The latter results point to a homologous-type of processing (such as SDSA) involved in a "a priori" non-homologous repair pathway. The underlying mechanisms for those events are not yet understood. Conversely, it was observed that repair processes initiated as HR can be terminated by NHEJ [87]. Interestingly, the involvement of homologies in the NHEJ repair process was mostly observed in tumour cells, suggesting a more active HR pathway compared to normal cells.

Table 3. Chromosomal junctions of non-homologous endjoining of DNA-dsb.

Type of change	Number
No change	6
1–2 bp loss	15
≤ 8bp loss	9
≤ 20 bp loss	10
> 20 bp loss	9

All analyzed products of non-homologous endjoining illustrate the high competence of retaining genomic integrity without major sequence alteration. The position of a radiation induced dsb is extremely rare in a coding or critical regulating DNA sequence since they amount to less than one percent of the human genome. Thus, a deletion of less than 50 bp, as observed here and by others [91, 65, 75, 97, 63], or an insert of moderate length should have no adverse effect towards cell function or mutagenesis.

This picture changes when regular NHEJ fails (see above). When unrejoined dsb are left until next mitosis cells are prone to mitotic death. In other cases, cells make use of a pathway deviating from classical NHEJ. This "backup system" allows a promiscuous endjoining such that one end of a dsb is not connected to the original partner but to a distant DNA end instead. This may be located either on the same or on a different chromosome. These erroneous repair events give rise to translocations, deletions and inversions. In contrast to NHEJ, HR is only rarely responsible for those chromosomal rearrangements [89, 103]. As outlined above, translocations in various forms can be one critical event for transformation if a combination of two genes is involved that are sensitive for tissue-specific carcinogenesis.

Techniques to monitor repair fidelity, mechanisms and genetic background

To investigate of repair effects several methods are available. Pulsed- and constant-field electrophoresis (PFGE and CFGE) are widely used to study dsb repair kinetics and the amount of non-repaired breaks, which have been shown to be a critical parameter for cellular sensitivity [110, 53, 24, 25]. These techniques are not suitable to measure misrepair. Erroneous repair events can either be monitored by conventional cytogenetic analysis or in more detail when they are combined with FISH-techniques. However, chromosomal aberrations of smaller size (< 1Mbp) escape from optical detection. A considerable progress was made by introduction a molecular method that combines PFGE and restriction fragment analysis based on Southern hybridization [64, 92, 32, 94, 28, 93]. This method allows detection of rearrangements as small as 100 kb including inter- and, as a major advantage over optical analysis, also intrachromosomal changes.

All the methods mentioned allow quantitative analysis but are not directly informative about the pathways involved in the repair process. Cell survival experiments had shown that cells defective defects in either homologous or non-homologous recombination (HR and NHEJ) are radiosensitive suggesting that both pathways are involved in the repair of radiation-induced damage, whereby HR defects compromise cell survival to a much lower extend [33, 46, 78, 29, 34, 42, 107] compared to NHEJ- defects [51, 12, 54, 58, 59, 24, 1]. Moreover, repair kinetics and the number of non-repaired dsb are largely altered only in NHEJ- [52, 39, 20, 57, 59, 5, 24, 77] but not in HR-mutants [4, 34, 107].

In order to obtain a more detailed insight into mechanisms, regulation, and genetic determinants of DNA repair a broad variety of functional assays have been developed that allow to study all particular pathways separately. These assays are based either on nuclear extracts where isolated proteins perform *in vitro* rejoining of plasmid DNA or oligonucleotides [22, 18, 30, 17, 76]. Alternatively, damaged plasmids can be introduced into the cell and repair of these extrachromosomal substrates is then performed within the nuclear- or cytoplasm [36, 96, 86, 15, 104, 113]. The least artificial and thus most relevant constructs are those that are integrated into the chromosomal structure. After dsb induction within this sequence cells process this kind of damage exactly like a dsb in the remainder "natural"genome. The constructs currently utilized, including those from our laboratory, are based on plasmids that are equipped with a marker gene by which recombination can be monitored. Those plasmids will be stably integrated into the genome. The marker genes typically confer resistance against antibiotics such as geneticin, puromcin, hygromycin or zeocin. Products from other genes, such as the gpt- or tk-gene, open a rescue pathway allowing the cells to survive otherwise toxic agents that interfere with the nucleotide metabolism (e.g. mycophenolic acid and the combination of hypoxanthine-aminopterine-thymidine). Intact gene functions commonly render cell resistant to the above mentioned compounds in the medium that are otherwise lethal. In the experimental repair settings those resistance genes are usually inactivated by a mutation that prevents translation of a functional protein. DNA damage and subsequent repair restitutes gene function which switches cells from sensitive to resistance status and can be selected for in a colony formation assay.

In HR constructs the mutation is usually an insertion within the coding sequence that inactivates antibiotic resistance gene. This insertion mutation consist of the recognition site for the rarely cutting restriction enzyme I-Sce I (Fig. 1.A). The I-SceI endonuclease, originating from yeast mitochondria [71], can then be used to introduce a single dsb into the target sequence without damaging other genomic sites [44]. This single dsb triggers the HR machinery. The HR pathway requires a homologous sequence as a template which is provided on the integrated plasmid as a second non-functional gene copy. The recombination process (gene conversion) removes the insertion and by that restitutes the gene function (similar constructs were employed in Rouet 1994, Moynahan and Jasin, 1997, Sargent 1997, Lin 1999, Lambert and Lopez, 2000, Delacote 2002). Those precisely controlled systems are taken as a model for radiation-induced damage although the latter might be more complex than dsb induced by a restriction endonuclease [40, 108, 76].

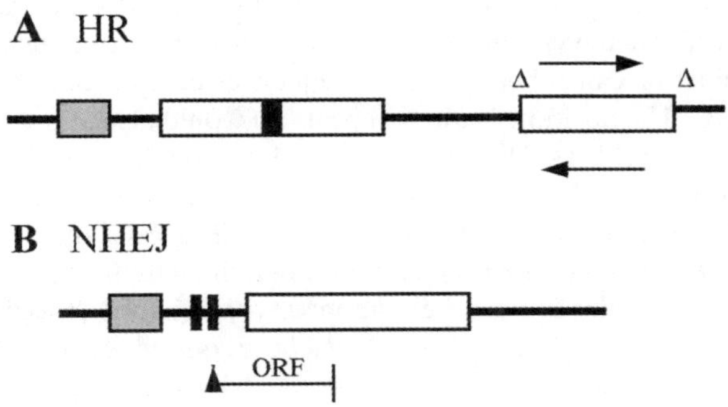

Fig. 1. Constructs for HR and NHEJ. (A) A viral promoter (grey box) controls the antibiotic resistance gene (abr, white box). The I-Sce insertion mutation (black bar) inactivates the gene. A second in-complete gene copy is placed downstream in direct or inverse orientation.
(B) the insertion mutation is placed between promoter and gene. It carries two Sce-sites (black bars) flanking a translational start codon. This triplet (ATG) opens a non-sense ORF that overlaps the original gene.

In order to study NHEJ, we designed a substrate that interrupts the functional unit of promoter and gene. (Fig. 1.B). The inserted sequence consists of two tandem copies of the I-Sce I recognition sequence flanking a start codon for an artificial open reading frame (ORF). This nonsense ORF overlaps the ORF of the antibiotic resistance gene, which, leads to its inactivation. Induction of a dsb at both I-Sce-sites removes the artificial start codon and shifts translation to the original reading frame. This NHEJ restitutes the translation of the antibiotic resistance gene. In both cases (HR and NHEJ), the number of resistant clones correspond to the number of recombination events. The NHEJ product that leads to cellular antibiotic resistance can be analyzed for each clone by direct sequencing. The NHEJ construct described here is specifically designed to detect small changes independent of the exact conformation of the repair junction. It allows to monitor repair events that are associated with a high degree of genomic conservation. Other NHEJ assays developed so far either select for large deletions only [97] or for only a single type of DNA junction i.e. 4-bp microhomology [60, 2, 63] and are thus less informative with respect to the spectrum of natural repair events.

The typical experimental procedure (see Fig. 2., page 128) includes (1) transfection of cells with the recombination substrate (plasmid) and allowance for stable chromosomal integration. (2) Single clones are transfected with I-SceI for dsb induction and subsequent recombination. (3) colony growth in selective medium. (4) DNA sequencing of the repair junction in single resistant clones.

Recently the green fluorescent protein (GFP) was introduced as alternative selectable marker. Those cells that restitute the GFP-gene show a green fluorescence that can be monitored by FACS-scanning. GFP-testing takes two days only and spares long-term cell culture.

With those recombination substrates versatile tools are at hand that can be easily tailored to investigate the various repair pathways with particular demands. Homologous recombination can either operate on sequences in direct or inverse orientation (Fig. 3.A and B). The position as direct repeat (A) would also allow to measure single-strand annealing (SSA). The length of homology and the space between the copies can be modified. Partner sequences for HR can be placed on different chromosomes (not shown). In the NHEJ substrate (Fig. 3.C) the distance between promoter and ORF can be modified and with that the length of possible deletions. The regions flanking the dsb can be equipped with microhomologies of different positions and length including extensive stretches required for SSA. Substrates can be designed that simultaneously allow for HR and NHEJ (Fig. 3.D) leaving the decision to the cell as to which pathway is to be used. Finally, plasmids can be constructed that in particular allows to select for various types of crossover events. These exchanges between two distant sites on the same or on different chromosomes can be achieved either by HR or, more importantly, by NHEJ (not shown). The latter repair events would correspond to chromosomal translocations and inversions.

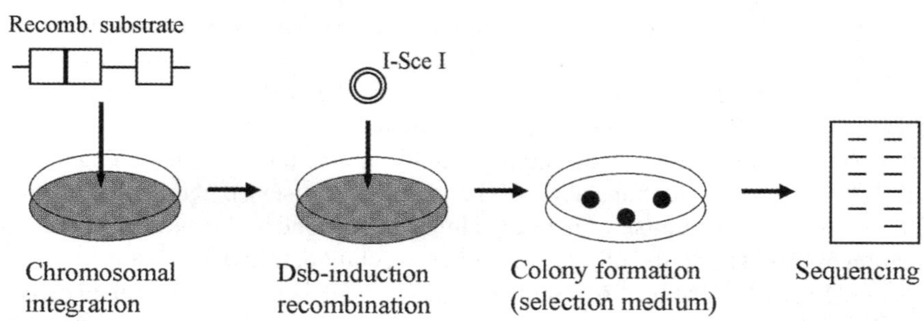

Fig. 2. Illustration of experimental principle of measuring recombination with a plasmid-based assay.

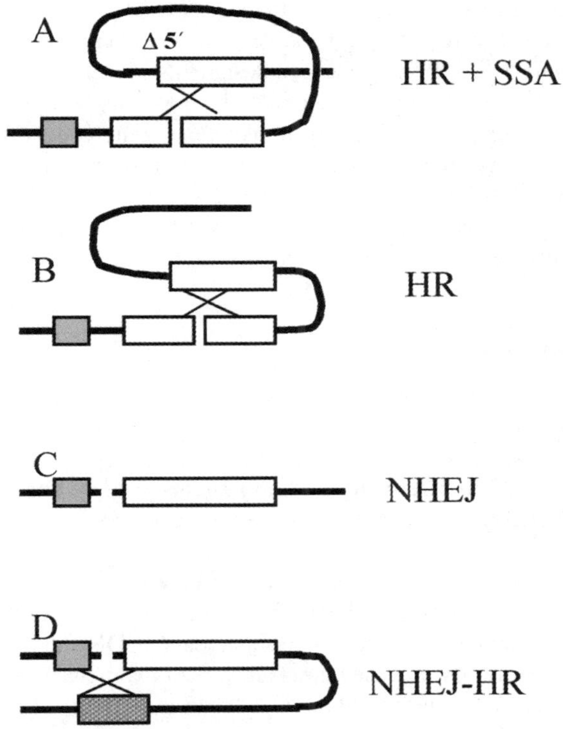

Fig. 3. Scheme of different constructs serving a substrates for HR and NHEJ.

All those constructs can now be used to investigate the general mechanism of repair, the regulation between , and the utitility of combinations of pathways. Several lines of evidence show, that a switch between NHEJ and HR occurs in dependence of the cell cycle progression [58]. It is assumed that HR is not active in G1 phase since no sister chromatid is available and interchromosomal HR should be a rare event [89, 88] and RAD51 the central gene for HR is not expressed in G1 [16, 105, 31]. In contrast, in S/G2 phase both HR and NHEJ should be active and eventually complement each other. Both, knock-out yeast and mouse cells have been shown particularly radiosensitive when both pathways were defective. For example, cell survival was reduced by some 3 orders of magnitude when Ku70, Rad54 and Nbs were knocked out inactivating NHEJ and HR simultaneously [67, 101, 115]. In parallel the frequency of G2 phase-aberrations of both chromatid and chromosome-type were synergistically enhanced in double knock out cells [115] Plasmid based genetic assays using rarely cutting restriction enzymes (e.g. I-SceI) can now be used to investigate the underlying mechanisms directly. The genetic players can be identified, some of which have already been described as controlling HR. Mutations in RAD51, RAD54, XRCC3,Nbs1 and BRCA1 and 2 were shown to suppress HR [85, 56, 27, 73, 102, 74, 106], while mutations of p53 [111, 70, 112, 2] and several genes of NHEJ pathway [84, 23] enhance the frequency of HR by 1 to 4 orders of magnitude. In contrast only few results are available of genetic control of NHEJ in a chromosomal context. [60, 73, 102]. The assays designed so far allow for only a limited number of various repair events (see above). This has now been overcome by the newly developed system which leads to translational gene activation upon NHEJ without restriction to certain junction types (see above). With this system p53 could be shown for the first time to control NHEJ. The number of NHEJ events was grately enhanced upon lose of p53wt function [20]. This result shows a new mechanism of how genomic stability can be compromised upon p53 dysfunction.

Future directions

The variety of recently developed genetic repair systems allows to investigate the genetic and chromosomal prerequisites necessary for an error free repair that best conserves genetic stability. Cellular "backup systems" can be characterized when classical HR or NHEJ fail such as repair processes depending on microhomologies, or longer homologous repeats at nearby or distant locations and other structural requirements for DNA joint formation. A major task will be to study the interaction and the mutual regulation of the various repair pathways. Combining those genetic studies with assays that monitor larger rearrangements on DNA at the chromosomal level would appreciably advance the knowledge about maintenance of genomic stability and mechanisms leading to radiation-induced instability, translocations, and carcinogenesis.

References

1 Adachi N, Ishino T, Ishii Y, Takeda S; Koyama (2001) H. DNA ligase IV-deficient cells are more resistant to ionizing radiation in the absence of Ku70: Implications for DNA double-strand break repair. Proc Natl Acad Sci US 98:12109–12113

2 Akyüz N, Boehden GS, Süsse S, Rimek A, Preuss U, Scheidtmann KH; Wiesmüller L (2002) DNA substrate dependence of p53-mediated regulation of double-strand break repair. Mol Cell Biol 22:6306–6317

3 Arai T, Nakano T, Fukuhisa K, Kasamatsu T, Tsunematsu R, Masubuchi K, Yamauchi H, Fukuda T, Noguchi H; Murata M (1991)second cancer after radiation therapy for cancer of the uterine cervix. Cancer 67:398–405

4 Asaad NA, Zeng ZC, Guan J; Thacker J (1994)IG. Homologous recombination as a potential target for caffeine radiosensitization in mammalian cells : reduced caffeine radiosensitization in XRCC2 and XRCC3 mutants. DNA Repair 30:1127–1132

5 Badie C, Iliakis G, Foray N, Alsbeih G, Pantellias GE, Okayasu R, Cheong N, Russell NS, Begg AC, Arlett CF; Malaise EP (1995) Defective Repair of DNA Double-Strand Breaks and Chromosome Damage in Fibroblasts from a Radiosensitive Leukemia Patient. Cancer Res 55:1232–1234

6 Balmain A, Gray J; Ponder B (2003) The genetics and genomics of cancer. Nature Genetics 33, suppl.:238–244
7 Bhatia S, Robinson LL, Oberlin O, Greenberg M, Bunin G, Fossati-Bellani F; Meadows AT (1996) Breast cancer and other second neoplasms after childhood hodgin´s disease. New Eng J Med 334:745–751
8 Blume-Jensen P; Hunter T (2001) Oncogenic kinase signalling. Nature 411:355–365
9 Boice JD, Land CE, Shore RE, Norman JE; Tokunaga MB (1979) Risk of breast cancer following low-dose radiation exposure. Radiology 131:589–597
10 Boice JD, Preston D, Davis FG; Monson RR (1991) Frequent chest X-ray fluoroscopy and breast cancer incidence among tuberculosis patients in Massachusetts. Radiat Res 125:214–222
11 Boice. JD, Engholm G, Kleinerman RA, Blettner M, Stovall M, Lisco H, Moloney WC, Austin DF, Bosch A, Cookfair DL, Krementz ET, Latourette H, Merrill JA, Peters LJ, Schulz MD, Storm HH, Björkholm E, Pettersson F, Bell CMJ, Coleman MP, Fraser P, Neal FE, Prior P, Choi NW; Hislop TG (1988) Radiation dose and second cancer risk in patients treated for cancer of the cervix. Radiat Res 116:3–55
12 Boubnov NV, Hall KT, Wills Z, Lee SE, He DM, Benjamin DM, Pulaski CR, Band H, Reeves W, Hendrickson EA; Weaver DT(1995) Complementation of the ionizing radiation sensitivity, DNA end binding, and V(D)J recombination defects of double-strand break repair mutants by the p86 Ku autoantigen. Proc Natl Acad Sci US 92:890–894
13 Brenner DJ, Miller RC; Huang YHEJ (1995) The biological effectiveness of radon-progeny alpha particles. III Quality factors. Rad Res 142:61–69
14 Brenner DJ, Curtis RE, Hall EJ; Ron E (2000) Second malignancies in prostate carcinoma patients after radiotherapy compared with surgery. Cancer 88:398–406
15 Britten RA, Liu D, Kuny S; Alalunis-Turner MJ (1997) Differential level of DSB repair fidelity effected by nuclear protein extracts derived from radiosensitive and radioresistant human tomour cells. Brit J Cancer 76:1440–1447
16 Chen F, Nastasi A, Shen Z, Brenneman M, Crissman H; Chen DJ (1997) Cell cycle-dependent protein expression of mammalian homologs of yeast DNA double-strand break repair genes Rad51 and Rad52. Mut Res 384:205–211
17 Chen S, Inamdar KV, Pfeiffer P, Feldmann E, Hannah MF, Yu Y, Lee JW, Zhou T, Lees-Miller SP; Povirk LF(2001) Accurate in vitro end joining of a double-strand break with partially cohesive 3´-overhangs and 3´-phosphoglycolate termini. J Biol Chem 276:24323–24330
18 Cheong N, Perrault AR, Wang H, Wachsberger P, Mammen P, Jackson I; Iliakis G (1999) DNA-PK-independent rejoining of DNA double-strand breaks in human cells extracts in vitro. Int J Radiat Biol 75:67–81
19 Court-Brown WM, Doll R; Hill AB (1960) The incidence of leukaemia following exposure to diagnostic radiation in utero. Brit Med J 2:1539
20 Dahm-Daphi J, Dikomey E; Pyttlik C (1994) Relationship between non-reparable strand breaks and cell survival studied in CHO, CHO K1 and the radiosensitive mutants xrs1 and xrs5. Int J Radiat Biol 65:657–663
21 Dahm-Daphi J, Hubbe P, El-Awady RA, Powell SN; Willers H (2003) p53 suppresses non-homologous Endjoining of chromosomal DNA double-strand breaks. Mol Cell Biol submitted:

22 Daza P, Reichenberger S, Göttlich B, Hagmann M, Feldmann E; Pfeiffer P (1996) Mechanisms of nonhomologous DNA end-joining in frogs, mice and men. Biol Chem 377:775–786
23 Delacote F, Han M, Stamato TD, Jasin M; Lopez BS (2002) An XRCC4 defect or Wortmannin stimulates homologous recombination specifically induced by double-strand breaks in mammalian cells. Nucl Acid Res 30:3454–3463
24 Dikomey E, Dahm-Daphi J, Brammer I, Martensen R; Kaina B (1998) Correlation between cellular radiosensitivity and non-repaired double-strand breaks studied in nine mammalian cell lines. Int J Radiat Biol 73:269–278
25 Dikomey E, Brammer I, Johansen J, Bentzen SM; Overgaard J (2000) Relationship between DNA double-strand breaks, cell killing, and fibrosis studied in confluent skin fibroblasts derived from breast cancer patients. Int J Radiat Oncol, Biol, Phys 46:481–490
26 Dörr W; Herrmann T (2002) Second primary tumours after radiotherapy for malignancies. Strahlentherapie Onkol 178:357–362
27 Dronkert MLG, Beverloo HB, Johnson RD, Hoeijmakers JHJ, Jasin M; Kanaar R (2000) Mouse RAD54 affects DNA double-strand break repair and sister chromatid exchange. Mol Cellr Biol 20:3147–3156
28 El-Awady RA, Dikomey E; Dahm-Daphi J (2001) Heat effects on DNA repair after ionizing radiation. Hyperthermia commonly increases the number of non-repaired double-strand breaks and structural rearrangements. Nucl Acid Res 29:1960–1966
29 Essers J, Hendriks RW, Swagemakers SMA, Troelstra C, de Wit J, Bootsma D, Hoeijmakers JHJ; Kanaar R (1997) Disruption of mouse RAD54 reduces ionizing radiation resistance and homologous recombination. Cell 89:195–204
30 Feldmann E, Schmiemann V, Goedecke W, Reichenberger S; Pfeiffer P (2000) DNA double-strand break repair in cell-free extracts from Ku80-deficient cells: implications for Ku serving as an alignment factor in non-homologous DNA end joining. Nuc Acids Res 28:2585–2596
31 Flygare J, Benson F; Hellgren D (1996) Expression of human RAD51 gene during the cell cycle in primary human peripheral blood lymphocytes. Biochim Biophys Acta 1312:231–236
32 Fouladi B, Waldren CA, Rydberg B; Cooper PK (2000) Comparison of repair of DNA double-strand breaks in identical sequences in primary human fibroblast and immortal hamster-human hybrid cells harbouring a single copy of human chromosome 11. Radiat Res 153:795–804
33 Fuller LF; Painter RB (1988) A Chinese hamster ovary cell line hypersensitive to ionizing radiation and deficient in repair replication. Mutat Res 193:109–121
34 Girard P-M, Foray N, Stumm M, Waugh A, Riballo E, Maser RS, Phillips WP, Petrini J, Arlett CF; Jeggo PA (2000) Radiosensitivity in nijmegen breakage syndrome cells is attributable to a repair defect and not cell cycle checkpoint defects. Cancer Res 60:4881–4888
35 Hall EJ; Wuu CS (2003) Radiation induced second cancers: The impact of 3D-CRT and IMRT. Int J Radiat Oncol Biol Phys 56:83–88
36 Hamilton AA; Thacker J (1987) Gene recombination in X-ray-sensitive hamster cells. Mol Cell Biol 7:1409–1414
37 Hamre MR, Severson RK, Chuba P, Lucas DR, Thomas RL; Mott MP (2002) Osteosarcoma as a second malignant neoplasm. Radiother Oncol 65:153–157
38 Hawkins MM (1990) Second primary tumours following radiotherapy for childhood cancer. Int J Radiat Oncol Biol Phys 19:1297–1301

39 Hendrickson EA, Qin XQ, Bump EA, Schatz DG, Oettinger M; Weaver DT (1991) A link between double-strand break-related repair and V(D)J recombination: the scid mutation. Proc Natl Acad Sci US 88:4061–4065
40 Henner WD, Grunberg SM; Haseltine WA (1982) Sites and structure of radiation-induced DNA strand breaks. J Biol Chem 257:11750–11754
41 Heyn R, Hoeberlen V, Newton WA, Ragab AH, Raney RB, Tefft M, Wharam M, Ensign LG; Maurer HM (1993) second malignant neoplasms in children treated for rhabdomyosarcoma. J Clin Oncol 11:262–270
42 Hsu HL, Yannone SM; Chen DJ (2002) Definig interactions between DNA-PK and ligase IV/XRCC4. DNA Repair 1:225–235
43 Jacob P, Goulko G, Heidenreich WF, Likhtarev I, Kairo I, Tronko ND, Bogdanova TI, Kenigsberg J, Buglova E, Drozdovitch V, Demidchik A, Balonov M, Znonova I; Beral V (1998) Thyroid cancer risk to children calculated. Nature 392:31–32
44 Jasin M (1996) Genetic manipulation of genomes with rare-cutting endonucleases. Trends Genet 12:224–228
45 Jhiang SM (2000) The RET proto-oncogene in human cancers. Oncogene 19:5590–5597
46 Jones NJ, Cox R; Thacker J (1987) Isolation and cross-sensitivity of X-ray-sensitive mutants of V79-4 hamster cells. Mutat Res 183:279–286
47 Kabotyanski EB, Gomelsky L, Han J-O, Stamato TD; Roth DB (1998) Double-strand break repair in Ku86- and XRCC4-deficient cells. Nucl Acids Res 26:5333–5342
48 Kanaar R, Hoejimakers JHJ; van Gent DC (1998) Molecular mechanisns of double-strand break repair. Trends Cell Biol 8:483–489
49 Kazakov VS, Demidchik EP, Astakhova LN, Baverstock K, Egloff B, Pinchera A, Ruchti C; Williams D (1992) Thyroid cancer after Chernobyl (Correspondence). Nature 359:21–22
50 Kemp CJ, Sun S; Gurley KE (2001) p53 Induction and Apoptosis in Response to Radio- and Chemotherapy in Vivo is Tumor-Type-dependent. Cancer Res 61:327–332
51 Kemp LM; Jeggo P (1986) Radiation-induced chromosome damage in X-ray-sensitive mutants (xrs) of the Chinese hamster ovary cell line. Mutat Res 166:255–263
52 Kemp LM, Sedgwick SG; Jeggo PA (1984) X-ray sensitive mutants of Chinese hamster ovary cells defective in double-strand break rejoining. Mutat Res 132:189–196
53 Kiltie AE, Orton CJ, Ryan AJ, Roberts SA, Marples B, Davidson SE, Hunter RD, Margison GP, West CML (1997) A correlation between residual DNA double-strand breaks and clonogenic measurements of radiosensitivity in fibroblasts from preradiotherapy cervix cancer patients. Int J Radiat Oncol Biol Phys 39:1137–1144
54 Kirchgessner CU, Patil CK, Evans JW, Cuomo CA, Fried LM, Carter T, Oettinger MA; Brown JM (1995) DNA-dependent kinase (p350) as a candidate gene for the murine SCID defect. Science:1178–1183
55 Lagarde F, Axelson G, Damber L, Mellander H, Nyberg F; Pershagen G (2001) Residential radon and lung cancer among never-smokers in Sweden. Epidemiology 12:396–404
56 Lambert S; Lopez BS (2000) Characterization of mammalian RAD51 double strand break repair using non-lethal dominant-negative forms. EMBO 19:3090–3099
57 Lee SE, Pulaski CR, He DM, Benjamin DM, Voss M, Um J; Hendrickson EA (1995) Isolation of mammalian cell mutants that are X-ray sensitive, impaired in DNA double-strand break repair and defective for V(D)J recombination. Mutat Res 336:279–291
58 Lee SE, Mitchell RA, Cheng A; Hendrickson EA (1997) Evidence for DNA-PK-dependent and -independent DNA double-strand break repair pathways in mammalian cells as a function of the cell cycle. Mol Cell Biol 17:1425–1433

59 Li Z, Otevrel T, Gao Y, Cheng H-L, Seed B, Stamato TD, Taccioli GE; Alt FW (1995) The XRCC4 gene encodes a novel protein involved in DNA double-strand break repair and V(D)J recombination. Cell 83:1079–1089
60 Liang F, Romanienko PJ, Weaver DT, Jeggo PA; Jasin M (1996) Chromosomal double-strand break repair in Ku-80-deficient cells. Proc Natl Acad Sci US 93:8929–8933
61 Likhtarev I, Sobolev BG, Kairo I, Tronko ND, Bogdanova TI, Oleinic VA, Epshtein EV; Beral V (1995) Thyriod cancer in the Ukraine. Nature 375:365
62 Lin Y, Lukacsovich T; Waldman A (1999) Multiple pathways for repair of DNA double-strand breaks in mammalian chromosoms. Mol Cell Biol 19:8353–8360
63 Lin Y, Waldman B. C.; Waldman AS (2003) Suppression of high-fidelity double-strand break repair in mammalian chromosomes by pifithrin-a, a chemical inhibitor of p53. DNA Repair 2:1–11
64 Löbrich M, Rydberg B (1995) Cooper PK. Repair of X-ray-induced DNA double-strand breaks in specific NotI restriction fragments in human fibroblasts: joining of correct and incorrect ends. Proc Natl Acad Sci US 92:12050–12054
65 Lukacsovich T, Yang D; Waldman AS (1994) Repair of a specific double-strand break generated within a mammalian chromosome by yeast endonuclease I-SceI. Nucl Acids Res 22:5649–5657
66 Lundell M, Hakulinen T; Holm LE (1994) Thyroid cancer after radiotherapy for skin hemangioma in infancy. Radiat Res 140:334–339
67 Mages GJ, Feldmann HM; Winnacker E-L (1996) Involvement of the saccharomyces cerevisiae HDF1 gene in DNA double-strand break repair and recombination. J Biol Chem 271:7910–7915
68 Matheson J,B, Burmeister BH, Smithers BM, Gotley D, Harvey JA; Doyle L (2002) Second primary oesophageal cancer following radiation for breast cancer. Radiother Oncol 65:159–163
69 Meadows AT (1988) Risk factors for second malignant neoplasms: report from the Late Effects Study Group. Bull Cancer 75:125–130
70 Mekeel KL, Tang W, Katchnic LA, Luo CM, DeFrank JS; Powell SN (1997) Inactivation of p53 results in high rates of homologous recombination. Oncogene 14:1847–1857
71 Monteilhet C, Perrin A,TA,CL; Dijon B (1990) Purification and characterization of the invitro activity of I-Sce I, a novel and highly specific endonuclease encoded by a group I intron. Nucl Acid Res 18:1407–1413
72 Moynahan ME; Jasin M (1997) Loss of heterozygosity induced by a chromosomal double-strand break. Proc Natl Acad Sci US 94:8988–8993
73 Moynahan ME, Chiu JW, Koller BH; Jasin M (1999) BRCA1 controls homology-directed DNA repair. Mol Cell 4:511–518
74 Moynahan ME, Pierce AJ; Jasin M (2001) BRCA2 is required for homology-directed repair of chromosomal breaks. Mol Cell 7:263–272
75 Nicolas AL; Young CS (1994) Characterization of DNA end joining in a mammalian cell nuclear extract: junction formation is accompanied by nucleotide loss, which is limited and uniform but not site specific. Mol Cellular Biol 14:170–80
76 Odersky A, Panyutin IV, Panyutin IG, Schunck C, Feldmann E, Goedecke W, Neumann RD, Obe G; Pfeiffer (2002). Repair of sequence-specific 125I-induced double-strand breaks by nonhomologous DNA end-joining in mammalian cell-free extracts. J Biol Chem: in press
77 O`Driscoll M, Cerosaletti KM, Girard PM, Gai Y, Stumm M, Kysela B, Hirsch B, Gennery A, Palmer SE, Seidel J, Gatti R, Neitzel H, Jeggo PA; Concannon P (2001) DNA

ligase IV mutations identified in patients exhibiting developmental delay and immunodeficiency. Mol Cell 8:1175–1185

78 Ohnishi T, Taki T, Hiraga S, Arita N; Morita T (1998) In vitro and in vivo potentiation of radiosensitivity of malignant gliomas by antisense inhibition of the RAD51 gene. Biochem Biophys Res Comm 245:319–324

79 Pedersen-Bjergaard J; Larsen SO (1982) incidence of acute nonlymphotic leukemia, preleukemia, and acute myeloproliferative syndrome up to 10 years after treatment of Hodgin´s disease. New Engl J Med 307:965–971

80 Perkel VS, Gail MH, Lubin J, Pee DY, Weinsein R, Shore-Freeman E; Schneider AB (1988) radiation-induced thyroid neoplasms: evidence for familial susceptibility factors. J Clin Endocrin Metabol 66:1316–1322

81 Peters L; McKay M (2001) Predictive Assays: Will they ever have a role in the Clinic?. Int J Radiat Oncol Biol Phys 49 (2):501–504

82 Pfeiffer P, Goedecke W; Obe G (2000) Mechanisms of DNA double-strand break repair and their potential to induce chromosomal aberrations. Mutagenesis 15:289–302

83 Pickles T; Phillips N (2002) The risk of second malignancy in men with prostate cancer treated with or without radiation in British Columbia. Radiother Oncol 65:145–151

84 Pierce AJ, Hu P, Han MEN; Jasin M (2001) Ku DNA end-binding protein modulates homologous repair of double-strand breaks in mammalian cells. Genes Developm 15:237–3242

85 Pierce AJ, Johnson RD, Thompson LH; Jasin M (1999) XRCC3 promotes homlogy-directed repair of DNA damage in mammalian cells. Genes Developm 13:2633–2638

86 Powell SN; McMillan TJ (1994) The repair fidelity of restriction enzyme-induced double-strand breaks in plasmid DNA correlates with radioresistance in human tumor cell lines. Int J Rad Oncol Biol Phys 29:1035–1040

87 Richardson C; Jasin M (2000a) Coupled homologous and nohomologous repair of a double-strand break preserves genomic integrity in mammalian cells. Mol Cell Biol 20:9068–9075

88 Richardson C; Jasin M (2000b) Frequent chromosomal translocations induced by DNA double-strand breaks. Nature 405:697–700

89 Richardson C, Mohynahan ME; Jasin M (1998) Double-strand break repair by interchromosomal recombination: suppression of chromosomal translocations. Genes Developm 12:3831–3842

90 Ron E, Modan B, Preston D, Alfandary E, Stovall M; Boice JD (1989) Thyreoid neoplasia following low-dose radiation in childhood. Radiat Res 120:516–531

91 Roth DB, Porter TN; Wilson JH (1985) Mechanisms of nonhomologous recombination in mammalian cells. Mol Cell Biol 5:2599–2607

92 Rothkamm K; Löbrich M (1999) Misrejoining of DNA double-strand breaks in primary and transformed human and rodent cells: a comparison between the HPRT region and other genomic locations. Mutat Res 433:193–205

93 Rothkamm K; Löbrich M (2002) Misrepair of radiation-induced DNA double-strand breaks and its relevance for tumorigenesis and cancer treatment (Review). Int J Oncol 21:in press

94 Rothkamm K, Kühne M, Jeggo PA; Löbrich M (2001) Radiation-induced genomic rearrangements formed by nonhomologous end-joining of DNA double-strand breaks. Cancer Res 61:3886–3893

95 Rouet P, Smith F; Jasin M (1994) Introduction of double-strand breaks into the genome of mouse cells by expression of a rare-cutting endonuclease. Mol Cell Biol 14:8096–8106

96 Rouet P, Smith F; Jasin M (1994) Expression of a site-specific endonuclease stimulates homologous recombination in mammalian cells. Proc Natl Acad Sci US 91:6064–6068

97 Sargent RG, Brenneman MA; Wilson JH (1997) Repair of site-specific double-strand breaks in a mammalian chromosome by homologous and illegitimate recombination. Mol Cell Biol 17:267–277

98 Shimizu Y, Kato H; Schull WJ (1990) Studies of the mortality of A-bomb survivors. Radiat Res 121:120–141

99 Shore RE, Woodard E, Hildreth, N., Dvoretsky P, Hempelmann L; Pasternack B (1985) Thyroid tumors following thymus irradiation. J Nat Cancer Inst 74:1177–1184

100 Shore RE, Hildreth N, Woodard E, Dovretsky P, Hemplemann L; Pasternack B (1986) Breast cancer among woman given X-ray therapy for acute postpartum mastitis. J Nat Cancer Inst 77:689–696

101 Siede W, Friedl AA, Dianova I, Eckardt-Schupp F; Friedberg EC (1996) The Saccharomyces cerivisiae Ku autoantigen homologue affects radiosensitivity only in the absence of homologous recombination. Genetics 142:91–102

102 Snouwaert JN, Gowen LC, Latour A, Mohn AR, Xiao A, DiBaise L; Koller BH (2000) BRCA1 deficient embrionic stem cells display a decreased homologous recombination frequency and an increased frequency of non-homologous recombination that is corrected by expression of a Brca1 transgene. Oncogene 18:7900–7907

103 Stark JM; Jasin M (2003) Extensive loss of heterozygosity is suppressed during homologous rapair of chromosomal breaks. Mol Cell Biol 23:733–743

104 Tang W, Willers H; Powell SN (1999) P53 directly enhances rejoining of DNA double-strand breaks with cohesive ends in g-irradiated mouse fibroblasts. Cancer Res 59:2562–2565

105 Tashiro S, Kotomura N, Shinohara A, Tanaka K, Kazuhiro U; Kamada N. S phase specific formation of the human Rad51 protein nuclear foci in lymphocytes. Oncogene 12:2165–2170

106 Tauchi H, Kobayashi J, Morishima KI, van Gent DC, Shiraishi T, Verkalk NS, van Heems D, Ito E, Nakamura A, Sonoda E, Takata M, Takeda S, Matsuura S; Komatsu K (2002) Nbs1 is essential for DNA repair by homologous recombination in higher vertebrates. Nature 420:93–98; United Nations Committee on the Effects of Atomic Radiation (Unscear)(2000), Annex I. Epidemiological evaluation of radiation-induced cancers, New York: United Nations, 297–450

107 Wang H, Zeng ZC, Bui T-A, DiBiase SJ, Qin W, Xia F, Powell SN; Iliakis G (2001) Nonhomologous end-joining of ionizing radiation-induced DNA double-stranded breaks in human tumor cells deficient in BRCA1 or BRCA2. Cancer Res 61:270–277

108 Ward JF (1985) Biochemistry of DNA lesions. Radiat Res 104:103–111

109 Weis HA, Darby SC; Doll R (1994) Cancer mortality following X-ray treatment for ankylosing spondylitis. Int J Cancer 59:327–338

110 Whitaker SJ, Ung YC; McMillan TJ (1995) DNA double-strand break induction and rejoining as determinants of human tumour cell radiosensitivity. A pulsed-field gel electrophoresis study. Int Journal of Radiat Biol 67:7–18

111 Wiesmüller L, Cammenga J; Deppert W (1996) In vivo assay of p53 function in homologous recombination between Simian Virus 40 chromosomes. J Virol 70:3737–744

112 Willers H, McCarthy E, Alberti W, Dahm-Daphi J; Powell SN (2000) Loss of wild-type p53 function is responsible for up-regulated homologous recombination rates in immortalized rodent fibroblasts. Int J Radiat Biol 76:1055–1062

113 Willers H, McCarthy EE, Hubbe P, Dahm-Daphi J; Powell SN (2001) Homologous recombination in extrachromosomal plasmid substrates is not suppressed by p53. Carcinogenesis in press

114 Wong FL, Boyse JD, Abramson DH, Tarone RE, Kleierman RA, Stovall MB, Goldman MB, Seddon JM, Tarbell N, Fraumeni JF; Li FP (1997) Cancer incidence after retinoblastoma. Radiation dose and sarcoma risk. J Amer Med Ass 278:1261–1267

115 Yamaguchi-Iwai Y, Sonoda E, Sasaki MS, Morrison C, Haraguchi T, Hiraoka Y, Yamashita YM, Yagi T, Takata M, Price C, Kakazu N; Takeda S (1999) Mre11 is essential for the maintenance of chromosomal DNA in vertebrate cells. EMBO 18:6619–6629

116 Zhong Q, Boyer TG, Chen PL; Lee WH (2002) Deficient nonhomologous end-joining activity in cell-free extracts from Brca-null fibroblasts. Cancer Res 62:3966–3970

117 Bongarzone I, Butti MG, Coronelli S, Borello MG, Santoro M, Mondellini P, Pilotti S, Fusco A, Della Porta G; Pierotti MA (1994) Frequent activation of ret protooncogene by fusion with a new activating gene in papillary thyroid carcinomas. Cancer Res 54(11):2979–85

UV-Radiation Biology as Part of Cancer Research

R. Greinert, E.W. Breitbart, B. Volkmer

Dermatologisches Zentrum Buxtehude, Abt. Mol. Zellbiologie, Germany
dermatologie@hospit.std.shuttle.de

Introduction

The existence of UV-radiation is known since about 200 years, starting with the pioneering work of Johann Wilhelm Ritter [1]. Since then an enormous amount of knowledge has been collected about this kind of non-ionizing radiation and its impacts on nearly every part of animated and in-animated material, because the main source of environmental UV is the sun. A picture is emerging now, that solar UV-radiation, although necessary in having driven the evolution of many biological systems [2], has to be seen as one of, if not the most prominent environmental challenge presented to many organisms on earth. UV-radiation (280–400 nm) can be absorbed by many molecular components in the cell. Energies of different parts of the solar UV-spectrum (UVC: 100–280 nm, UVB: 280–315 nm, UVA: 315–400 nm) are therefore able to excite molecules to reactive electronic states which are photophysical precursors of a plenitude of photochemical reactions, depending on the UV-wavelength region. This may lead to changes in molecular structure or even create new molecular species with high and often hazardous reactivity against important molecular and genetic components in the cell. In this way cells might be damaged in an UV-dependent way. The main target of UV-induced damage in the cell is DNA. This is because DNA is able, both, to absorb UV-photons directly or to interact with UV-induced cellular species (e.g. reactive oxygen species, ROS). Both reaction pathways may introduce lesions into DNA, which if not, not fully or misrepaired may give rise to mutations with cancerogenic potential.

The tissue with the highest cancer risk with respect to UV-radiation is naturally skin. UV induces specific DNA-damages in different cell-types of human epidermis (e.g. keratinocytes, melanocytes), which, left un- or mis-repaired, might cause mutations in important oncogenes or tumor-suppressor genes which are necessary for cell-cycle control or coordination of DNA-repair processes. If these functions are deficient or disrupted due to UV-induced DNA-damage, skin cancers like basal cell carcinoma (BCC), squamous cell carcinoma (SCC) or malignant melanoma (MM) might develop.

Deficiencies in nucleotide excision repair (NER), the main DNA repair-pathway responsible for the removal of UV-induced lesions or other "bulky DNA-lesions", seem to be of special importance for the development of non-melanocytic skin cancers (BCC, SCC and MM). This has been learned from patients suffering from

Xeroderma Pigmentosum (XP), a syndrome chararacterized by deficiencies in those genes which code for proteins involved in NER. XP-patients, therefore, carry a 1000 fold higher risk than the normal population to develop skin cancers [3]. Furthermore, even in the normal population, repair capacity with respect to the removal of UV-induced DNA-lesions shows a high degree of inter-individual variability, which manifests itself in individual repair kinetics for the removal of photoproducts and the amount of persisting UV-damage which cannot be removed from the genome at all. For the latter case there exists new evidence that a special sub-population of epidermal keratinocytes in the basal layer of human epidermis (probably epidermal stem cells) is not able to repair UV-induced DNA-lesions (CPD) and that these cells might be involved in (non-melanocytic) skin cancer development.

There exists also evidence that induction of UV-induced DNA lesions and activities of enzymatic (mis-)repair of photoproducts produce some kind of genomic instability that may be responsible to render cells more sensitive in later stages of replication or proliferation to the actions of tumor promoting agents (like e.g UV-radiation itself), which then, finally induce pre-neoplastic transformation and carcinogenesis in skin. This might be especially important for the development of malignant melanoma of the skin, where increasing levels of genomic instability at a chromosomal level (visualized by the appearance of chromosomal aberrations) have been reported during successive stages of melonoma genesis.

UV-induced damage and mutations

UV-radiation is classically subdivided into 3 spectral bands:
UVC (100–280 nm), UVB (280–315 nm) and UVA (315–400 nm). Only UVB and UVA are biologically relevant at earth's ground level, because the UVC-region of the solar spectrum is effectively shielded by an intact stratospheric ozone layer [4]. About 95% of UV-radiation reaching the earth comprises UVA. However, the remaining 5% of UVB seem to be responsible for the majority of biologically harmful effects of UV-radiation because of the high potential of this type of radiation to induce specific DNA-damage. DNA is the main target for interaction with UV-radiation and might be damaged via direct or indirect effects.

Direct effects are caused by photon absorption of DNA and photochemical conversion of absorbed energy into photodimerization between adjacent pyrimidine bases. According to the action spectrum UVB is about 1000 times more effective than UVA to induce this kind of photodimerization [5]. UVB also induces, at a much lower frequency, inter- and intrastrand crosslinks, protein-DNA crosslinks and rare base aducts (for a review on DNA photochemistry see e.g. Cadet and Vigny) [6]. Photodimerizations between adjacent pyrimidine bases are by far the most prevalent photoreactions resulting from the direct action of UV-radiation. Two major photoproducts are formed via this reaction pathway: the cyclobutyl pyrimidine dimer (CPD) and the pyrimidine(6-4)pyrimidone photoproduct (6-4PP). CPDs are formed between adjacent pyrimidines linked by a cyclobutyl ring between the 5 and 6 carbons of adjacent thymine (Thy) or cytosine (Cyt) bases. The 6-4PP links the 6 carbon on a 3' Cyt or Thy with the 4 carbon of a 5' Cyt. However, using the same wavelength of excitation in the UVB, 6-4PPs are induced only at 15–30 % the rate of CPDs [7]. The possible reasons for this difference in induction frequency of CPD and 6-4PP are not well understood but it is speculated that 6-4PP are mainly induced in functionally active regions of the genome [8]. Using fluorescently labeled monoclonal antibodies against CPD and a calibration with the radio immuno assay (RIA) [5] we were recently able to show that a UVB dose of 300 J/m² is able to induce several hundred thousands of CPDs per genome per cell [9]. Both CPDs and 6-4 PP can inhibit DNA synthesis and gene transcription. However, because of structural differences, 6-4 PP, even though induced at lower frequency, may be responsible for most of the *lethal* effects of UVB radiation in human cells. This could be deduced from studies which show that CPDs and 6-4 PP distort DNA to significantly different degrees; that is, the CPD induces a 7° bend whereas 6-4 PP bend the phosphodiester backbone of DNA by 44° [10], the latter being responsible for a more efficient blocking of the progression of DNA polymerases during DNA synthesis and transcription. On the other hand the high degree of DNA bending introduced by 6-4 PP might help to recognize this kind of photo damage easier by the cellular repair machinery, which could explain the more rapid removal of 6-4 PP from the genome compared to CPD (see „DNA repair and skin cancer", page 143). In line with this, the small bending angle introduced by CPDs may be responsible for the observation that CPDs are considerable more likely than 6-4 PPs to allow replication bypass (including base mis-pairing) because the small amount of DNA distortion due to CPDs is not as well recognized by repair enzymes as the 44° bend introduced by 6-4 PP. This implies that CPDs may be considerably more mutagenic than 6-4 PPs and are a likely candidate for UV-induced signature mutations [10] (see below).

When CPDs are not repaired (see § III), these DNA lesions can lead to mutations in the DNA sequence. Mutations are in the form of C → T and CC → TT transitions and are known as "UV-signature mutations". A so called "A-rule" has be proposed to explain how these signature mutations arise [11, 12]. According to the A-rule, DNA polymerase inserts adenine (A) residues by default opposite to lesions that it cannot interpret. A mutation is then created upon DNA replication of the strand containing base pair changes. At a CC cyclobutane dimer, therefore, a CC to TT transition occurs because two A residues are placed opposite the dimer by default in place of two guanine (G) residues. In a similar default mechanism 6-4 PPs give rise to C → T transitions [12].

UV-signature mutations have been detected in a number of tumor suppressor genes and oncogenes (including e.g. *patched, p16, ras* and *p53)* in human SCC, BCC and malignant melanoma [12]. Because the important biological function of p53 originates from transcriptional activation of a large number of genes involved in fundamental cellular processes like cell cycle control, apoptosis, DNA replication, repair, genome instability and senescence [e.g. 12, 13, 14], it is not surprising, that genetic alterations of the p53 gene have most frequently been found in human tumors [15] and especially in skin cancers like SCC, BCC and precancerous actinic keratoses [16, 17, 18, 19]. Convincing models for the initial steps of UV-carcinogenesis, especially for SCC, already exist [20, 21], based on findings that p53 mutations in non-melanoma skin cancers are detected at higher frequency (50–90 %) than those in internal malignancy and that the predominant alterations are C → T and CC → TT transitions at dipyrimidine sites [16].

Since UVA is only poorly absorbed by DNA its possible harmful effect on a cellular level has long been underestimated. However, it is clear now that genotoxic stress induced via UVA can be attributed to **indirect** photosensitizing reactions, which take place when cellular chromophores (photosensitizers like e.g. riboflavins, nicotinamid coenzym) are excited by the absorption of UVA-photons. Photophysical and photochemical reaction pathways then produce reactive molecular species which are able to interact with DNA (indirect DNA damage) [22]. At the level of DNA, photosensitization induces oxidative damage either by charge transfer from excited endogenous chromophores or by the reaction with reactive oxygen species (ROS) like e.g. superoxide anions, hydroxy radicals or singlet oxygen that are generated at UVA wavelengths. Indeed, photo-induced oxidative DNA-modifications, such as single-strand breaks; DNA-protein crosslinks, alkali labile sites and 8-oxo-7,8-dihydro-7'-deoxyguanosine (8-oxo-Guanin) have been observed (for a review see Cadet [23]), [24]. 8-oxo-Guanin, as the most prominent oxidative damage, has been detected in cells exposed e.g. to radiation from broad spectrum sun lamps [25], simulated solar light, SSL [26], to UVA [24, 27, 28, 29] and even visible light in the range of 400–450 nm [24, 30]. The mutagenic potential of 8-oxo-Guanin is well established and it has been shown that this type of photo-lesion mainly induces T → G transversions in the genome [31, 32, 33, 34]. However, this type of oxidative-damage induced mutation seems to represent only a minor fraction of UV-induced mutations, even in the UVA region [35, 36]. On the other hand it was found that the number of T → G transversions was higher in human skin cancers than in UVB-induced tumors in mice [37] and that the action spectrum of photocarcinogenesis in hairless mice is at least compatible with a role of UVA-induced non-dimer DNA modifications relevant for skin cancer induction [38]. Furthermore, there exists good evidence for a specific role of UVA-induced DNA lesions and mutations in the development of malignant melanoma in animal models [39].

It should also be kept in mind, that 8-oxo-Guanin, although this lesion represents the most prominent oxidative damage induced by UVA, is produced even more effectively by UVB radiation. A 1000fold higher yield of 8-oxo-Guanin induction has been reported in rodent cells [26]. This indicates that 8-oxo-dependent T → G transversions are also induced more effectively by UVB than UVA. However, recent investigations seem to rule out an involvement of 8-oxo-guanin in cell killing and demonstrate only a minor contribution of this type of oxidative photodamage to the overall solar UV mutagenic process [26]. Nevertheless, this does not speak against an important role of, especially, UVA for the induction of other types of mutations or genomic instabilities which might contribute to UV-carcinogenesis. This can be deduced e.g. from the fact that UVA is also able to induce CPDs. Production of CPDs upon UVA irradiation has been a matter of controversy for long time. However, it has been observed by a number of laboratories [27, 40, 41] and has now been accepted [42]. UVA induces CPDs with an approximately 1000fold lower yield than UVB [26] according to the absorption spectrum of DNA [43] and to an action spectrum published by Setlow already in 1974 [44]. Thus compared to UVC or to UVB, UVA doses have to be 10^5 [42] or 10^2 times higher [42, own results 2002] to produce the same amount of CPDs (in vitro about 1000 kJ/m^2 of UVA are needed to induce 0.25 CPD/kbp [42]). Despite this differences, the UVA doses are still relevant, since they correspond to a few hours of sunlight exposure at zenith in summer [41] and to significantly lower exposure times if irradiation takes place e.g. in solaria, which use UVA exposures up to 3–5 times higher than zenith equator sun. Recent data furthermore show that about 10 % of CPDs induced by simulated solar light were formed by the UVA component of this light source used [42]. Thus, UVA is ably to contribute significantly to the induction of CPD-dependent UV-signature mutations which have been shown to be one of the most important initial steps in development of skin cancer (see above).

Recent research also highlights other UVA-induced premutagenic DNA-lesions to be important for mechanisms possibly involved in UV-carcinogenesis. For example, if DNA single strand breaks (ssb), which can be induced as part of UVA-dependent oxidative damage, are not repaired in time within the cell cycle (i.e. in G1-phase) by base excision repair [45, 46], they are able to cause DNA double-strand breaks (dsb) in a S-phase specific repair pathway [47] and, thus result in chromosomal rearrangements and instability [48]. Molecular and genetic analysis of repair deficient strains from various organisms, therefore suggest, that ssb are one of the major risk factors for genome instability induced by oxidative DNA damage [49].

DNA repair and skin cancer

Skin cancer is one of the most common cancers worldwide and its incidence is increasing [50]. There is no doubt that UV-irradiation plays a major role in the induction of all three forms of skin cancer [51] including squamous and basal cell carcinomas as well as melanoma. For squamous cell carcinoma (SCC) a model exists in which UV-irradiation is effective in the initiation, promotion and progression step of tumorigenesis [20] including as an early step a mutation of the tumor suppressor gene p53. For SCC the cumulative UV dose enhances the risk. To the contrary, basal cell carcinoma (BCC) arises from intermittent as well as cumulative UV exposure and malignant melanoma (MM) is associated with intermittent UV exposure [51], especially through sun burns in childhood [52, 53, 54].

Evidence strongly suggests that DNA photoproducts formed at adjacent pyrimidine bases by direct absorption of solar UVB (290–320 nm) are the predominant, premutagenic events that initiate human BCC and SCC. The main photoproducts occurring are (CPD) and P(6-4)PP (see „UV-induced damage and mutations", page 140). Nucleotide excision repair (NER) removes most of these DNA photoproducts very effectively in cells. However, defects in NER, as found in Xeroderma pigmentosum, a rare, genetically recessive disease, increases the skin cancer risk to a factor of 1000 and more [3, 55] indicating that unrepaired or misrepaired UV-induced DNA lesions play a central role in the induction of skin cancer.

Characterization of the dose-dependence of photoproduct induction and repair has therefore become an important topic in research on UV mutagenesis and carcinogenesis, and the knowledge about essential processes is continuously increasing. In mammalian cells CPDs seem to be repaired much slower than (6-4)-PP [7, 56], and the repair rates for both CDP and (6-4)-PP are considered to be heterogeneous along the mammalian genome. Preferential repair of transcriptionally active genes compared to inactive ones has been demonstrated, as well as faster repair of the transcribing strand of active genes compared to the non-transcribing strand [56, 57, 58, 59]. Furthermore, repair rates of UV-induced CPDs are found to vary in the same gene sequence, even between adjacent base positions, and it has been shown that some mutational hotspots (e.g., in the p53 gene) can be considered as "repair coldspots" [60, 61, 62, 63, 64, 65]. There is also increasing evidence that the specific chromatin environment of UV-induced DNA lesions may influence the recognition of damage and its repair [66, 67, 68, 69, 70, 71]. Beside these genetic- and sequence-specific differences in repair efficiency, we have recently shown (using immunofluorescently labeled CPDs) that overall repair capacity in primary human keratinocytes is characterized by a large inter-individual variability with respect to repair kinetics and residual damage which cannot be removed at all [72] and that repair capacity of human keratinocytes can be saturated in a dose-dependent way [9].

In a recent study [73] we treated Skh-1 hairless mice with sub-erythemal chronic UVB exposures for 40 days and analysed the amount and distribution of CPDs. It was found that CPDs accumulated as a result of chronic irradiation and the damage persisted in the epidermis and the dermis over several weeks. While these persistent CPDs were evenly distributed in the dermis we found a small number of heavily-damaged cells in the basal layer of the epidermis up to 50 days after the chronic treatment was terminated. CPD retaining basal cells (CRBC) are not apoptotic and disappeared when the skin cells were induced to proliferate by treatment with a tumor promotor (TPA) or a sunburn dose of UVB [73, 74]. At the same time we found a slight increase of p53 mutated cell clones in the skin. In a preliminary study we could show, that the occurrence of CRBCs is not limited to mouse skin since similarly heavily damaged cells were found in human skin [74] (see Fig. 1., page 145)

In current models, skin cancer induction is described as a multistage process including initiation, promotion and progression induced by different chemical carcinogens in combination with UV irradiation or by UV alone. The initiation stage seems to be irreversible and tumor development by a promotor can be observed up to one year after the initial treatment [75]. Since it has been shown that a small number of skin cells with low proliferation activity are damage retaining, the occurrence of persistent damage retaining cells represents a tumor initiation step [76]. Work from other laboratories [76, 77] and our own group [78] suggests that these persisting cells are epidermal stem cells. In a recent study, we investigated the correlation between CRBCs in human skin and sun exposure history by quantifying CRBCs in skin biopsies taken from parts of the body with different sun exposure histories. A good correlation between sun exposure of skin areas where the biopsies had been taken from and the number of CRBCs in the basal layer of epidermis could be documented [Volkmer 2003].

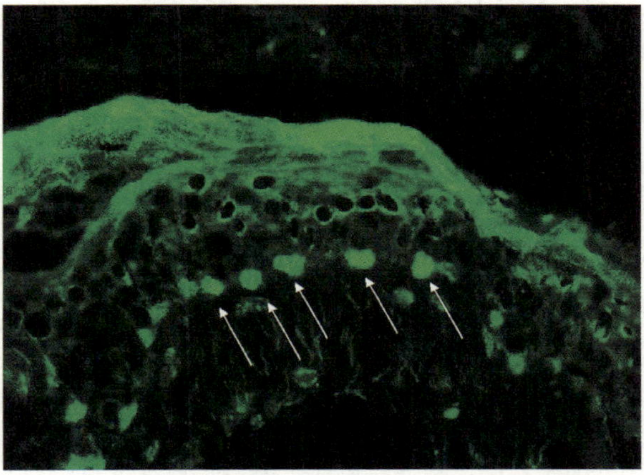

Fig. 1. Fluorescence-immunological detection of CRBCs (using monoclonal antibodies against CPDs) in a section of human skin. Biopsies were taken from an skin area which had been sun exposed in the past.

These findings suggest that the CRBCs we observe in mouse and human epidermis are possible skin cancer precursor cells and can be used as biomarkers for assessing skin cancer risk in the human population.

Genomic instability and skin cancer

Malignant melanoma of the skin represent that kind of skin cancer with the poorest prognosis. In contrast to many other solid tumors, melanoma may also affect a younger population and is the leading case of death due to malignancy in females between the age 20 and 30, e.g. in USA [79]. Overall mortality due to malignant melanoma is reported to be in the range of 20–25 % [80] of all incident cases (e.g in Germany). A number of epidemiological studies as well as in-vitro and in-vivo work on animal and human material show that UV is the main risk factor in the etiology of melanoma focussing on the UVA-range, which might be of special importance [39, 81, 82, 83, 84, 85].

UV-induced mutations have been found in several tumor suppressor gene and oncogenes [86], and especially those induced in the CDKN2A-gene (encoding the tumor suppressor $p16^{INK4A}$) and in other components of the $p16^{INK4A}$/CDK4/

CyclinD/pRB-pathway (which controls the cell cycle) have been shown to have important ramifications in melanocyte transformation [86]. Nearly all melanomas have been found to have defects in this pathway [87, 88]. Interestingly, (probably) UV-induced genomic instability seems to play an important role in melanoma formation because nonrandom deletions, losses of heterozygosity and rearrangements have been found frequently in chromosomal regions including e.g. 1p, 7q, 9p, 10q and 11q [89, 90, 91]. Future research has now to be focussed to answer the question how and when these and perhaps other chromosomal changes are introduced in melanocyte's genome by the action of UV-radiation at certain time points of melanoma development.

Some kind of genomic instability, which manifests itself by the appearance of chromosomal aberrations, might be introduced via non-repaired, persistent UV-induced photoproducts in the DNA of skin cells. The ability of UV-radiation to produce chromosome aberrations was already described about 30 years ago [92]. Later on Kaufmann and Wilson demonstrated that the greatest number of chromosomal aberrations occurred when normal cells were UV-irradiated at the G1/S border of the cell cycle and that A-T fibroblasts were hypersensitive to the induction of chromosomal aberrations by UV [93]. The fact that UV-radiation mainly induces chromatid-type aberrations further suggests that they arise after DNA replication has occurred [94] and that they are most probably the result of errors or deficiencies in post-replicative repair mechanisms [92, 94]. These models have been refined recently to include new information on the mechanisms of post-replicative repair. When DNA replication occurs while UV-photoproducts still remain in the DNA, some of these lesions block the progression of DNA replication forks [95]. If this blockage is persistent, because, e.g. it could not be bypassed by the action of certain polymerases (e.g. polymerase η and ξ) [96, 97], aberrant DNA replication intermediates are produced that contain long patches of single-stranded DNA and DNA double-strand breaks accumulate [94, 98, 99], which are known to be precursor lesions of chromosomal aberrations.

The aberrant replication intermediates that accumulate after UV-irradiation are thought to be resolved through recombinational mechanisms between sister chromatids [92, 94, 100, 101]. The enzymology and signalling pathways responsible for this process are not yet fully understood. However, recently it could be shown that the p34-subunit of the single-stranded DNA-binding protein (replication protein A, RPA), which is a key protein in the cellular processes of NER, DNA recombination and DNA double-strand break repair [102, 103], becomes hyperphosphorylated in response to UV(C)-radiation, depending on the expression of ATM, the kinase mutated in the human genetic disorder ataxia-teleangiectasia (A-T). Induction of this DNA UV-damage response was further shown to be dependent on replication [104]. It is postulated that this pathway is triggered by the accumulation of aberrant replication intermediates (probably DNA-dsb) resulting from DNA replication fork blockage by UV-photoproducts. It is suggested that RPA-p34 is hyperphosphorylated as a participant in the recombinational postreplication repair of these replication products. Successful resolution of these replication intermediates reduces the accumulation of chromosomal aberrations that would otherwise occur as a consequence of UV-radiation.

If DNA double-strand breaks that accumulate after UV-radiation are the inducing signals, this would suggest similarities in the signalling pathways responsible for the processing of UV-radiation and ionizing-radiation (IR) induced DNA damage, since it is well known that IR-induced DNA-dsb induce ATM-kinase activity towards a number of important checkpoint- and repair-proteins involved in DNA-dsb repair [105]. Recently, Limoli and coworkers [106] furthermore demonstrated co-localization of proliferating cell nuclear antigen, PCNA (involved in DNA replication and repair) and the MRE11 complex (Mre11/Rad50/Nbs1, involved in recombinatinal DNA-dsb repair after ionizing radiation) in nuclear foci 4–8 h after UV-irradiation of XPV human fibroblast (deficient in polymerase η). This further indicates that common or related signalling and repair pathways are shared after UV-radiation and ionizing radiation and implicates an important role of the MRE11 complex also in recognition and/or resolution of DNA replication intermediates (most probably DNA-dsb) after UV-radiation which could be precursors of chromosomal aberrations. The involvement of DNA-dsb in a resolution of UV-induced replication fork arrests has furthermore clearly been presented very recently by Limoli's group [107] which detected γ-H2AX formation in Xerpderma Pigmentosum variant (XPV-) cell lines after UVC irradiation. γ-H2AX represents the phosphorylated form of the H2AX sub-population of H2A histone in the nucleosomal structure of chromatin. ATM-dependent phosphorylation of H2AX at places of (ionizing radiation-induced) dsb has been shown to be the earliest detectable dsb-dependent signal, rendering dsb-damaged chromatin susceptible for the dsb repair machinary [108, 109].

Using immunofluorescently labeled antibodies against γ-H2AX, we (see Fig. 2., page 148) and others [110] have very recently be able to show that UVA is able to produce DNA-dsb directly in the genome of human keratinocytes. This is the first direct demonstration of a causal link between UV(A)-induced DNA-lesions and a possible later appearance of genomic instability on a chromosomal level, which has been shown to be important in skin cancer development, especially malignant melanoma (see above). In this context, it is interesting to see that melanoma cell lines developed from melanoma metastases (the last step of melanoma development) are characterized by a high level of chromosomal aberrations (of both numerical and structural type) involving nearly all chromosomes of the human genome, as could be shown by us recently using spectral karyotyping (SKY, multi-colour FISH) in metaphase preparations of human melanoma cell lines [72] (see Fig. 3., page 148)

Future research will therefore be needed to show, how and when these chromosomal instability will be produced by UV(A)-radiation at certain stages of skin cancer (especially melanoma) development in dependence of radiation quality and exposure conditions, and, how certain chromosomal alterations might be used as diagnosis, prognosis or risk markers for skin cancer. Identification of molecular (DNA-repair/misrepair) processes and signalling pathways involved might also influence therapeutic strategies in skin cancer treatment.

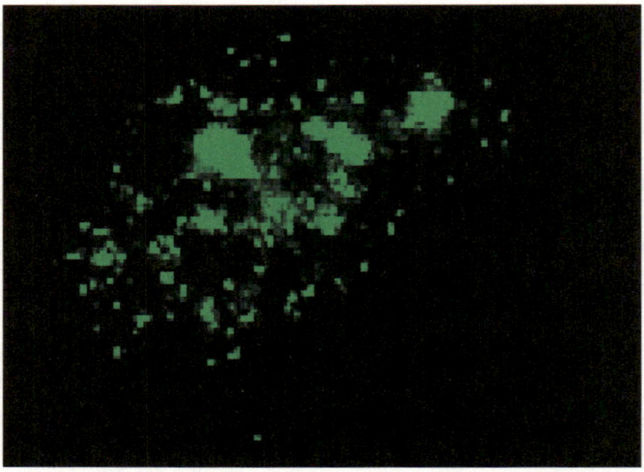

Fig. 2. Visualization of H2AX-phosporylation-foci (indicative of dsb-induction [108, 109] induced by UVA (λmax = 365 nm) using fluorescently labeled monoclonal antibodies against γ-H2AX (the phophorylated form of H2AX) in a skin fibroblast.

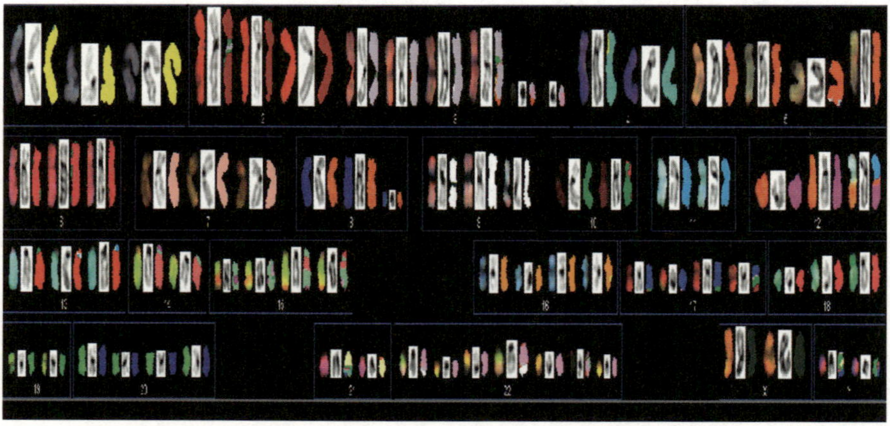

Fig. 3. Spectral karyotyping (SKY) of a human melanoma metastasis cell line showing a high degree of aneuploidy (e.g. # 1 and # 20) complex aberrations (e.g. # 15) and reciprocal translocations (e.g. # 12) (from [72])

Outlook

This review could only focus on a selected part of UV-radiation biology in skin cancer research. DNA-repair mechanisms and their involvement in skin cancer development have been discussed in more detail and other important mechanisms, like e.g. UV-induced apoptosis and immunosuppression could not be described and their essential roles in skin carcinogenesis could (unfortunately) not be touched.

However, even for the latter mechanisms, DNA-repair processes like NER and BER, their deficiencies and interindividual variability in repair capacities will surely be of predominant importance, because it is widely accepted now that unrepaired UV-induced DNA damages are those primary signals which cause apoptotic events and immuno-modulation.

Further work has also to be focused on the problem how modulations of nuclear chromatin structure by epigenetic regulation via methylation, chromatin remodelling and chromatin condensation, brought about by acetylation and deacetylation of core histone tails [111] influence processes of DNA-damage induction and DNA repair activities, and how these processes are, generally, determined by the over all dynamics of interphase chromatin [112]. The influence of different radiation qualities (e.g. UVA vs UVB) and different exposure regimes (chronic vs acute) have to be studied, against the above mentioned background, in more detail. Especially the induction of genomic instabilities, probably induced by UVA, seems to be an important field of future research, because it seems clear now that UVA produces DNA-dsbs [110, Greinert, see this review] which are well known to be precursor lesions of chromosomal aberrations. Work should also concentrate on primary cell culture from human skin and on epidermal stem cells. Modern genetic, molecular biological and cytogenetical experimental approaches are available (and will be developed in the future) to show that UV-radiation biology (as part of radiation biology) has played an important part of (skin) cancer research and will do so in the future.

References

1 Ritter JW (1803) Versuche über das Sonnenlicht. Gilberts Ann Physik 12:409–415 (München)

2 Cleaver JE, Mitchell DL (1997) Ultraviolet radiation carcinogenesis. In: Holland JF, Bast Jr. RC, Morton DL, Frei III E, Kufe DW, Weichselbaum RR (eds) Cancer Medicine Vol. 1, Willams and Wilkens, Baltimore:307–318.

3 Cleaver JE, Thompson LH, Richardson AS, States JC (1999) A summary of mutations in UV, sensitive disorders: Xeroderma pigmentosum, Cockayne syndrome, and Trichothiodystrophy. Human Mutat 14:19–22

4 Slaper H, Velders GS, Daniel JS, deGruijl FR, van der Leund JC (1996) Estimation of ozone depletion and skin cancer incidence to examine the Vienna convention achievements. Nature 21:256–258.

5 Rosenstein BS, Mitchell DL (1987) Action spectra for the induction of pyrimidine(6-4)pyrimidone photoproducts and cyclobutane dimers in normal human skin fibroblasts. Photochem Photobiol 45:775–781

6 Cadet J, Vigny P (1990) The photochemistry of nucleic acids. In: Morrison H (ed) Bioorganic Photochemistry: Photochemistry and the nucleic acids Vol. 1 J. Wiley and Sons, New York, NY: 1–272

7 Mitchell DL, Brash DE, Nairn RS (1990) rapid repair of pyrimidine (6-4) pyrimodone photoproducts in human cells does not result from change in epitope conformation. Nucl Acids Res 18:963–971

8 Mitchell DL, Karentz D (1993) The iduction and repair of DNA photodamage in the environment. In: Young AR, Björn LO, Moan J, Nultsch W (eds) Environmental UV Photobiology Plenum Press, New York and London:345–377
9 Greinert R, Boguhn O, Harder D, Breitbart EW, Mitchell DL, Volkmer B (2000b) The dose dependence of cyclobutane dimer induction and repair in UVB-irradiated human keratinocytes. Photochem Photobiol 72:701–708
10 Mitchell DL, Pederson M, Gosh R (1996) UV-induced DNA damage and skin cancer. In: Volkmer B, Heller H (eds) Environmental UV-radiation, risk of skin cancer and primary prevention Fischer Verlag, Stuttgart, Jena, Lübeck, Ulm:117–136
11 Tressmann I, Liu SK, Kennedy MA (1992) Mechanism of SOS muatgenesis of UV-irradiated DNA: mostly error-free processing of deaminated cytosin. Proc Natl Acad Sci USA 89:1159–1163
12 Matsumura Y, Ananthaswamy HN (2000) Molecular mechanisms of photocarcinogenesis Front. Bioscience 7 765–783
13 Harris CC (1996) Structure and function of the p53 tumor suppressor gene: clues for rational cancer therapeutic strategies. J Natl Cancer Inst 88:1442–1455
14 Prives C, Hall PA (1999) The p53 pathway. J Pathol 187:112–126
15 Hollstein, M. Sidransky, D, Vogelstein, B, Harris, CC (1991) p53 mutation in human cancer. Science 253:49–53
16 Ziegler A, Leffell DJ, Kunala S, Sharma HW, Gailani M, Simon JA, Halperin AJ, Baden HP, Shapiro PE, Bale AE, Brash DE (1994) Mutation hotspots due to sunlight in the p53 gene of nonmelanoma skin cancer. Proc Natl Acad Sci USA 90:4216–4220
17 Brash DE, Rudolph JA, Simon JA, Liu A, McKenna GJ, Baden HP, Halperin AJ, Ponten J (1991) A role for sunlight in skin cancer: UV-induced p53 mutations in squamous cell carcinoma. Proc Natl Acad Sci USA 88:10124–10128
18 van der Riet P, Karp D, Farmer E, Wei Q, Grossmann L, Tokino K, Ruppert JM, Sidransky D (1994) Progression of basal cell carcinoma through loss of chromosome 9q and inactivation of a single p53 allele. Cancer Res 54:25–27
19 Nelson MA, Einspahr JG, Alberts DS, Balfour CA, Wymer JA, Welch KL, Salasche SJ, Bangert JL, Grogan TM, Bozzo PO (1994) Analysis of p53 gene in human precancerous actinic keratosis lesions and squamous cell carcinoma. Cancer Lett 85:23–29
20 Brash DE (1997) Sunlight and the onset of skin cancer. Trends Genet 13:420–413
21 Fusenig N, Boukamp P (1998) Multiple steps of genetic alterations in immortalization, malignant transformation and tumor progression in human skin keratinocytes. Mol Carcinogen 23:144–158
22 Cadet J, Aselmino C, Douki, T Voituriez L (1992) Photochemistry of nucleic acids in cells. J Photochem Photobiol B 15:277–298
23 Cadet J, Berger M, Douki T, Ravanat J-L (1997) Oxidative damage to DNA: formation, measurement and biological significance. Rev Physiol Biochem Pharmacol 137:1–87
24 Kielbassa C, Roza L, EPE, B (1997) Wavelength dependence of oxidative DNA damage induce by UV and visible light. Carcinogenesis 18:811–816
25 Fischer-Nielson A, Loft S, Jensen KG (1993) Effect of ascorbate and 5-aminosalicylic acid on light-induced 8-oxo-guanosone formation in V79 hamster cells. Carcinogenesis 14:2431–2433
26 Douki T, Perdiz D, Grof P, Moustacchi E, Cadet J, Sage E (1999) Oxidation of guanin in cellular DNA by solar UV-radiation: Biological role. Photochem Photobiol 70:184–190

27 Zhang XS, Rosenstein BS, Lebwohl M, Mitchel DL, Wei HC (1997) Induction of 8-oxo-7,8 dihydro-2'-deoxyguanosine by ultraviolet radiation in calf thymus DNA and HeLa cells. Photochem Photobiol 65:119–124

28 Kvam E, Tyrell RM (1997) Induction of oxidative DNA damage in human skin cells by UV and visible radiation. Carcinogenesis 18:2379–2384

29 Wamer WG, Wei RR (1997) In-vitro photo-oxidation of nucleic acids by ultraviolet A radiation. Photochem Photobiol 65:560–563

30 Pflaum M, Kielbassa C, Garmyn M, Epe B (1998) Oxidative DNA damage induced by visible light in mammalian cell: extent, inhibition by antioxidants and genotoxic effects. Mutat Res 408:137–146

31 LePage F, Margat A, Grollmann AP, Sarasin A, Gentil A (1995) Mutagenicity of a unique 8-oxoguanine in human a Ha-ras sequence in mammalian cells. Carcinogenesis 16:2779–2784

32 Grollmann AP, Moriya M (1993) Mutagenesis by 8-oxo-guanine: an enemy within. Trends Genet 9:246–249

33 Moriya M (1993) Single-stranded shuttle plasmid for mutagenesis studies in mammalian cells: 8-oxo-guanine in DNA induces targeted GC→TA transversions in simian kidney cells. Proc Natl Acad Sci, USA 90:1122–1126

34 Kamiya H, Murata-Kamiya S, Koizume S, Nishimura S, Ohtsuka E (1995) 8-hydroxyguanine (7,8-dihydroxy-8-oxoguanine) in hot-spots of the c-Ha-ras gene: effects of sequence contexts in mutation spectra. Carcinogenesis 16:883–889

35 Dobretsky EA, Turcotte J, Chateauneuf A (1995) A role of ultraviolet A in solar mutagenesis. Proc Natl Acad Sci USA 92:2350–235

36 Roberts C, Muel B, Benoit A, Dubetret A, Sarasin A, Stary A (1996) Cell survival and shuttle vector mutagenesis induce by ultraviolet A and ultraviolet B radiation in a human cell line. J Invest Dermatol 106:721–728

37 Duma N, van Kranen HJ, de Vries A, Berg RJW, Wester CF, van Kreijl CF, Sarasin A, Daya-Grosjean L, deGruijl FR (1997) The role of UVB light in skin carcinogenesis through the analysis of p53 mutations in squamous cell carcinoma of hairless mice. Carcinogenesis 18:897–904

38 deGruijl FR, Sterenborg HJ, Forbes PD, Davies RE, Cole C, Kelfkens G, van Weelden H, Slapers H, van der Leun JC (1993) Wavelength dependence of skin cancer induction by ultaviolet irradiation of albino hairless mice. Cancer Res 53:53–60

39 Ley RD (2002) Animal models of ultraviolet radiation (UVR)-induced cutaneous melanoma. Front Biosci 7:1531–153

40 Tyrell RM (1973) Induction of pyrimidine dimers in bacterial DNA by 365 nm radiation. Photochem Photobiol 17:69–73

41 Kuluncics, Z, Perditz, D, Brulay, E, Muel, B and Sage, E (1999) Wavelength dependence of ultraviolet-induced DNA damage distribution: involvement of direct and indirect mechanisms and possible artefacts. J Photochem Photobiol B: Biol. 49:71–80

42 Perdiz D, Grof P, Mezzina M, Nikaido O, Moustacchi E, Sage E (2000) Distribution and repair of bipyrimidine photoproducts in solar UV-irradiated mammalian cells. J Biol Chem 35:26732–26742

43 Sutherland JC, Griffin KP (1981) Absorption spectrum of DNA for wavelength greater than 360 nm. Radiat Res 86:399–409

44 Setlow RB (1974) The wavelength in sunlight effective in producing skin cancer: a theoretical analysis. Proc Natl Acad Sci USA 71:3363–3366

45 Cabelof DC, Raffoil JJ, Yanamadala S, Guo Z, Heydari AR (2002) Induction of polymerase beta-dependent by excision repair in response to oxidative stress in vivo. Carcinogenesis 23:1419–1425
46 Mitra S, Izumi,T, Boldogh I, Bhakkat KK, Hill JW, Hazra TH (2002) Choreography of oxidative damage repair in mammalian genomes. Free Radic Biol Med 33:15–28
47 Taylor RM, Moore DJ, Whitehouse J, Johnson P, Caldecott KW (2000) A cell cycle-specific requirement for XRCC1 BRCT II domain during mammalian strand break repair. Mol Cell Biol 20:735–740
48 Paulovich AG, Toczyski DP, Hartwell LH (1997) When checkpoints fail. Cell 88:315–321
49 Okano S, Kanno S, Nakajima S, Yasui A (2000) Cellular response and repair of single-strand breaks introduced by UV damage endonuclease in mammalian cells. J Biol Chem 42:32635–32641
50 IARC: Solar and Ultraviolet Radiation (1992) IARC Monographs on the Evaluation of carcinogenic risks to humans. IARC Scientific Publication 55:73–138
51 Armstrong BK, Kricker A (2001) The epidemiology of UV induced skin cancer. J Photochem Photobiol B: Biology 63:8–18
52 Berwick M, Halpern A (1997) Melanoma epidemiology. Curr Opin Oncol 9:178–182
53 Breitbart M, Garbe C, Büttner P et al. (1997) Ultraviolet light exposure, pigmentary traits and the development of melanocytic naevi and cutaneous melanoma. A case-control study of the German Central Malignant Melanoma Registry. Acta Derm Venerol 77:374–378, 11
54 Dulon M, Weichenthal M, Blettner M, Breitbart M, Hetzer M, Greinert R, Baumgardt-Elms C, Breitbart EW (2002) Sun exposure and number of nevi in 5- to 6-year old European children, J Clin Epidemiol 55:1075–1081
55 Kraemer KH, Lee MM, Andrews AD, Lambert WC (1994) The role of sunlight and DNA repair in melanoma and nonmelanoma skin cancer. The xeroderma pigmentosum paradigm. Arch Dermatol 130:1018–1021
56 van Hoffen A., Venema J, Meschini R, van Zeeland AA, Mullenders LHF (1995) Transcription coupled repair removes both cyclobutane pyrimidine dimers and 6-4 photoproducts with equal efficiency and in a sequential way from transcribed DNA in xeroderma pigmentosum group C fibroblasts. EMBO J 14:360–367.
57 Bohr VA, Smith CA, Okumotu DS, Hanawalt PC (1985) DNA repair in an active gene: Removal of pyrimidine dimers from the DHFR gene of CHO cells is much more efficient than in the genome overall. Cell 40:359–369
58 Mellon I, Spivak G, Hanawalt PC (1987) Selective removal of transcription blocking DNA damage from the transcribed strand of mammalian DHFR gene. Cell: 51:241–249.
59 Selby CP, Sancar A (1993) Molecular mechanism of transcription-repair coupling. Science 260:53–58.
60 Kunala S, Brash DE (1992) Excision repair at individual bases of the Escherichia coli lacI gene: Relation to mutation hot spots and transcription coupling activity. Proc Natl Acad Sci: USA 89:1031–11035.
61 Gao S, Drouin R, Holmquist GP (1994) DNA repair rates mapped along a human gene at nucleotide resolution. Science 263:1438–1440.
62 Tornaletti S, Pfeiffer GP (1994) Slow repair of pyrimidine dimers at p53 mutation hot spots in skin cancer. Science 263:1436–1438.
63 Tu Y, Tornaletti S, Pfeiffer GP (1996) DNA repair domains within a human gene: Selective repair sequences near the transcription initiation site. EMBO J 15:5–683.

64 Tijsterman M, Tasseron-de Jong JG, van de Putte P, Bouxer J (1996) Transcription-coupled and global genome repair in Saccharomyces cerevesiae RPB2 gene at nucleotide resolution. Nucleic Acids Res 24:3499–3505.
65 Dammann R, Pfeiffer GP (1997) Lack of gene- and strand-specific DNA repair in RNA polymerase III-transcribed human tRNA genes. Mol Cell Biol 17:219–229.
66 Pfeiffer GP (1997) Formation and processing of UV photoproducts: effects of DNA sequence and chromatin environment. Photochem Photobiol 65:805–819
67 Smerdon MJ, Thomae F (1990) Site-specific DNA repair at the nucleosome level in a yeast minichromosome. Cell 61:675–684.
68 Schieferstein U, Thoma F (1998) Site-specific repair of cyclobutane pyrimidine dimer formation in a positioned nucleosome by photolyase and T4 endonuclease V in vitro. EMBO J 17:306–316.
69 Wang Z, Wu X, Friedberg EC (1991) Nucleotide excision repair of DNA by human cell extracts is suppressed in reconstituted nucleosomes. J Biol Chem 266:22472–22478
70 Suquet C, Mitchell DL, Smerdon MJ (1995) Repair of UV-induced (6-4) photoproducts in nucleosome core DNA. J Biol Chem 270: 16507–16509
71 Schieferstein U, Thoma F (1996) Modulation of cyclobutane pyrimidine dimer formation in a positioned nucleosome containing poly(dA·dT) tracts. Biochem 35:7705–7714.
72 Greinert R, Breitbart EW, Mitchell DL, Smida J, Volkmer B (2000a) Characterization of human keratinocytes by measuring cellular repair capacity of UVB-induced DNA damage and monitoring of cytogenetic changes in melanoma cell lines. Radiat Protect Dosim: 91,41–45
73 Mitchell DL, Greinert R, De Gruijl FR, Guikers KLH., Breitbart EW, Byrom M., Gallmeier M, Lowery M, Volkmer B (1999) Effects of chronic low-dose ultraviolet B radiation on DNA damage and repair in mouse skin. Cancer Res 59:2875–2884
74]Mitchell DL, Volkmer B, Breitbart EW, Byrom M., Lowery MG, Greinert R (2001) Identification of a non-dividing subpopulation of mouse and human epidermal cells exhibiting high levels of persistent UV photodamage. J Invest Dermatol:117, 590–595
75 Slaga TJ, Budunova IV, Gimenez-Conti IB, Aldaz M (1996) The mouse skin carcinogenesis model. J Invest Dermatol Symposium Proceedings 1:151–156
76 Morris RJ (2000) Keratinocyte stem cells: targets for cutaneous carcinogens. The Journal of Clinical Investigation 106:3–8
77 Jones PH, Harper S, Watt FM (1995) Stem cell patterning and fate in human epidermis. Cell 80:83–93
78 Volkmer B, Mitchell DL, Breitbart EW, Greinert R (2003): Immunofluorescent detection of persistent heavily damaged cell in the basal layer of sun exposed human skin. J Invest Dermatol, submitted
79 Johnson TM, Dolan OM, Hamilton TA, Lu MC, Swanson NA, Lowe L (1998) Clinical and histological trends in melanoma. J Am Acad Dermatol 38:731–744
80 Becker N, Warendorf J (1998) Krebsatlas der Bundesrepublik Deutschland, 1981–1990. 3. Auflage, Springer Verlag.
81 Moan J, Dahlback A, Setlow RB (1999) Epidemiological support for a hypothesis for melanoma induction indicating a role for UVA radiation. Photochem. Photobiol 70:243–247
82 Rünger T (1999) Role of UVA in pathways of melanoma and nonmelanoma skin cancer. Photodermatol Photoimmunol Photomed 15:212–216

83 Robinson ES, Hill RJ, Kripke ML, Setlow RB (2000) The monodelphis melanoma model: initial report on large ultraviolet A exposure of suckling young. Photochem Photobiol 71:743–746
84 Wang SQ, Setlow R, Berwick M, Polsky D, Marykoob AA, Kopf AW, Bart RS (2000) Ultraviolet A and melanoma: a review. J Am Acad Dermatol 146:837–846
85 Ortonne JP (2002) Photobiology and genetics of malignant melanoma. Br J Dermatol 146:11–16
86 Walker GJ, Hayward NK (2002) Pathways to melanoma development: lessons from the mouse. J Invest Dermatol 119:783–792.
87 Castello M, Pollock PM, Walters MK et al. (1997) CDKN2A/p16 is inactivated in most melanoma cell lines. Cancer Res 57:4868–4874
88 Walker, GJ, Flores, JF, Glendening, JM, Lin, AH, Markl, ID, Fountain, JW (1998): Virtual 100% of melanoma cell lines harbour alterations at the DNA level within CDKN2A, CDKN2B or one of their downstream targets. Genes Chrom Cancer 22: 157–163
89 Dracopoli NC, Fountain JW (1996) CDKN2A mutations and melanoma. Cancer Surveys 26:115–132
90 Swedley D, Sidhar S, Birdsall S, Bennett D, Herlyn M, Cooper C, Shipley J (2000) Characterization of chromosome 1 abnormalities in malignant melanomas. Genes Chrom Cancer 28:121–125
91 Sargent LM, Nelson MA, Lowry DT, Senft JR, Jefferson AM, Ariza ME, Reynolds SH (2001) Detection of three novel translocations and specific common chromosomal break sites in malignant melanoma by spectral karyotyping. Genes Chrom Cancer 32:18–25
92 Bender MA, Griggs HG, Walker PL (1973) Mechanisms of chromosomal aberration production. I. Aberration induction by ultraviolet light. Mutat Res 20:387–402
93 Kaufmann WK, Wilson SJ (1994) G1 arrest and cell-cycle dependent clastogenesis in UV irradiated human fibroblasts. Mutat Res 314:67–76
94 Kaufmann WK (1989) Pathways of human cell post-replication repair. Carcinogenesis 10:1–11
95 Park SD, Cleaver JE (1979) Recovery of DNA synthesis after ultraviolet irradiation of xeroderma pigmentosum cells depends on excision repair and is blocked by caffein. Nucleic Acids Res 6:1151–1159
96 Gibbs PE, McGregor WG, Maher VM, Nisson P, Lawrence CW (1998) A human homolog of sacharomyces cerevisiae REV3 gene, which encodes the catalytic subunit of polymerase eta. Proc Natl Acad Sci USA 95:6876–6880
97 Masutani C, Kusamoto R, Yamada A, Dohmae N, Yokoi M, Yuasa M, Araki M, Iwai S, Takio K, Hanaoka F (1999) The XPV (xeroderma pigmentosum variant) gene encodes human DNA polymerase eta. Nature 399:700–704
98 Wang TC, Smith KC (1986) Postreplication repair in ultraviolet-irradiated human fibroblasts: formation and repair of DNA double-strand breaks. Carcinogenesis 7:389–392
99 Michel B, Ehrlich SD, Uzest M (1997): DNA double-strand breaks caused by replication arrest. EMBO J 16:430–438
100 Haber JE (2000) Recombination: a frank view of exchanges and vice versa. Curr Opin Cell Biol 12:286–292
101 Lowndes NF, Mugia JR (2000) Sensing and responding to DNA damage. Curr. Opin Genet Dev 10:17–25
102 Wold MS (1997) Replication protein A: a heterodimeric, single-stranded DNA-binding protein required for eukaryontic DNA metabolism. Annu Rev Biochem 66: 61–92

102 Iftode, C, Daniely, Y and Borowiec, JA (1999): Replication protein A (RPA): the eukaryotic SSB. Crit. Rev. Biochem Mol Biol 34:141–180
104 Oakley GG, Loberg LI, Yao J, Risinger MA, Yunker RL, Zernik-Kobak M, Kanna KK, Lavin MF, Carty MP, Dixon K (2001) UV-induced hyperphosphorylation of replication protein A on DNA replication and expression of ATM protein. Mol Biol Cell 12:1199–1213
105 Abraham RT (2001) Cell cycle checkpoint signalling through ATM and ATR kinases. Genes Develop 15:2177–2196
106 Limoli CL, Giedzinski E, Morgan WF, Cleaver JE (2000) Inaugural article: polymerase eta deficiency in xeroderma pigmentosum variant uncovers an overlap between S-phase checkpoint and double-strand break repair. Proc Natl Acad Sci USA 97:7939–7946
107 Limoli CL, Giedzinski Bonner WM, Cleaver JE (2002) UV-induced replication arrest in Xeroderma Pigmentosum variant leads to DNA double-strand breaks, γ-H2AX formation, and Mre11 recombination. Proc Natl Acad Sci USA, 8:233–238
108 Burma S, Chen BP, Murphy M, Kurimasa A, Chen DJ (2001) ATM phosphorylates histone H2AX in response to DNA double-strand breaks. J Biol Chem 276:42462–42467
109 Paull TT, Rogakou EP, Yamozaki V, Kirchgessner CU, Gellert M, Bonner WM (2000) A critical role for histone H2AX in recruitment of repair factors to nuclear foci after DNA damage. Curr Biol 10:886–895
110 Rapp A, Dittmar H, Hausmann M, Greulich KO (2002) UVA irradiation produces DNA double strand breaks. Abstract. In: Proceedings of 6. Jahrestagung der Gesellschaft für Strahlenforschung, Sept. 25.–27., Göttingen, Germany
111 Rakyan VK, Preis J, Morgan HD, Whielaw E(2001) The mark, mechanisms and memory of epigenetic states in mammals. Biochem J 356:1–10
112 Roix J, Misteli T (2002) Genomes, proteomes, and dynamic networks in the cell nucleus. Histochem. Cell Biol 118:105–116

Contribution of Radiation Research to Human Space Exploration: Approaches to mitigate Radiation Health Risk in Spaceflight

Marco Durante

Dipartimento di Scienze Fisiche, Università "Federico II", Monte S. Angelo, Via Cintia, 80126 Napoli, Italy
marco.durante@na.infn.it

Abstract.

It is likely that the 21st century will be characterized by human space exploration and colonization. The International Space Station (ISS) already provides a living environment for humans outside the Earth; a manned mission to Mars is planned for 2025; exploration of other planets (e.g. Jupiter) are also planned; space tourism may soon become a reality. A major area of concern is the possible detrimental effects on health (including cancer, cataracts, hereditary effects and neurological disorders) caused by the exposure to galactic cosmic rays (GCR) and solar particle events (SPE). Uncertainties in space radiation risk for interplanetary missions are substantially high (400–1500% according to NASA estimates), and international scientific cooperation is needed to reduce these uncertainties before manned exploration missions start. Even if uncertainties in risk assessment will be reduced in the next few years, there is little doubt that appropriate countermeasures have to be taken to reduce the exposure or the biological damage produced by cosmic radiation. In addition, it is necessary to develop appropriate countermeasures against SPE, which can produce acute effects, even life threatening, for inadequately protected crews. Strategies that may prove to be effective in reducing exposure, or the effects of the irradiation, include shielding, administration of drugs or dietary supplements to reduce the radiation effects, crew selection based on a screening of individual radiation sensitivity. It is foreseeable that research in passive and active radiation shielding, radioprotective chemicals, and individual susceptibility will boost in the next years to provide efficient countermeasures to the space radiation threat.

Introduction

When human space exploration started over 40 years ago, it was soon recognized that space radiation represented a potential hazard for the crews [13]. As a matter of fact, astronauts are exposed in space to cosmic radiation at a dose rate substantially higher than on Earth. The absorbed dose in space depends on mission duration, altitude, and inclination, as outlined in Table 1. Current plans for manned space exploration require longer and longer sojourns of astronauts in space (see Fig. 1., page 159). For long-term interplanetary missions, such as a Mars mission, doses can exceed maximum recommended limits [7].

NASA has carefully studied the current uncertainties on space radiation risk estimates. Estimates of the errors associated with values for the major contributors to a calculation of the risk from high-LET radiation have been calculated by Curtis et al. [4]:

- 10 to 15% uncertainty in the initial charged-particle spectra;
- 50% uncertainty in the radiation transport calculation;
- 200 to 300% uncertainty in the risk coefficients for low-LET radiation;
- 200 to 500% uncertainty in the risk coefficients for high-LET radiation;
- 400 to 1500% uncertainty in the overall risk.

It is noted that the major uncertainties are associated to the poor knowledge of the biological effects of charged particles, with a smaller contribution coming from the characterization of the space radiation field and its primary interactions. The goal of the NASA Space Radiation Health Program is to reduce these uncertainties to less than ±50% within the year 2023 [17]. This ambitious goal can be fulfilled only by an international cooperative effort in field of radiation physics, biology, and medicine.

Table 1. Average radiation dose equivalents for different missions. Data are from radiation measurements, after correction for quality factors, except for ISS and Mars, where we present the best estimates. Maximum values are reported in parenthesis. Data courtesy of Dr. F. Cucinotta.

Program	Average altitude [km]	Inclination [degrees]	Crew-members	Dose-rate [mSv/day]	Total dose [mSv]
Gemini	454 (1370)	30	20	0.87 (4.7)	0.53 (4.7)
Apollo	–	–	33	1.3 (3.9)	12.2 (33)
Skylab	381 (435)	50	9	1.2 (2.1)	72 (170)
STS(alt.>450 km)	570 570	28.5	85	3.2 (7.7)	26.5 (78)
STS(alt.<450 km)	337 337	28.5	207	0.23 (0.4)	2.1 (7.1)
STS/Mir	341 (355)	51.6	4	0.72 (1.0)	100 (140)
ISS	360 (450)	51.6	280	0.5 (1.0)	80 (180)
Mars	–	–	4(8)	1.5 (2.0)	400 (1200)

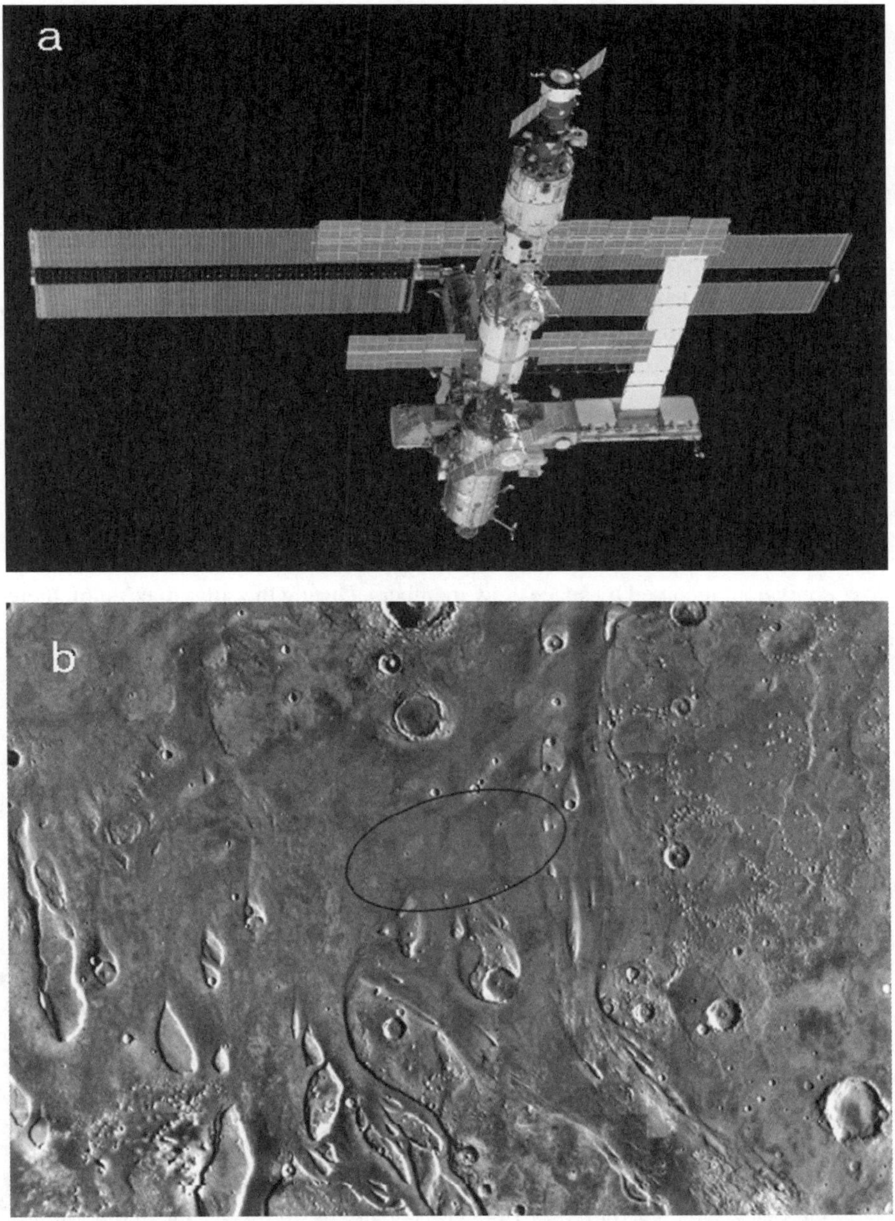

Fig. 1. Cosmic radiation is a cause of concern for the International Space Station (ISS) and the Mars exploration programs. (a) ISS in a photograph from the Space Shuttle Atlantis in October 2002 (STS-112 mission) following the undocking of the two spacecraft. (b) The Mars Pathfinder (August 1997) landing site, Ares Vallis, in the Chryse Planitia region in the northern hemisphere of Mars. Pictures from NASA, Educational Resource Department.

Even if risk uncertainties will be reduced in the next few years, there is little doubt that appropriate countermeasures have to be taken to reduce the exposure or the biological damage produced by radiation. In all basic radioprotection textbooks, it is clearly stated that there are three means to reduce exposure to ionising radiation: increasing the *distance* from the radiation source, reducing the exposure *time*, and by *shielding*. Distance is not an issue in space, space radiation beeing isotropic. Time in space should be increased rather than decreased according to the plans of exploration and colonization. Shielding remains as the only feasible countermeasure. Other strategies can be effective in reducing exposure, or the effects of the irradiation, in space. These strategies include the choice of an appropriate time of flight, administration of drugs or dietary supplements to reduce the radiation effects, and crew selection.

Space weather forecast

Because of the long period of the solar cycle (typically 11 years), it may be possible with a judicious choice of starting date to significantly reduce the radiation dose equivalent received by space crew members. During the solar maximum, there are increased numbers of SPE with a concurrent reduction in fluence rate for the GCR, in comparison with the rate during solar minimum. The National Academy of Sciences has proposed that the optimal time for a planetary mission is during solar minimum, despite the increased total fluence and dose equivalent associated with the increased galactic fluence [16]. This recommendation is based on dose estimates for different flight scenarios, and on the importance of minimizing the probability of large SPE.

A number of empirical and semiempirical models are currently used to predict the radiation environment at ISS during the scheduled missions [18]. Data-based, neural net, nonlinear filtering models appear to be close to producing operational-quality forecasts of the radiation field on ISS. However, our ability to predict the occurrence of SPE due to flares is poor, and the accuracy of the prediction varies inversely with the length of the advance warning desired. In addition, the probability of SPE reaching Earth depends on the location of the flare on the sun's surface. The magnetic field line that connects Earth to the sun has its origin on the sun at location W55. Thus the closer the flare event is to this location, the shorter the time it takes the solar flare particles to reach Earth. With increasing distances from location W55 on the sun, the protons take longer to reach Earth or may not be connected by magnetic field lines to Earth. Hence, such SPE would not be observed on Earth. The US National Space Weather started in 1995 specifically to fill these gaps in our knowledge.

Radioprotective agents

The search for efficient radioprotectors is a major goal of research in radiation protection and therapy. Protection from deleterious radiation effects by antioxidants was demonstrated about half a century ago. It was soon recognised that both radiation injury and oxygen poisoning occur through the formation of reactive oxygen species, and therefore antioxidants could be efficiently used to prevent the damage.

Phosphorothioates and other aminothiols, which are usually administered shortly before irradiation, have been hypothesized to act by one or by a combination of effects: scavenging of radiation-induced free radicals before their reaction with biomolecules; induced hypoxia; formation of mixed disulfides; scavenging of metals; repair of DNA through hydrogen donation to carbon-centered radicals; and genome stabilization. Thiols may have further effects at the physiological and immunological levels that may not be readily predicted by considering only free-radical events [23]. Not surprisingly, research in antioxidants compounds has been very intense for potential military applications in case of a nuclear conflict. The Walter Reed Army Institute of Research has developed a number of phosphorothioate agents, designated as WR compounds, with radioprotective ability. WR-2721 (amifostine, or Ethyol®) is probably the most important compound synthesized at the Walter Reed Institute. Ethyol is the first broad-spectrum cytoprotectant approved in many countries for clinical use during chemotherapy and radiotherapy cancer treatments [22]. Amifostine seems to be able to protect not only from acute deterministic radiation effects, but also from stochastic effects, in particular mutations and carcinogenesis [8]. In addition, amifostine and its active metabolite (WR-1065) appear to be protective also against exposure to neutrons [3] and heavy ions [9].

Unfortunately, WR-2721 and other thiols have significant side effects, including nausea, vomiting, vasodilatation, and hypotension. Therefore, it is not possible to use these compounds in space flights for protection against chronic exposure to cosmic radiation. However, the use of WR aminothiols or similar drugs on a short-term basis following acute radiation exposure associated with a SPE is worth consideration.

Natural occurring antioxidants are less effective than phosphorothioate agents in protection against high-dose acute radiation burden. However, nutritional antioxidants have a low toxicity, can be used for prolonged time, and they seem to play a key role in the prevention of cancer. A comprehensive list of natural compounds displaying radioprotective ability can be found in Weiss and Landauer [23]. Retinoids and vitamins (A, C, and E) are probably the most well-known and studied natural radioprotectors, but hormones (e.g. melatonin), gluthatione, superoxide dismutase, phytochemicals from plant extracts (including green tea and cruciferous vegetables), and metals (especially selenium, zinc, and copper salts) are also under study as dietary supplements for individuals overexposed to radiation, including astronauts.

Effectiveness of antioxidants in space is further complicated by the presence of HZE particles. In principle, antioxidants should provide reduced or no protection against densely ionizing radiation, because the direct effect is more important than free radicals-mediated indirect radiation damage at high-LET. However, recent experiments suggest that an efficient radioprotection by dietary supplements can be achieved even in case of exposure to high-LET radiation. Ascorbic acid suppresses *in vitro* transformation of mouse fibroblasts exposed to fission-spectrum neutrons [21]. Preliminary results show that retinyl acetate, a naturally-occurring retinoid, strongly reduces the induction of fibroma in rats exposed to swift Fe ions [1]. Following exposure of human cells to accelerated high-LET carbon ions, it has been shown that even beer reduces the yield of chromosomal aberrations [15]. Most of the detrimental effects of heavy ions on the CNS of rats are suppressed by dietary supplements of strawberries [11].

It appears from these data that the effects of diet on human radiosensitivity have been overlooked in the past. Astronauts might be able to reduce significantly the deleterious effects of radiation, and perhaps weightlessness, using nutritional supplements increasing the level of endogenous antioxidants. According to Joseph *et al.* [11], "fruits and vegetables may become as important on space-going vessels as limes were on the sea-going vessels of old".

Individual susceptibility

There is little doubt that genetic predisposition plays a decisive role in cancer development. It is also possible that genetic factors determine human radiosensitivity. Radiation risk in space might be reduced if astronauts are genetically radioresistant. On the other hand, hypersensitive individuals will be at higher risk in long-term space missions.

Most of our knowledge on individual susceptibility comes from clinical experience in radiotherapy. Therapeutic misadventures involving *ataxia-telangiectasia* (AT) patients have dramatically demonstrated the importance of genetic susceptibility to radiation damage in cancer treatment. *Ataxia-telangectasia* is a rare (approximately one case per 40,000 live births) autosomal recessive disorder characterized by high radiosensitivity, immune deficiency, recurrent respiratory infections, and predisposition to childhood lymphoma. During the 60's, AT patients suffered severe or fatal radiation injuries when treated with a standard course of radiotherapy. An explanation of the clinically observed severe radiation injuries came from *in vitro* studies of AT patients' cultured fibroblasts, showing that AT cells are up to three times more radiosensitive than wild type [5]. The cloning of the human AT mutated gene (ATM) [19] has shown that the increased radiosensitivity in AT cells is probably not caused by a defect in DNA repair but results from an altered cell cycle control, and perhaps an inability to activate damage-inducible DNA repair [14].

ATM homozygotes only represent a small fraction of the radiosensitive patients, although they appear to be the most sensitive. Clinical experience indicates that there is a heterogenous radiosensitivity in the adult population: approximately 15% of the radiotherapy patients can be classified as "sensitive", and approximately 5%

of patients are injured during the radiotherapy course [2]. ATM heterozygotes, who are also cancer-prone, are suspected to represent a large fraction of the extremely radiosensitive patients. They represent about 1% of the population but about 5% of the cancer patients in USA. The coincidence of 5% of cancer patients being ATM heterozygotes and 5% of patients suffering severe side effects in radiotherapy is striking, but there is no direct evidence that ATM heterozygotes are hypersensitive to radiation, although fibroblasts cultured from ATM heterozygotes have radiosensitivity between that of normal and AT patients [5].

A predictive assay able to identify radiation hypersensitive or cancer-prone subjects could be useful in crew selection for long-term space flights. However, this assay is not available at the moment. Given the impact of environmental and dietary factors on radiosensitivity, it is not even sure that such a test will ever be feasible. Genetic diseases characterized by enhanced radiosensitivity are very rare and phenotipically obvious. Thus, the normal process for crew selection should generally be adequate for screening unusually susceptible individuals. Genetic screening would be necessary to pick ATM heterozygotes, that are less rare (around 1% of the US population), healthy, but may be at increased risk of radiation-induced stochastic or nonstochastic effects.

Shielding

For terrestrial radiation workers, additional protection against radiation exposure is usually provided through increased shielding. Unfortunately, shielding in space is problematic, especially when GCR is considered. High-energy radiation is very penetrating: a thin or moderate shielding is generally efficient in reducing the equivalent dose, but as the thickness increases, shield effectiveness drops. This is the result of the production of a large number of secondary particles, including neutrons, caused by nuclear interactions of the GCR with the shield. These particles have generally lower energy, but can have higher quality factors than incident cosmic primary particle.

NASA is using the analytical code HZETRN coupled to NUCFRG2 [25], developed at the Langley Research Centre, to estimate GCR doses in different shielding configurations for the ISS and Mars mission. HZETRN is based on the solution of the *Boltzmann transport equation* for the flux density $\phi j(x,\Omega,E)$ of the particle j in the spatial location x, moving in the direction Ω with energy E:

$$\Omega \cdot \nabla\phi_{\nabla j}(x, \Omega, E) = \Sigma\int\sigma_{jk}(\Omega, \Omega', E, E')\phi_k(x, \Omega', E')d\Omega' dE - \sigma_j(E)\phi_j(x, \Omega, E) \qquad (1)$$

here $\sigma_j(E)$ is the macroscopic cross section (in cm^2/g) for removal of j particles of energy E, and $\sigma_{jk}(\Omega,\Omega',E,E')$ are the double differential particle production and fragmentation cross sections for the processes adding j particles of energy E in direction Ω, including spontaneous disintegration. Since the atomic, nuclear elastic and reactive cross sections are ordered as $1{:}10^{-5}{:}10^{-8}$, solutions to Boltzmann equation are expanded as a sum of physical perturbations, under the approximation that all secondary particles produced by nuclear collisions move in the same direction Ω as the incident particle. The HZETRN code calculates the straight-ahead particle transport along 512 different rays converging on the point x, and provides as output the energy spectra of the charged particles in the point x within the ISS module or the astronauts' body.

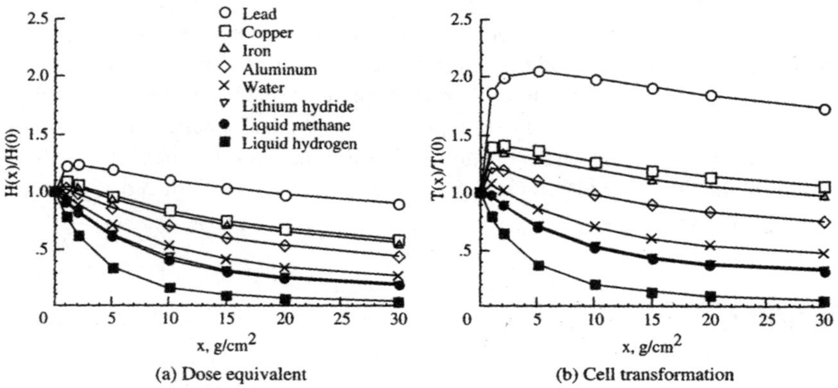

Fig. 2. Attenuation of dose equivalent and cell transformation frequency in one-year exposure at the solar minimum behind several shield materials. GCR transport calculated by HZETRN. (a) Equivalent dose, calculation based on ICRP-60; (b) neoplastic cell transformation, calculation based on track-structure Katz's model. Plot courtesy of Dr. F. Cucinotta.

The relative attenuation of the dose equivalent H(x) produced by GCR at solar minimum as a function of the thickness x (in g/cm^2) of different shield materials as calculated by HZETRN is given in Fig. 2.a [24], assuming one year exposure. Low-Z shields, including water, provide a good protection, although clearly the shield effectiveness increases slowly for thickness exceeding 10 g/cm^2. On the contrary, for high-Z material shield, thin shields provide no protection. Lead appears to be the worst shield material. The differential shield effectiveness in Fig. 2. is a consequence of the relative attenuation of different LET components in the GCR by different materials. The spectrum of the GCR transmitted through the shield is generally shifted towards lower LET values than in the free space: high-LET values are attenuated, and low-LET values amplified. However, the absolute LET values that are attenuated or amplified depend on the shield material. For liquid hydrogen, the region below 1 keV/μm is amplified, and GCR with LET> 1 keV/μm is attenuated. After water shielding, the amplified region is shifted to around 20 keV/mm and to more than 50 keV/μm after lead shielding. As a result, liquid hydrogen attenuates the LET component with higher quality factors (Q), while lead amplifies the region around the peak of the Q(LET) curve (ICRP 1991). As a result, the dose equivalent is increased by a lead shield and decreased by a liquid hydrogen shield.

Liquid hydrogen appears to have the maximum performance as shield material. Of course hydrogen is not a practical shield material, being a low temperature liquid. Hydrogen storage in grafite nanofibers or lithium hydride (^{6}LiH) may have a large impact in space shield design. So far, it appears that polyethylene can be a good compromise. NASA has recently replaced aluminium panels with lighter polyethylene slabs in the crew sleeping quarters on ISS. Dose measurements were consistent with the expected dose reduction of around 30% in the shielded area. Quite clearly, research in the field of shielding is likely to have quick applications on ISS.

The results of the transport calculations can be also used to predict any type of biological effect in a given point of the space vessel or in the astronauts' body. To this purpose, the solution of equation (1) is written in terms of the Green's function G(ρ,σ,x) as [20]:

$$\phi(\rho,x) = G(\rho, \sigma, x)\phi_B(\Gamma) \tag{2}$$

where $\phi(\rho,x)$ is the particle fields at the position x within the material of compositional density ρ, and $\phi_B(\Gamma)$ is the space radiation field at the boundary Γ. The net effect $I_{se}(x,\rho)$ for the endpoint e observed at the point x of the system s can then be written as:

$$I_{se}(\rho,x) = \Sigma R_{se} G(\rho, \sigma, x)\phi_B(\Gamma) \tag{3}$$

where Rse is the probability of the effect e in the system s, and can be estimated using a biophysical model which predicts the biological response in each point given the radiation field. Wilson and coworkers [24] have used the track-structure repair model originally developed by Katz and coworkers [12] to evaluate the effect of shielding on the risk of neoplastic transformation. The results of this calculation are given in (see Fig. 2.b, page 164) as the ratio of transformation frequency T(x) at a depth x to the transformation T(0) of unshielded cells. The attenuation characteristics are qualitatively similar to the results shown in (see Fig. 2.a, page 164), where the ICRP (1991)[10] quality factor approach had been used. However, there are clearly large quantitative differences. Results of these simulations suggest that the estimates of the shield effectiveness depend on the biological endpoint under consideration, and on the biophysical model used for the calculations.

Unfortunately, very few biological experiments are available to validate these models. Recently, a large international collaboration involving Italian, US, and Japanese research groups has started with the aim of measuring genetic damage (chromosomal aberrations, lethal mutations, and DNA double-strand breakage) in human cells behind shields [6]. The project has so far produced preliminary data relative to high-energy iron ions. Fig. 3 shows the results for the induction of chromosomal aberrations by 500 MeV/n Fe ions shielded with PMMA (lucite), a common low-Z plastic target material, and aluminium. The frequency of aberrations, scored by FISH in prematurely condensed chromosomes, is plotted either as a function of the dose at the sample position, or of the fluence of primary ions incident on the shield. No significant effects are introduced by shielding when data are plotted vs. the absorbed dose, but the effectiveness per incident ion is clearly increased by the shield. This appears to be caused by the increased dose-rate behind the shield. The shield increases the dose-average LET for the 500 MeV/n beam: the slow-down of the primary iron beam produces an increase in the LET, which is not compensated by the loss of primary ions and production of lighter fragments. Therefore, for a given number of incident 500 MeV/n Fe ions, the dose is higher behind 6–8 g/cm^2 shield than behind 0 cm shield. Results on DNA double-strand-breakage obtained with the 1 GeV/n Fe beam (accelerated at the AGS of the Brookhaven National Laboratory) shielded with 19 cm PMMA show an opposite trend. Here the shield efficiently removes primary ions from the beam, a large amount of low-LET fragments is produced, and as a result both the dose-rate and the DNA breakage per incident ion are reduced behind the shield. These preliminary results provide support to the prediction of Wilson and coworkers that shield can increase rather than decrease the effectiveness of GCR. Shield effectiveness depends on shield material and thickness, incident ion charge and velocity, biological target and endpoint. Hopefully, results from these experiments will provide soon indications about the reliability of the models described above.

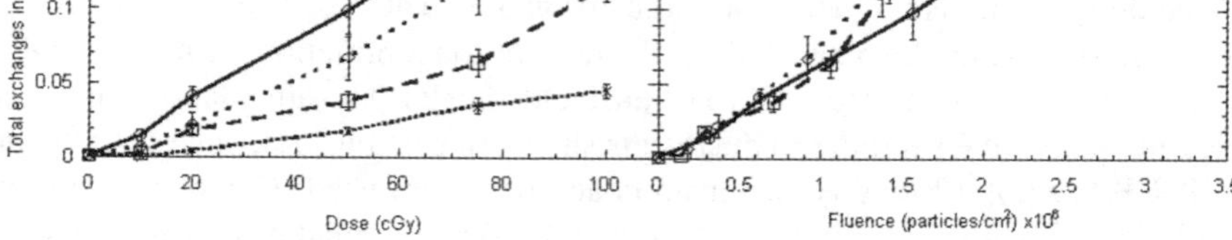

Fig. 3. Frequency of total exchanges (simple-type plus complex-type, both reciprocal and incomplete) involving prematurely condensed human chromosomes 1, 2, and 4 in blood lymphocytes exposed to accelerated iron ions, plotted vs. a) the dose at the sample position; and b) the fluence of iron ions incident on the shield. The shield thickness is 0 mm for (○) and (□), or 3 cm Al for (◇). X-ray data are shown for comparison in panel (a). Bars are standard errors of the mean values.

Acknowledgements

The author is grateful to Dr. Francis A. Cucinotta for valuable suggestions and discussions. I also thank Dr. Walter Schimmerling and Prof. Giancarlo Gialanella for support and encouragement during these studies.

References

1 Burns FJ, Zhao P, Xu G, Roy , Loomis C (2001) Fibroma induction in rat skin following single or multiple doses of 1.0 GeV/nucleon 56Fe ions from the Brookhaven Alternating Gradient Synchrotron (AGS). Phys Med 17:S194–195
2 Busch D (1994) Genetic susceptibility to radiation and chemotherapy injury: diagnosis and management. Int J Radiat Oncol Biol Phys 15, 30(4):997–1002
3 Carnes BA, Grdina DJ (1992) In vivo protection by the aminothiol WR-2721 against neutron-induced carcinogenesis. Int J Radiat Biol 61:567–576
4 Curtis SB, Nealy JE, Wilson JW (1995) Risk cross sections and their application to risk estimation in the galactic cosmic-ray environment. Radiat Res 141:57–65
5 Deschavanne PJ, Debieu D, Fertil B, Malaise EP (1986) Re-evaluation of in vitro radiosensitivity of human fibroblasts of different genetic origins. Int J Radiat Biol 50:279–293
6 Durante M (2001) Influence of the shielding on the space radiation biological effectiveness. Phys Med 17:S269–271
7 Durante M (2002) Biological effects of cosmic radiation in low-Earth orbit. Int J Mod Phys A17:1713–1721
8 Grdina DJ, Shigematsu N, Dale P, Newton GL, Aguilera JA, Fahey RC (1995) Thiol and disulfide metabolites of the radiation protector and potential chemopreventive agent WR-2721 are linked to both its anti-cytotoxic and anti-mutagenic mechanisms of action. Carcinogenesis 16:767–774
9 Hei TK, Piao CQ, Wu LJ, Willey JC, Hall EJ (1998) Genomic instability and tumorigenic induction in immortalized human bronchial epithelial cells by heavy ions. Adv Space Res 22:1699–1707
10 ICRP (1991) Recommendations of the International Commission on Radiological Protection. Publication 60. Annals of the ICRP 21 (Pergamon Press, London)
11 Joseph JA, Shukitt-Hale B, McEwen J, Rabin BM (2000) CNS-induced deficits of heavy particle irradiation in space: the aging connection. Adv Space Res 25:2057–2064
12 Katz R, Ackerson B, Homayoonfar M, Sharma SC (1971) Inactivation of cells by heavy ion bombardment. Radiat Res 47:402–425
13 Kiefer J (2001) Space radiation research in the new millenium – from where we come and where we go. Phys Med 17:S1–4
14 Meyn MS (1995) Ataxia-telangiectasia and cellular responses to DNA damage. Cancer Res 55:5991–6001
15 Monobe M, Ando K (2002) Drinking beer reduces radiation-induced chromosome aberrations in human lymphocytes. J Radiat Res 43:237–245

16 NAS, US National Academy of Sciences (1997) Radiation Hazards to Crews on Interplanetary Missions. Report of the Task Group on the Biological Effects of Space Radiation (National Academy Press, Washington DC)
17 NASA, Life Sciences Division (1998) Strategic Program Plan for Space Radiation Health Research (NASA, Washington DC)
18 NRC, US National Research Council (2000) Radiation and the International Space Station (National Academy Press, Washington DC)
19 Savitsky K, Bar-Shira A, Gilad S, Rotman G, Ziv Y, Vanagaite L, Tagle DA, Smith S, Uziel T, Sfez S (1995) A single ataxia telangiectasia gene with a product similar to PI-3 kinase. Science 268:1749–1753
20 Shinn JL, Wilson JW, Schimmerling W, Shavers MR, Miller J, Benton EV, Frank AL, Badavi FF (1995) A Green's function method for heavy ion beam transport. Radiat Environm Biophys 34:155–159
21 Tauchi H, Sawada S (1993) Suppression of gamma- and neutron-induced neoplastic transformation by ascorbic acid in Balb/c 3T3 cells. Int J Radiat Biol 63:369–374
22 Wasserman T (1999) Radioprotective effects of amifostine. Semin Oncol 26:S89–94
23 Weiss JF, Landauer MR, (2000) Radioprotection by antioxidants. Ann NY Acad Sci 899:44–60
24 Wilson JW, Kim M, Schimmerling W, Badavi FF, Thibeault SA, Cucinotta FA, Shinn JL, Kiefer R (1995) Issues in space radiation protection: galactic cosmic rays. Health Phys 68:50–58
25 Wilson JW, Miller J, Konradi A, Cucinotta FA. (eds.) (1997) Shielding Strategies for Human Space Exploration. NASA CP 3360, Langley Research Center

Repair

Radiation Damage and Human Cells: You have to know one to understand the other

Markus Löbrich

Fachrichtung Biophysik, Medizinische Fakultät der Universität des Saarlandes,
66421 Homburg, Germany
markus.loebrich@uniklinik-saarland.de

Abstract.

Ionizing radiation causes various lesions in human cells of which damage to DNA is no doubt biologically most significant. In particular, DNA double-strand breaks (DSBs) pose a problem for the cells and the organism as a whole. The faithful repair of DSBs is probably the most critical challenge for a cell in order to maintain its genomic integrity. If repaired incorrectly, these lesions can lead to genomic rearrangements and may initiate tumorigenesis. Also, loss of sequence information at the break site may be an important biological consequence of unfaithful DSB repair. Human cells can repair DSBs either by homology-dependent or by non-homologous repair mechanisms. In this contribution, I discuss the role of these two pathways for repairing DSBs produced by ionizing radiation. Particular emphasis is given to the comparison between high and low radiation doses, and a concept is developed that speculates about cellular responses to DNA damage introduced by very low radiation doses (a few mGy).

DNA damage and the unkindest cut of all

The harmful effects of ionizing radiation are derived from the potential of highly energetic photons to generate charged particles, in particular electrons, that can break chemical bonds in biological macromolecules. Upon exposure of human cells, basically all molecules of the cell's machinery are affected. A radiation dose of 1 Gy, known to cause acute radiation effects in humans, produces many thousand ionizations in a single cell, and various kinds of proteins, lipids and sugar molecules will irreversibly be destroyed. Yet it is not the damage to these molecules that represents the greatest harm of ionizing irradiation. Rather, it is the DNA, packaged into 46 chromosomes, which is the most critical target of radiation and determines most, if not all, biological consequences of radiation exposure. The reason is as simple as obvious: except for our genetic information, all biological macromolecules can be replaced, no matter how complicated and diverse they are. The instructions on how to rebuild them is stored in the DNA, and it will only be a question of time until a damaged and nonfunctional enzyme is replaced by a newer one. There is simply no need to repair molecules such as enzymes that could as well be removed and regenerated. This concept of reversing or healing radiation damage works very efficiently, at least in theory, for almost every molecule – except for the DNA itself. If this molecule suffers irreversible change, it has to take care of itself as there is simply no other molecule with the knowledge of how to rebuild the original DNA molecule. This is probably the reason why a human cell has such a fascinating variety of DNA repair ensembles – different repair mechanisms for the different kinds of damages that can arise in the DNA molecule upon radiation exposure.

Also within the class of DNA damages, there exists something similar to a hierarchy, and not all lesions are repaired by the same strategy or concept. As the two strands of a DNA molecule carry the same, although complementary, genetic information, it makes a big difference whether only one strand is damaged or whether lesions occur in close proximity on both strands. If the damage is limited to one strand, the cell can easily remove the lesion, e.g. a severely modified base, and fill the gap with a new base, actually with a new nucleotide carrying the base. This is a simple repair pathway, basically because the information of which base to include is available on the second strand from the base pair rule A:T and G:C. Even if longer stretches of a single strand are damaged or erroneous, a repair pathway can remove the altered nucleotides, fill-in the single-stranded gap and through this mechanism restore the original genetic information. All this works quite well with high efficiency and little error as long as the damage is limited to one strand. It is much more problematic, however, if both strands are damaged at nearly the same position so that one cannot serve as a template for the other's repair. It may be argued that this must be a very rare event – given the chances that out of the billions of bases in each strand of a human cell just the base opposite of a lesion is damaged. Indeed, if ionizations and chemical modifications were always single and independent events, it would be highly unusual to observe such double-stranded lesions. As a matter of fact, DNA damage that arises from thermal forces or is produced by chemical species originating from normal oxidative processes constitutes almost always a single lesion and is thus confined to one of the two DNA strands. However, it is a hallmark of ionizing radiation to produce clusters of ionizations, particularly at the end of electron tracks, which can cause two or more lesions in close proximity and, not all that infrequently, also double-stranded lesions. One form of double-stranded lesions are double-strand breaks (DSBs) where the sugar-phosphate backbone is interrupted at nearly the same position in both strands of the DNA molecule. This is a severe form of DNA damage and cells should be equipped with some kind of machinery to respond to it.

The cell's dilemma in repairing DNA double-strand breaks

A DSB disrupts the physical integrity of the DNA molecule and, if left unrepaired, leads to death during cell division when the DNA is disproportionately distributed to the progeny and large parts of the genomic material are lost. Although not obvious for many years, cells have the potential to repair DSBs and survive even this kind of DNA insult. However, some multicellular organisms and also human beings do this at the risk of inducing a tumor, suggesting that the process of DSB repair can go awry. This may not be surprising given that the simple repair concepts employed for single-stranded lesions are obviously not applicable to the repair of DSBs. The most simple DSB repair concept imaginable is the direct pasting of the broken ends without concern about base alterations at the break site. In this scenario, base-sequence alterations of a few base pairs would occur, but at least the overall structural integrity of the genome would be retained. Then, the cell will only die if important genes for cell viability were affected by these small sequence alterations. This may be expected to be highly unlikely as only a tiny fraction of the billions of base pairs are needed for the viability of a given cell type. So for most of the cells in our body the risk of cell death seems to be acceptable and should not represent any major hazard for the organism. What happens, however, if some stem cells of our body suffer base-sequence changes from DSB repair following radiation exposure in particular genes that affect cell growth or growth control regulation? This faulty information can be passed on to daughter cells and may eventually initiate tumorigenesis. Also this is unlikely, but it should be kept in mind that a single genetic change in a single cell of our body could theoretically be sufficient to start or accelerate tumor development.

From this point of view one may expect that the cell takes much effort in avoiding any change to its genetic material. If the information is lost on both strands of a DNA molecule, as will most often be the case for a DSB, it would be highly desirable to retrieve the lost information from an undamaged region of the genome. This would require that genetic information is redundantly available, which is indeed the situation in organisms with diploid chromosomal content. Except for the germ line, most of the cells' genetic information is available in two copies on two homologous chromosomes. The two homologs represent our mother's and father's heritage and carry essentially the same genetic information. It should therefore be possible to restore a DSB on one chromosome by retrieving information from the homologous partner chromosome. Such a mechanism is termed homologous recombination and scientists have begun to appreciate the significance of this process for maintaining genomic stability in human cells. However, it is not usually the homolog that serves as a template for repair but the sister chromatid. After a cell has replicated its DNA but before it has devided into daughters, each chromosome consists of two chromatids that represent identical copies (in addition to the two homologs with nearly the same information). Because the two sister chromatids are connected to each other and thus are localized in close proximity, they can easily serve as the information donor during the homologous recombination process of DSB repair. As DSBs also occur during DNA replication, this mechanism is a very clever way of repairing these spontaneous lesions without the unacceptable loss of genetic information. It is therefore reasonable to assume that DSBs produced by ionizing radiation are also repaired by chromatid-mediated homologous recombination in situations when chromatids are available. True, but what happens to a DSB introduced before DNA replication when no sister chromatids are available? Does a cell then use the homolog as a template for repair? The answer is: it depends! Yes, if the genome of the cell is small and simple as it is for most unicellular organisms; and No for human cells with a highly diversified genome. Although theoretically possible, a DSB in a human cell is hardly ever repaired by homologue-mediated homologous recombination. Instead, the simple paste-and-risk-the-base-sequence-error pathway (more scientifically, this pathway is abbreviated NHEJ for non-homologous end joining) is employed to remove a DSB that is induced when human cells are irradiated before their DNA is replicated.

Being by yourself isn't all that bad

To grasp the idea behind the cells' decision to rely on NHEJ instead of homologous recombination for repairing radiation-induced DSBs, one has to look at the kind of events that are associated with erroneous DSB repair. In addition to errors on the base-sequence level, incorrect DSB repair events can involve the joining of formerly unconnected parts of the genome. Such erroneous processes, which I refer to as misrejoining events, generate genomic rearrangements that encompass chromosomal translocations, deletions, inversions, or insertions of huge parts of the genome. Importantly, the same events are intrinsically connected to tumor induction. Having said this, the most important task of a cell should be to avoid any genomic rearrangement resulting from incorrect DSB repair. The quality of a particular DSB repair pathway or concept is hence not only based on its ability to restore the original base-sequence information but – possibly more so – on its error-rate with regard to the production of genomic rearrangements. Taking the philosophical standpoint that a cell is doing what is best for itself or the organism, homologue-dependent homologous recombination may be more erroneous than NHEJ for causing rearrangements in the genome. It may be possible that the process of retrieving sequence information from a homologous partner chromosome, which at some point must involve a physical connection between the donor and the acceptor of genetic information, is highly prone to interfere with genomic stability. This is particularly so because many pieces of our genetic information are not only found in two copies on the two homologous chromosomes, but sometimes appear with near-identical sequence at several places in our genome. As long as a sister chromatid with the exact homologous sequence is located next to a DSB, there is no worry about near-identical sequences elsewhere in the genome. However, if the cell has to search through the entire genome to find the homologous chromosome, it may be at risk to interact with an inappropriate partner that carries a near-identical sequence, and may possibly produce a genomic rearrangement.

Despite all the efforts of a cell to repair DSBs as well and faithfully as possible, genomic rearrangements do occur upon ionizing irradiation and humans can develop radiation-associated tumors. So what is going wrong and who is the culprit? A possible mechanism for the generation of genomic rearrangements involves the interaction of two lesions, in particular two DSBs. The probability with which ends from different DSBs are connected may be expected to depend on the number of other nearby break ends and, correspondingly, on the radiation dose deposited. The higher the dose, the more break ends occur in a cell, leading to a higher probability for rearrangements formed by pasting incorrect break ends. Additionally, a separation of DSBs in space and time can be expected to considerably reduce the probability for DSB misrejoining as break ends from different DSBs have to coincide in time in order to interact and form genomic rearrangements. Radiobiologists have experimentally confirmed these ideas by showing that a reduction in dose, a decrease in dose rate or a dose fractionation protocol all reduce the probability with which a DSB leads to genomic rearrangements – at least for photon irradiation. However, in contrast to sparsely ionizing photon radiation, fractionation experiments with α-particles have demonstrated that the probability of DSB misrejoining is nearly independent of dose and that there is no reduction factor for low dose rates or for dose fractionation in case of densely ionizing particles. Hence, a separation of densely ionizing particle tracks in time and space does not appreciably reduce the probability for DSB misrejoining. This again is consistent with a mechanism for the generation of genomic rearrangements by the interaction of incorrect break ends from DSBs in close proximity. Whereas randomly distributed DSBs from a low dose of sparsely ionizing radiation are not likely to be induced in close proximity, a single densely ionizing α-particle can produce several breaks that will be distributed along the track of the particle and, therefore, may occur in sufficient proximity to generate DSB misrejoining with high probability. The suggestion that the traversal of a single α-particle through a human cell nucleus will produce an appreciable level of genomic rearrangements is particularly striking when one considers fact that α-particles emitted from radon and radon daughters contribute substantially to the annual exposure of the human population to ionizing radiation. It is probably no coincidence that some of the best documented cases of radiation-induced tumors were observed in mine-workers exposed to appreciable levels of radon gas.

The simpler the better? Pros and cons of NHEJ

The culprit is unmasked! If genomic rearrangements are formed by pasting incorrect DSB ends, it must be the above mentioned non-homologous end joining (NHEJ) pathway that causes all this trouble. Unfortunately(?), NHEJ is the major pathway for repairing radiation-induced DSBs in human cells and so most of the breaks are simply repaired by connecting broken DNA ends with little or no requirement for sequence homology – and sometimes even by connecting ends from different DSBs at different locations. So what went wrong during evolution? Why are we stuck with such a disastrous way of repairing DSBs? Would we not benefit tremendously by abolishing NHEJ from the functional repertoire of our cells? If we still believe that natural processes evolved to our best interest, the answer should be No. And indeed, scientists have shown with genetically engineered mice that inactivation of the NHEJ pathway leads to an increase in chromosomal rearrangements instead of a decrease. Also tumorigenic processes are not slowed but accelerated if an organism lives without NHEJ. Obviously, these experiments demonstrate that NHEJ has an important role as a caretaker of genomic integrity and tell us that it should contribute substantially to correct repair of radiation-induced DSBs. Indeed, low dose-rate experiments with sparsely ionizing photon radiation, in which cells rarely encounter multiple DSBs simultaneously, have revealed that NHEJ can repair radiation-induced DSBs with high fidelity and that it suppresses the occurrence of misrejoining events. However, the faithful nature of NHEJ is transformed into an error-prone character when experiments with high acute doses are performed. In this case, multiple breaks occur in close proximity and NHEJ frequently forms genomic rearrangements. So in addition to its caretaker role, NHEJ also has the potential to effect genomic rearrangements. Given that even under low dose and low dose-rate conditions multiple DSBs may occasionally coincide within a critical interaction distance, the NHEJ pathway is probably responsible for the generation of at least some genomic rearrangements that participate in tumorigenesis in normal individuals. From this, it can also be argued that even the lowest doses have a certain probability of producing two DSBs in a single cell, which – again with a measurable probability – can form a genomic rearrangement and possibly cause a tumor.

To die or not to die

The provocative statement that any radiation dose, no matter how small, can theoretically cause a tumor, but also the reasoning that the probability for DSB misrejoining and tumorigenesis should be marginal for very low doses are both derived from our knowledge gained through experiments at much higher doses than those typically encountered during daily life. What scientists – including myself – really know about the biological effects of extremely low doses is minimal. Each cell of our body is traversed – on average – by a single electron once a year. The probability that one electron causes a DSB somewhere in the genome of a cell is approximately 5% (assuming that the dose deposited per cell is ~1 mGy and that 1 Gy produces ~50 DSBs per cell). So what we really want to know is how a cell handles a single break induced by ionizing radiation. Are the enzymatic processes that operate at these low doses the same as those at much higher doses? NHEJ should be a good candidate for repairing a single DSB since we have learned that this pathway is very efficient at low doses and low dose-rate conditions. At least it should not cause genomic rearrangements as long as there is not more than a single break in a single cell. This sounds reasonable and it may be hard to argue that repair pathways could operate differently at very low doses. There is, however, the problem of base-sequence alterations caused by NHEJ of a single DSB. These changes will be confined to the point in the genome where the DSB was introduced and they will certainly be much less harmful for a cell than a genomic rearrangement. But nevertheless, they do represent an error in the genetic information and – if they happen to occur in genes controlling cell growth – might pose a tumorigenic risk. Given that we have billions of cells in our body which potentially can form a tumor if damaged at particular genomic locations, and given that per year one out of twenty cells will suffer a base-sequence alteration, the paste-and-don't-worry-too-much NHEJ pathway may not be our best bet. We may want to remember what a cell is doing if an enzyme gets destroyed by radiation – it throws it away and makes a new one. For DNA, this concept did not work as the DNA is the depository of our genetic information. That is why a cell has all these different DNA repair pathways. However, upon reconsideration the same concept may indeed work also for DNA damage – the DNA cannot be replaced but the cell can. At a high dose, most of the cells of an organism have several DSBs and will suffer some kind of genetic change – so repair is essential. But at the very low doses of daily life, there will only be a tiny fraction of our cells hit at any given time by an electron. If the electron causes a DSB in a particular cell, the neighboring cells will likely not have a break. So why repair the DSB of the hit cell? This damage to the organism can easily and without risk be removed if the hit cell is eliminated and replaced by a new one derived from the division of an undamaged neighboring cell. If this concept has any rationale, DSBs produced by a few mGy of ionizing radiation should not be repaired. The persistent DSB could then represent a signal to have this particular cell destroyed. An organism would not suffer from the cell loss because a sufficient number of undamaged cells is available to have the missing cell replaced. This would imply that DSBs induced by radiation doses of a few mGy may not pose any tumorigenic risk for the organism, and that a treshold may exist for the harmful effects of ionizing radiation. Whether all this is only fantasy or has meaning – nobody knows. But one thing is certain: radiation research is alive!

I would like to thank all the members of my group who have contributed substantially to the development of the presented ideas.

Different Means to an End: DNA Double-Strand Break Repair

Kai Rothkamm

Fachrichtung Biophysik, Universität des Saarlandes, 66421 Homburg, Germany
kai.rothkamm@uniklinik-saarland.de

Abstract.

One of the current hot topics in radiation biology is the field of DNA double-strand break repair. Double-strand breaks are not only important for determining the biological effects of ionising radiation but they are also formed during normal DNA metabolism in eukaryotic cells. At least two different repair pathways have evolved which contribute differentially to the removal of double-strand breaks, depending on the cell cycle phase, the type of lesion induced and the species under investigation. The elucidation of the cellular machinery that is involved in repairing DNA double-strand breaks is crucial for the improvement of cancer treatment, for gaining insight into the mechanisms of tumourigenesis and for precise assessments of cancer risk emanating from radiation and other clastogenic agents.

The many faces of radiation research – some already gone

What is the situation of radiation research at the beginning of the 21st century? This extremely heterogeneous field (see Figure 1) has seen a glorious past; many fundamental and far-reaching findings which had enormous implications for other scientific areas have been made by radiation researchers. A wealth of disciplines within this field, like radiation ecology, radiobiology, radiation toxicology, radiation physics, radiation chemistry, radiation hygiene, etc. have contributed to our current understanding of radiation action, unravelled important mysteries of general scientific interest and, last not least, have helped to ensure public safety in times of nuclear power stations, intercontinental air traffic and computed tomography. Over the last decades, however, several traditional disciplines of radiation research have been reduced to mere skeletons, others have been completely closed down – at least in Germany. This raises the question whether these areas are still needed today: not only for radiation research itself but also for the general scientific community and for the society. A lot has been learned about radiation action, and in several disciplines a basic understanding of the most demanding problems has been established.

Therefore one might think that, given that financial resources are limited, priorities – and money – have to be withdrawn from areas of reduced significance in favour of disciplines with potentially better prospects. In principle, one cannot argue against such a strategy. If, however, the consequences go as far as a complete disappearance of disciplines, then new aspects have to be taken into account, such as the question, whether the total lack of experts with a detailed knowledge in the field is compatible with public safety, disaster control and the sustainable development of a country.

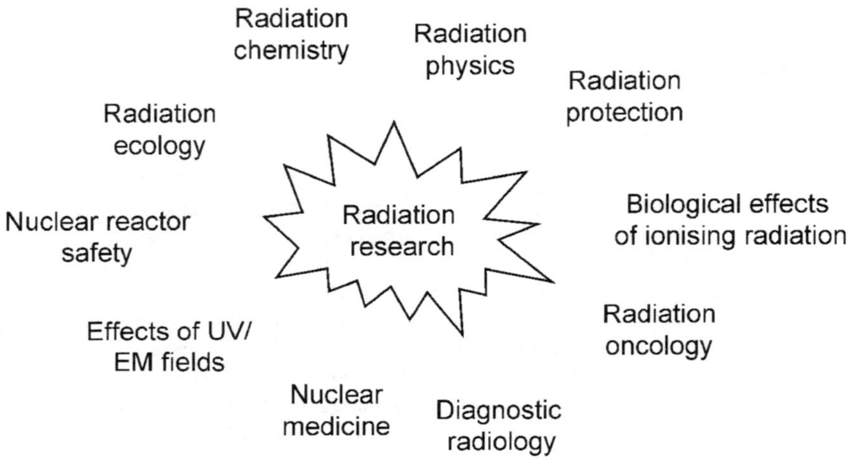

Fig. 1. Radiation research is a very heterogeneous field. Only a few important disciplines are quoted in this illustration.

In the next sections, the scientific and societal importance of one particular area of research within the discipline of radiation biology – research on DNA double-strand breaks and their repair – will be outlined.

DNA double-strand break repair: a critical cellular response to ionising radiation

DNA is under constant attack from endogenous agents, such as free radicals generated during oxidative metabolism and exogenous agents, including many carcinogens. The DNA double-strand break which is produced by ionising radiation and radiomimetic drugs but also arises endogenously during DNA replication, by topoisomerase failure and as an initiator of physiological recombination processes like V(D)J recombination and meiosis, is probably the most severe threat to the maintenance of genomic integrity. If left unrepaired, double-strand breaks can result in loss of genomic material and cell death, whereas incorrect repair of this lesion leads to genomic rearrangements, chromosome aberrations and mutations. Accordingly, the processing of double-strand breaks is not only a fascinating issue of basic research (see Figure 2, lower part) but also directly affects important societal issues like cancer treatment and radiation risk assessment (Figure 2, upper part).

In mammalian cells, at least two genetically separable pathways exist for repairing double-strand breaks, non-homologous end joining and homologous recombination. The former repairs broken DNA ends with little or no requirement for sequence homology. It involves the XRCC4 gene product, DNA ligase IV, and the DNA-PK holo-enzyme, consisting of the DNA end-binding heterodimer Ku70/Ku80 and the catalytic subunit DNA-PKcs (reviewed in [13,29]). Inactivation of non-homologous end joining sensitises cells to ionising radiation, impairs the repair of DNA double-strand breaks, leads to an increase in chromosomal rearrangements and accelerates carcinogenesis [1,3–5,8,12,14–16,38]. Notably, non-homologous end joining has been shown to serve as an efficient pathway for correctly repairing DNA double-strand breaks in situations of separated breaks but it generates genomic rearrangements if multiple double-strand breaks are close in time and space [24]. Therefore, non-homologous end joining may be responsible for the generation of at least some genomic rearrangements that participate in human carcinogenesis [20].

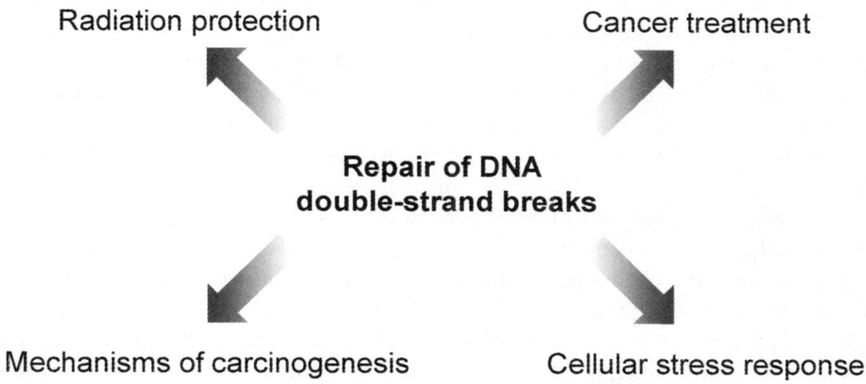

Fig. 2. Impact of research in the area of DNA double-strand break repair on other fields of basic research (lower part) and on issues relevant for public health (upper part).

Homologous recombination can also mediate the repair of DNA double-strand breaks but appears to be less important than non-homologous end joining for the repair of ionising radiation-induced breaks in higher eukaryotes [10,33–35] (in contrast to the situation in lower eukaryotes like yeast, where homologous recombination is the predominant pathway for repairing radiation-induced DNA double-strand breaks). It requires for the repair process a template of homologous sequence and is apparently restricted to situations where a sister chromatid is available to serve as template [31]. Its primary function is probably the repair of double-strand breaks that are associated to DNA replication [9], and cells totally defective in homologous recombination accumulate double-strand breaks spontaneously and die eventually [30,36]. Importantly, homologous recombination is thought to be almost error-free with respect to the formation of genomic rearrangements. In tumour cells, however, a deregulation of the homologous recombination pathway has often been observed, and some tumours contain mutations in factors that participate in homologous recombination [2,35]. These alterations may lead to non-conservative recombination events which probably contribute to loss of heterozygosity in tumour cells.

Cancer treatment and DNA double-strand breaks: a close connection

Cell killing determines the therapeutic success as well as the occurrence of acute side effects during radiation therapy. As unrepaired DNA double-strand breaks are lethal for a cell, the quantification of residual double-strand breaks following ionising irradiation and repair incubation is being utilised for understanding, predicting and optimising the clinical effectiveness of radiotherapy [17]. From the studies carried out to date, it has become clear that genetic differences among individuals might account for much of the so far unpredictable severity of early or late tissue reactions in a subset of patients undergoing radiotherapy. A growing number of genetic loci implicated in the repair of DNA double-strand breaks and in cell cycle checkpoint regulation has been shown to influence radiation sensitivity and cancer predisposition in humans [11] but to date no reliable technique applicable to the screening is available for patients who will possibly over-respond to radiation treatment.

Whereas the importance of double-strand breaks for the therapeutic success of radiation therapy has been established many years ago, accumulating evidence now suggests that many chemotherapeutic drugs like inhibitors of topoisimerases I and II, alkylating agents and nitrogen mustards also seem to mediate their cytotoxic effect via the induction of double-strand breaks [7]. Although the precise mechanisms of formation and repair of double-strand breaks induced by chemotherapeutic toxins have not yet been elucidated, recent evidence suggests that homologous recombination as well as non-homologous end joining are involved in processing these lesions and that enhanced repair confers resistance to chemotherapeutics in tumour cells (see for example [19,26,39]). Since drug resistance is one of the crucial problems in chemotherapy, it is of cardinal importance to elucidate the mechanisms underlying this phenomenon. Investigating the repair of DNA double-strand breaks may provide new clues for the development of anti-resistance strategies.

Currently, therapeutic strategies are being developed for selectively and efficiently sensitising tumour cells to the action of ionising radiation or chemotherapeutic drugs [7,22]. Chemical inhibitors of DNA repair and signal transduction pathways that impair the normal cellular response to DNA damage are promising candidates for sensitising cancer cells to radio- or chemotherapy. However, since this strategy is likely to reduce genome stability, it may have the potential to enhance the incidence of secondary malignancies. Therefore, a detailed knowledge of the involvement of repair processes in modulating the carcinogenic potential of ionising radiation and other agents that induce DNA double-strand breaks is crucial for optimizing cancer treatment.

Tumourigenesis and DNA double-strand breaks: implications (not only) for radiation risk assessment

One of the major goals in the field of radiation biology is to identify the physical and biological factors that affect the faithfulness of DNA repair at a damaged site. When the damage is repaired incorrectly, faulty information may be passed to future cells. Moreover, chromosomal rearrangements resulting from erroneous rejoining of double-strand breaks are thought to be the principal alterations involved in (radiation) carcinogenesis [6,18,23], and erroneous double-strand break repair events may be involved in genomic instability, a phenomenon that is closely related to the development of tumours. Therefore, understanding the mechanisms of DNA misrepair is of utmost importance for understanding how cells respond to DNA damage and for exploring the road to cancer. A cell can encounter two conceptually different types of errors during the repair of a double-strand break. First, small scale sequence alterations can occur at the breakpoint, generated by imprecise pasting of broken ends (reviewed by Pfeiffer [21]). These may only rarely result in a deleterious mutation as the target for such an event, the coding sequences of actively transcribed genes, is very small. The other type of misrepair involves the mixing up of DNA ends originating from distinct double-strand breaks, which results in the formation of large-scale genomic rearrangements (reviewed in [25]. These events, encompassing translocations, deletions, inversions and insertions, affect megabase pair regions of the genome and thus may be much more effective in triggering mutagenesis and cancer induction. In fact, the spectrum of ionising radiation-induced mutations that lead to gene inactivation, reflects the predominant role of large-scale rearrangements in mutagenesis [27,28,32,37].

Recent evidence suggests that the danger that emanates from a double-strand break may differ between breaks that arise endogenously during replication and radiation-induced breaks. Since replication-associated breaks appear to be exclusively repaired by homologous recombination which operates mainly error-free, these breaks constitute a much lower risk than radiation-induced breaks that are processed by the error-prone non-homologous end joining machinery (Rothkamm et al., submitted). This implies that radiation-induced double-strand breaks may be much more effective in triggering carcinogenesis than endogenous breaks that are formed during replication in every proliferating cell.

References

1 Biedermann KA, Sun JR, Giaccia AJ, Tosto LM, Brown JM (1991) scid mutation in mice confers hypersensitivity to ionizing radiation and a deficiency in DNA double-strand break repair. Proc Natl Acad Sci USA 88:1394–1397
2 Bishop AJ, Schiestl RH (2001) Homologous recombination as a mechanism of carcinogenesis. Biochim Biophys Acta 1471:M109–M121
3 Chang C, Biederman KA, Mezzina M, Brown JM (1993) Characterization of the DNA double-strand break repair defect in scid mice. Cancer Res 53:1244–1248
4 Difilippantonio MJ, Zhu J, Chen HT, Meffre E, Nussenzweig MC, Maxx EE, Ried T, Nussenzweig A (2000) DNA repair protein Ku80 suppresses chromosomal aberrations and malignant transformation. Nature 404:510–514
5 Dikomey E, Dahm-Daphi J, Brammer I, Martensen R, Kaina B (1998) Correlation between cellular radiosensitivity and non-repaired double-strand breaks studied in nine mammalian cell lines. Int J Radiat Biol 73:269–278
6 Elliott B, Jasin M (2002) Double-strand breaks and translocations in cancer. Cell Mol Life Sci 59:373–385
7 Fojo T (2001) Cancer, DNA repair mechanisms, and resistance to chemotherapy. J Natl Cancer Inst 93:1434–1436
8 Gao Y, Ferguson DO, Xie W, Manis JP, Sekiguchi J, Frank KM, Chaudhuri J, Horner J, De Pinho RA, Alt FW (2000) Interplay of p53 and DNA repair protein XRCC4 in tumorigenesis, genomic stability and development. Nature 404:897–900
9 Haber JE (1999) DNA recombination: the replication connection. Trends Biochem Sci 24:271–275
10 Haber JE (2000) Partners and pathways repairing a double-strand break. Trends Genet 16:259–264
11 Hall J, Angèle S (1999) Radiation, DNA damage and cancer. Mol Med Today 5:157–164
12 Hendrickson EA, Qin XQ, Bump EA, Schatz DG, Oettinger M, Weaver DT (1991) A link between double-strand break-related repair and V(D)J recombination: the scid mutation. Proc Natl Acad Sci USA 88:4061–4065
13 Jeggo PA (1998) DNA breakage and repair. Adv Genet 38:185–218
14 Johnston PJ, MacPhail SH, Stamato TD, Kirchgessner CU, Olive PL (1998) Higher-order chromatin structure-dependent repair of DNA double-strand breaks: involvement of the V(D)J recombination double-strand break repair pathway. Radiat Res 149:455–462
15 Kurimasa A, Kumano S, Boubnov NV, Story MD, Tung CS, Peterson SR, Chen DJ (1999) Requirement for the kinase activity of human DNA-dependent protein kinase catalytic subunit in DNA strand break rejoining. Mol Cell Biol 19:3877–3884
16 Lees-Miller SP, Godbout R, Chan DW, Weinfeld M,, Day RS III, Barron GM, Allalunis-Turner J (1995) Absence of p350 subunit of DNA-activated protein kinase from a radiosensitive human cell line. Science 267:1183–1185
17 McMillan TJ, Tobi S, Mateos S, Lemon C (2001) The use of DNA double-strand break quantification in radiotherapy. Int J Radiat Oncol Biol Phys 49:373–377
18 Mitelman F (2000) Recurrent chromosome aberrations in cancer. Mutat Res 462:247–253
19 Muller C, Christodoulopoulos G, Salles B, Panasci L (1998) DNA-Dependent protein kinase activity correlates with clinical and in vitro sensitivity of chronic lymphocytic leukemia lymphocytes to nitrogen mustards. Blood 92:2213–2219

20 Nikiforova MN, Stringer JR, Blough R, Medvedovic M, Fagin JA, Nikiforov YE (2000) Proximity of chromosomal loci that participate in radiation-induced rearrangements in human cells. Science 290:138–141
21 Pfeiffer P (1998) The mutagenic potential of DNA double-strand break repair. Toxicol Lett 96–97:119–129
22 Poggi MM, Coleman CN, Mitchell JB (2001) Sensitizers and protectors of radiation and chemotherapy. Curr Probl Cancer 25:334–411
23 Rabbitts TH (1994) Chromosomal translocations in human cancer. Nature 372:143–149
24 Rothkamm K, Kühne M, Jeggo PA, Löbrich M (2001) Radiation-induced genomic rearrangements formed by non-homologous end-joining of DNA double-strand breaks. Cancer Res 61:3886–3893
25 Rothkamm K, Löbrich M (2002) Misrepair of radiation-induced DNA double-strand breaks and ist relevance for tumorigenesis and cancer treatment (Review). Int J Oncol 21:433–440
26 Shen H, Schultz M, Kruh GD, Tew KD (1998) Increased expression of DNA-dependent protein kinase confers resistance to adriamycin. Biochim Biophys Acta. 1381:131–138
27 Simpson P, Morris T, Savage J, Thacker J (1993) High-resolution cytogenetic analysis of X-ray induced mutations of the HPRT gene of primary human fibroblasts. Cytogenet Cell Genet. 64:39–45
28 Singleton BK, Griffin CS, Thacker J (2002) Clustered DNA damage leads to complex genetic changes in irradiated human cells. Cancer Res 62:6263–6269.
29 Smith GC, Jackson SP (1999) The DNA-dependent protein kinase. Genes Dev 13:916–934
30 Sonoda E, Sasaki MS, Buerstedde JM, Bezzubova O, Shinohara A, Ogawa H, Takata M, Yamaguchi-Iwai Y, Takeda S (1998) Rad51-deficient vertebrate cells accumulate chromosomal breaks prior to cell death. EMBO J 17:598–608
31 Takata M, Sasaki MS, Sonoda E Morrison C, Hashimoto M, Utsumi H, Yamaguchi-Iwai Y, Shinohara A, Takeda S (1998) Homologous recombination and non-homologous end-joining pathways of DNA double-strand break repair have overlapping roles in the maintenance of chromosomal integrity in vertebrate cells. EMBO J 17:5497–5508
32 Thacker J (1986) The nature of mutants induced by ionising radiation in cultured hamster cells. III. Molecular characterization of HPRT-deficient mutants induced by gamma-rays or alpha-particles showing that the majority have deletions of all or part of the hprt gene. Mutat Res 160:267–275
33 Thompson LH, Schild D (1999) The contribution of homologous recombination in preserving genome integrity in mammalian cells. Biochimie 81:87–105
34 Thompson LH, Schild D (2001) Homologous recombinational repair of DNA ensures mammalian chromosome stability. Mutat Res 477:131–153
35 Thompson LH, Schild D (2002) Recombinational DNA repair and human disease. Mutat Res 509:49–78
36 Tsuzuki T, Fujii Y, Sakumi K, Tominaga Y, Nakao K, Sekiguchi M, Matsushiro A, Yoshimura Y, MoritaT (1996) Targeted disruption of the Rad51 gene leads to lethality in embryonic mice. Proc Natl Acad Sci USA 93:6236–6240
37 Vrieling H, Simons JW, Arwert F, Natarajan AT, van Zeeland AA (1985) Mutations induced by X-rays at the HPRT locus in cultured Chinese hamster cells are mostly large deletions. Mutat Res 144:281–286

38 Wachsberger PR, Li WH, GuoM, Chen D, Cheong N, Ling CC, Li G, Iliakis G (1999) Rejoining of DNA double-strand breaks in Ku80-deficient mouse fibroblasts. Radiat Res 151:398–407

39 Wang ZM, Chen ZP, Xu ZY, Christodoulopoulos G, Bello V, Mohr G, Aloyz R, Panasci LC (2001) In vitro evidence for homologous recombinational repair in resistance to melphalan. J Natl Cancer Inst. 93:1473–1478

Impact of Physical Developments

Electrostatistics of DNA Complexes

Roland R. Netz

Sektion Physik, LMU Munich, Theresienstr. 37, 80333 Munich, Germany
netz@theorie.physik.uni-muenchen.de

Introduction

DNA, the carrier of genetic information, is a strongly charged stiff polymer. Its structure and properties are of central importance in microbiology and gene- or biotechnological applications. For the biophysicist, it is characterized by a few parameters, such as its length, its bending rigidity or its linear charge density. Because of the straightforward preparation of well-defined DNA samples and the readily available techniques for their manipulation and characterization, DNA is now widely studied by experimental and theoretical physicists as a model system to understand the intricate behavior of charged biopolymers. In this contribution it is demonstrated that many effects that are obtained with complexes of DNA and oppositely charged objects can be rationalized using simple statistical-mechanics and electrostatic concepts.

Important for the following are the mechanic and electric properties of DNA: First of all, DNA is a rather thin macromolecule which consists of stacked base pairs connected by a double helix; it has a diameter of roughly *2 nm* and a length of up to few centimeters.Therefore it can be visualized with the atomic force microscope. The molecule is ratherflexible and at room temperatures is bent by thermal forces. Its bending energy can be described by linear elasticity theory. The persistence length characterizes the typical radius of curvature of DNA, which at room temperatures and in aqueous solution is roughly $l_0 \approx 50\ nm$. A DNA strand which is much longer than this length scale will form a random coil. Secondly, DNA is highly charged. At normal pH conditions, two protons dissociate from each base pair, so that DNA carries two negative charges per base pair. It is this high negative charge which lends DNA many of its salient structural properties. It is important to note that the so-called counterions are always close to the DNA and surround it in the form of a loosely bound counterion atmosphere. In order to understand the electrostatic properties of DNA and its interactionswith other charged objects, it is important to take the counterion clouds into account as well.

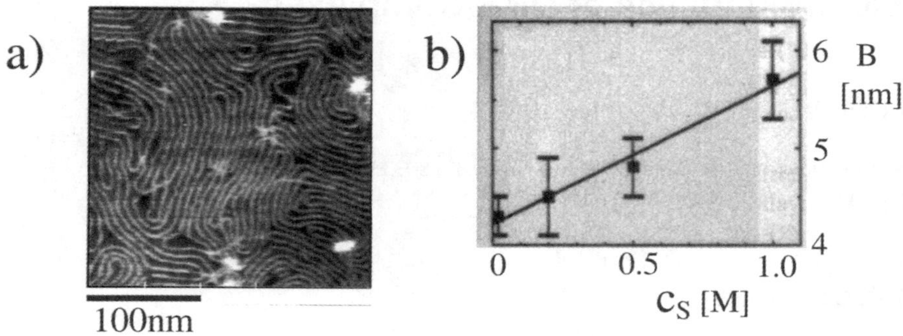

Fig. 1. a) AFM picture of a DNA layer adsorbed at an oppositely charged substrate at a salt concentration c_s = *1 M*. The average lateral distance between DNA strands is $B \cong 6$ *nm*. b) Mean lateral distance between DNA molecules, B, as a function of the salt concentration (reproduced after [5]).

DNA adsorption on planar substrates

For many applications, it is important to adsorb DNA or other highly charged polymers in a controlled way on planar substrates, for example for the production of DNA chips [1] or the fabrication of charge-oscillating multilayers [2–4]. Various experiments have been performed with DNA [5, 6] and synthetic polymers with comparable charge density [7, 8]. Important issues are the structure of the adsorbed layer or the amount of adsorbed material at a given set of parameter values (such as salt concentration of the ambient solution, polymer concentration, etc.). Fig.1a shows atomic-force-microscope pictures of an adsorbed DNA layer on a positively charged substrate, obtained at relatively high salt concentration of *1 M* [5]. The analysis of the AFM pictures shows that the adsorbed layer is extremely thin, which is in contrast to the rather diffuse layers that are obtained with neutral polymers. The individual DNA strands have a rather well-defined mutual distance of $B \approx 6$ *nm* at a salt concentration c_s = *1 M*, which is larger than the DNA diameter of $D \approx 2$ *nm*[1]. At length scales above *100 nm* the DNA strands change their orientation, the structure resembles a finger print. The lateral distance between DNA strands grows with increasing salt concentration, see Fig. 1.b [5]. All these findings can be theoretically explained by considering the competition between electrostratic attraction to the substrate and electrostatic and entropic repulsion between neighboring DNA strands [9], as will be shown in the following.

[1] It is important to note that the DNA layer shown in Fig. 1.a has been prepared at a salt concentration of c_s = *1 M* but imaged at a much smaller salt concentration (presumably without changing its structure), since at high salt the layer becomes extremely fuzzy and is impossible to image with an AFM.

A polyelectrolyte (PE) characterized by a linear charge density τ, is subject to an electrostatic potential created by σ_s, the homogeneous surface charge density (per unit area) on the substrate. Because this potential is attractive for an oppositely charged substrate (which is the situation that was considered in all above-mentioned experiments), it is the driving force for the adsorption and we neglect complications due to additional interactions or effects. One example for the latter are interactions due to the dielectric discontinuity at the substrate surface[2] and to the impenetrability of the substrate for salt ions [11]. Within the linearized Debye-Hückel (DH) theory, the electrostatic attractive force acting on a PE section at a distance δ from the homogeneously charged plane is in units of k_BT and per PE unit length

$$f_{att}(\delta) = -4\pi l_B \sigma_s \tau e^{-\kappa\delta} \tag{1}$$

The screening length κ^{-1} depends on the salt concentration c_s and ion valency q and is defined via $\kappa^2 = 8\pi q^2 l_B c_s$, where $l_B \equiv e^2/4\pi\varepsilon\varepsilon_0 k_BT$ is the Bjerrum length (e is the elementary charge, ε is the relative dielectric constant). In water, one typically has $l_B \approx 0.7\ nm$ .Assuming that the polymer is adsorbed over a layer of width δ smaller than the screening length κ^{-1}, the electrostatic attraction force per PE unit length can bewritten as

$$f_{att} \cong -4\pi l_B \sigma_s \tau \tag{2}$$

For simplicity, we neglect non-linear effects due to counter-ion condensation on the PE (as obtained by the Manning counterion-condensation argument [12]) and on the surface (as obtained within the Gouy-Chapman theory). Although these effects are important for highly charged systems [13], most of the important features of single PE adsorption already appear on the linearized Debye-Hückel level.

Because of the confinement in the adsorbed layer, the polymer feels an entropic repulsion. If the layer thickness δ is much smaller than the effective persistence length of the polymer, l_{eff}, as depicted in the side view of Fig.2a, a new length scale, the so-called deflection length λ, enters the description of the polymer statistics. The deflection length λ measures the average distance between two contact points of the polymer chain with the substrate. As shown by Odijk, the deflection length scales as $\lambda \sim d^{2/3} l_{eff}^{1/3}$ and is larger than the layer thickness δ but smaller than the persistence length l_{eff} [14]. The entropic repulsion follows in a simple manner from the

[2] An ion in solution has a repulsive interaction from the surface when the solution dielectric constant is higher than that of the substrate. This effect can lead to desorption for highly charged PE chains. On the contrary, when the substrate is a metal there is a possibility to induce PE adsorption on non-charged substrates or on substrates bearing charges of the same sign as the PE. See Ref. [10] for more details.

deflection length by assuming that the polymer loses roughly an entropy of one k_BT per deflection length.

On the other hand, if $\delta > l_{eff}$, the polymer forms a random coil with many loops within the adsorbed layer. The chain can be viewed as an assembly of decorrelated blobs, each containing a subchain of length $L \sim \delta^2/l_{eff}$, within which the polymer obeys Gaussian statistics. The decorrelation into blobs has an entropic cost of roughly one k_BT per blob. The entropic repulsion force per polymer unit length is thus [14]

$$f_{rep} \sim \begin{cases} \delta^{-5/3} l_{eff}^{-1/3} & for\ \delta \ll l_{eff} \\ l_{eff} \delta^{-3} & for\ \delta \gg l_{eff} \end{cases} \tag{3}$$

where we neglected a logarithmic correction factor which is not important for our scaling arguments.

The equilibrium layer thickness follows from equating the attractive and repulsive forces, Eqs. (2) and (3). For rather stiff polymers and small layer thickness, $\delta < \kappa^{-1} < l_{eff}$, we obtain

$$\delta \sim (l_B \sigma_s \tau l_{eff}^{1/3})^{-3/5} \tag{4}$$

For a layer thickness corresponding to the screening length, $\delta \approx \kappa^{-1}$, scaling arguments predict a rather abrupt desorption transition [9]. Setting $\delta \sim \kappa^{-1}$ in Eq. (4), we obtain an expression for the adsorption threshold (valid for $\kappa l_{eff} > 1$)

$$\sigma_s{}^* \sim \frac{\kappa^{5/3}}{\tau l_B l_{eff}^{1/3}} \tag{5}$$

For $\sigma_s > \sigma_s{}^*$ the polymer is adsorbed and localized over a layer with a width smaller than or comparable to the screening length (and with the condition $l_{eff} > \kappa^{-1}$ also satisfying $\delta < l_{eff}$, indicative of a flat layer). As σ_s is decreased, the polymer abruptly desorbs at the threshold $\sigma_s \cong \sigma_s{}^*$.

From Eq. (4) we see that the layer thickness δ is of the same order as l_{eff} for $l_B \sigma_s \tau l_{eff}^2 \sim 1$, at which point the condition $\delta \ll l_{eff}$ used in deriving Eq. (4) breaks down. Let us now consider the opposite limit, $l_{eff} < \kappa^{-1}$. If the layer thickness is larger than the persistence length but smaller than the screening length, $l_{eff} < \delta < \kappa^{-1}$, the prediction for δ obtained from balancing Eqs. (2) and (3) becomes

$$\delta \sim \left(\frac{l_{eff}}{l_B \sigma_s \tau}\right)^{1/3} \tag{6}$$

From the expression Eq. (6) we see that δ has the same size as the screening length κ^{-1} for

$$\sigma_s^* \sim \frac{l_{eff}\kappa^3}{\tau l_B} \tag{7}$$

This in fact denotes the location of a continuous adsorption transition at which the layer grows to infinity [9]. The scaling results for the adsorption behavior of a flexible polymer, Eqs. (6)–(7), are in agreement with previous results [15].

Now we generalize the discussion of the single PE chain adsorption and consider the effect of interactions between different adsorbed polymers on a simple scaling level. In order to do so, we assume that the adsorption energy is strong enough such that the polymers essentially lie flat on the substrate, which is the relevant case for describing the experiments shown in Fig. 1.a. Chain crossings are disfavored, and lateral chain correlations are large enough to locally induce the polymers to form some type of locally ordered lattice. The formation of this two-dimensionally correlated adsorbed layer is accompanied by a loss of energy and entropy, which will turn out to be important to understand the density of adsorbed chains on the substrate. We follow here the original ideas of Ref. [9], which were subsequently extended by Nguyen et al. [16]. To understand the idea, consider the top view in Fig. 2.a, where a *lamellar* phase is shown where different polymer strands are parallel locally, characterized by an average lamellar spacing B. The lamellar phase is stabilized either by steric or by electrostatic repulsions between neighboring polymers; we will in fact encounter both stabilization mechanisms for different values of the parameters. We calculate the free energy and other characteristics of the adsorbed lamellar phase, assuming that we are inside the adsorbed regime of a single polymer. We will also assume, later on, that the desorption transition obtained for the single-chain case also applies to the case of many-chain adsorption. As was shown in Ref. [9], to obtain the complete phase diagram it is sufficient to consider the lamellar phase depicted in Fig. 2.a, since other possible phase morphologies are metastable or degenerate. We assume that the distance between neighboring polymer strands, B, is much smaller than the effective persistence length, $B < l_{eff}$ (which can be checked self-consistently). Since the possible conformations of the adsorbed polymers are severely restricted in the lateral directions, we have to include, in addition to the electrostatic interactions, a repulsive free energy contribution coming from steric interactions between stiff polymers [14]. This is the same type of entropic repulsion that was used before to estimate the pressure inducing desorption from a substrate. The total free energy density per unit area and per k_BT in the lamellar phase is given by

$$F_{lam} \cong -\frac{2\pi l_B \sigma_s \tau}{B\kappa} + \frac{1}{l_{eff}^{1/3} B^{5/3}} \ln\left(\frac{l_{eff}}{B}\right) + F_{rep} \tag{8}$$

where the first term comes from the electrostatic attraction to the oppositely charged surface (which for consistency is taken to be penetrable to ions), the second term is the Odijk entropic repulsion between PEs [14] and F_{rep} is the electrostatic repulsion within the lamellar array.

To obtain the electrostatic repulsive energy, we first note that the reduced potential created by a charged line with line charge density τ is at a distance B within the Debye-Hückel approximation given by

$$V_{line}(B) = \tau \int_{-\infty}^{\infty} v_{DH} \sqrt{B^2 + s^2}\, ds = 2 l_B \tau K_0[\kappa B] \tag{9}$$

with the Debye-Hückel potential $v_{DH}(r) = l_B e^{-kr}/r$ and where K_0 denotes the modified Bessel function. The repulsive electrostatic free energy density of an array of parallel lines with a nearest-neighbor distance of B and line charge density τ can thus be written as

$$F_{rep} = \frac{2 l_B \tau^2}{B} \sum_{j=1}^{\infty} K_0 [jB\kappa] \tag{10}$$

This expression is also accurate for rods of finite radius D as long as $D \ll B$ holds. In the limit $B\kappa \ll 1$, when the distance between strands is much smaller than the screening length, the sum can be transformed into an integral and we obtain

$$F_{rep} \cong \frac{2 l_B \tau^2}{B} \int_0^{\infty} K_0[sB\kappa]\, ds = \frac{\pi l_B \tau^2}{B^2 \kappa} \tag{11}$$

This expression neglects effects due to the presence of a solid substrate. For example, and as discussed in Ref. [10], for a low-dielectric substrate the electrostatic interactions are enhanced by a factor of two close to the substrate surface, a rather small effect which will be neglected in the following. Since the average adsorbed surface charge density is given by $\sigma_{ads} = \tau/B$, it follows that the self energy Eq. (11) in the limit $B\kappa \ll 1$ is given by $F_{rep} \cong \pi l_B \sigma_{ads}^2 \kappa^{-1}$ and thus is identical to the self energy of a totally smeared-out charge distribution [9]. In this case, lateral correlations therefore do not matter. In the opposite limit, $B\kappa \gg 1$, when the polymers are much farther apart than the screening length, the sum in Eq. (10) is dominated by the first term and (using the asymptotic expansion of the Bessel function) the free energy density (in units of $k_B T$) becomes

$$F_{rep} \cong \frac{\sqrt{2\pi} l_B \tau^2 e^{-B\kappa}}{B^{3/2}\kappa^{1/2}} \tag{12}$$

In this limit, it is important to note that the smeared-out repulsive energy Eq. (11) is much larger and thus considerably overestimates the actual electrostatic repulsion between polymer strands. Conversely, this reduction of the electrostatic repulsion between polymers results in an enormous overcharging of the substrate, as we will see shortly.

In order to determine the equilibrium distance between the polymer strands, we balance the electrostatic attraction term, the first term in Eq. (8), with the appropriate repulsion term. There are three choices to do this. For $D < k^{-1} < B^* < B$ (with some crossover length B^* to be determined later on), the electrostatic repulsion between the polymers is irrelevant (i.e. the last term in Eq. (8) can be neglected), and the lamellar phase is *sterically* stabilized in this case. The equilibrium lamellar spacing is given by

$$B \sim \left[\frac{\kappa}{\tau\sigma_s l_B l_{eff}^{1/3}} \ln\left(\frac{\tau\sigma_s l_B l_{eff}}{\kappa}\right)\right]^{3/2} \tag{13}$$

In all what follows, we neglect the logarithmic cofactor.

For $D < k^{-1} < B < B^*$, the steric repulsion between the polymers is irrelevant (i.e. the second term in Eq. (8) can be neglected). The free energy is minimized by balancing the electrostatic adsorption term , the first term in Eq. (8), with the electrostatic repulsion term appropriate for the case $B\kappa > 1$, Eq. (12), which leads to the *electrostatically* stabilized lamellar spacing

$$B \sim \kappa^{-1} \ln\left[\frac{\tau\kappa}{\sigma_s}\right] \tag{14}$$

The adsorbed charge density then follows from $\sigma_{ads} \sim \tau/B$ as

$$\sigma_{ads} \sim \sigma_s \frac{\tau\kappa\sigma_s^{-1}}{\ln(\tau\kappa\sigma_s^{-1})} \tag{15}$$

Therefore, the electrostatically stabilized lamellar phase shows always strong charge reversal, since the polymer spacing B is larger than the screening length and thus $\tau\kappa\sigma_s^{-1} > 1$. This can be seen from comparing the two equations (14) and (15). The crossover between the sterically stabilized lamellar phase, described by Eq. (13), and the lamellar phase which is stabilized by electrostatic repulsion, Eq. (14), occurs when the predictions for B become simultaneously equal to the crossover spacing B^*, leading to a crossover for a surface charge density of (without logarithmic cofactors)

$$\sigma_s^* \sim \frac{\kappa^{5/3}}{\tau l_{eff}^{1/3} l_B} \tag{16}$$

For σ_s larger than the crossover value in Eq. (16) the distance between neighboring polymer strands is smaller than B^* and the electrostatic stabilization mechanism is at work, for σ_s smaller than the crossover value in Eq. (16) the lamellar spacing B is larger than the characteristic crossover length B^* and the Odijk entropic repulsion dominates. One notes that the transition Eq. (16) is, on the scaling level, the same as the adsorption threshold in Eq. (5) and it is therefore not clear a priori whether the sterically stabilized lamellar phase exists. However, we note that additional non-electrostatic adsorption forces will stabilize the sterically stabilized lamellar phase which should therefore occur in a finite range of parameters [9]. We obtain the interesting result that in the sterically stabilized adsorbed phase the strand separation increases with increasing salt concentration, see Eq. (13), while in the electrostatically stabilized phase the strand separation decreases with increasing salt, see Eq. (14). The intuitive reason for this is clear: in the sterically stabilized phase adding salt diminishes the electrostatic attraction to the substrate, while in the electrostatically stabilized phase the predominant effect of salt is to weaken the repulsion between neighboring PE strands.

The electrostatically stabilized lamellar phase crosses over to the *charge-compensated* phase when B as given by Eq. (14) becomes of the order of the screening length κ^{-1}. In the charge-compensated phase, the lamellar spacing is obtained by balancing the electrostatic adsorption energy with the repulsion in the smeared-out limit Eq. (11) and is given by

$$B \cong \frac{\tau}{\sigma_s} \tag{17}$$

In this case the adsorbed surface charge density $\sigma_{ads} = \tau/B$ exactly neutralizes the substrate charge density,

$$\sigma_{ads} \sim \sigma_s \tag{18}$$

The crossover between the charged-reversed phase and charge-compensated phase is obtained by matching Eqs. (14) and (17), leading to a threshold surface charge density of

$$\sigma_s^* \sim \tau\kappa \tag{19}$$

Finally, taking into account that the polymers have some width D, there is an upper limit for the amount of polymer that can be adsorbed in a single layer. Clearly, the lateral distance between polymers in the *full* phase is given by

$$B \cong D \tag{20}$$

and thus the adsorbed surface charge density in the full phase reads

$$\sigma_{ads} = \frac{\tau}{D} \tag{21}$$

The crossover between the full phase and the compensated phase is obtained by comparing Eqs. (17) and (20), leading to

$$\sigma_s^* \sim \tau/D \tag{22}$$

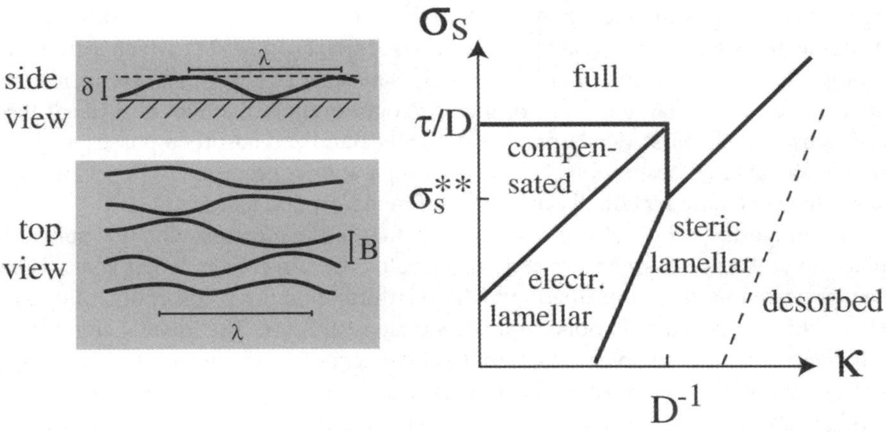

Fig. 2. a) Schematic structure of the adsorbed DNA layer in a side view and in a top view, exhibiting the deflection length λ, the adsorbed layer height δ, and the lateral distance between adsorbed DNA strands, *B*. b) Scaling diagram of the adsorption behavior of a highly charged polymer using logarithmic axes.

In Fig. 2.b we show the adsorption diagram, for strongly charged polymers (defined by $\tau\sqrt{l_B l_{eff}} > 1$) as a function of the substrate charge density σ_s and the inverse screening length κ. The electrostatically stabilized lamellar phase shows strong charge reversal as described by Eq. (15). At slightly larger surface charge densities we predict a charge-compensated phase which is not full (i.e. $B < D$) for a range of surface charge densities as determined by Eqs. (19) and (22). At even larger substrate charge density, the adsorbed polymer phase becomes close packed, i.e. $B \approx D$. We note that since the full phase is not charge reversed, the full phase can consist of a second adsorbed layer (or even more layers). It should however be clear that close to charge compensation the effective substrate charge density an additional layer feels is so small that the condition for adsorption is not necessarily met. At low substrate charge densities the distance between adsorbed polymer strands becomes so large that the entropic repulsion between polymers dominates the electrostatic repulsion, and finally, at even lower charge densities, the polymers desorb. One notes that the transition between the electrostatically and sterically stabilized adsorbed phases, Eq. (16), has the same scaling form (disregarding logarithmic factors) as the desorption transition of semi-flexible polymers, Eq. (5). We have shifted the desorption transition to the right, though, because typically there are attractive non-electrostatic interactions as well, which tend to stabilize adsorbed phases. This is also motivated by the fact that the sterically stabilized phase has been seen in experiments on DNA adsorption, as will be discussed below. The critical charge density σ_s where the full phase, the electrostatically and the sterically stabilized phases meet at one point, is given by $\sigma_s^{**} \sim 1/(D^{5/3} l_{eff}^{1/3} \tau l_B)$. In the phase diagram we have assumed that the charge density threshold for the full phase Eq. (22), $\sigma_s^{*} \sim \tau/D$, satisfies the inequality $\tau/D > \sigma_s^{**}$, which for a fully charged PE at the Manning threshold, $\tau = 1/l_B$, amounts to the condition $l_{eff} > l_B^3/D^2$, which is true for a large class of PE's (especially stiff biopolymers such as DNA). As one increases at fixed substrate charge density σ_s the salt concentration, one moves through the compensated, electrostatically and the sterically stabilized adsorbed phases, before one finally induces desorption. The polymer separation is predicted to first stay constant, then decrease and finally increase, before desorption takes place.

One important result of our discussion is that in the electrostatically stabilized phase the substrate charge is strongly reversed by the adsorbed polymer layer. This can give rise to a charge-oscillating multilayer formation if the adsorption of oppositely charged polymer is done in a second step [2, 3, 4]. The general trend that emerges is that charge reversal is more likely to occur for intermediate salt concentrations and rather low substrate charge density. For too high salt concentration and too low substrate charge density, on the other hand, the polymer does not adsorb at all. In essence, the salt concentration and the substrate charge density have to be tuned to intermediate values in order to create charge multilayers.

In experiments on DNA adsorbed on oppositely charged substrates one typically observes a lamellar phase [5, 6]. In one experiment, the spacing between DNA strands was found to increase with increasing salt concentration [5]. One theoretical explanation invokes an effective interaction between neighboring DNA strands mediated by elastic deformations of the membrane, which forms the substrate in these experiments [17]. In the sterically stabilized regime, the distance between adsorbed polymers increases as $B \sim \kappa^{3/2}$ with the salt concentration, see Eq. (13), which offers an alternative explanation for the experimental findings. It would be interesting to redo DNA adsorption experiments on rigid substrates, where the elastic coupling to the membrane is absent. For high enough substrate charge densities and by varying the salt concentration one should be able to see the crossover from the electrostatically stabilized phase, Eq. (14), where the DNA spacing decreases with added salt, to the sterically stabilized phase, Eq. (13), where the DNA spacing increases with added salt.

DNA adsorption on oppositely charged spheres

Although the DNA is a highly negatively charged polymer, although it is quite stiff and resists bending, the *2m* of human DNA is packaged in the cell nucleus which only has a diameter of a few micrometers. The mechanism for this folding is ingeneous: The DNA is wrapped around a large number of small, highly positively charged almost spherical proteins (called histones), it is thereby partially neutralized and greatly compactified.

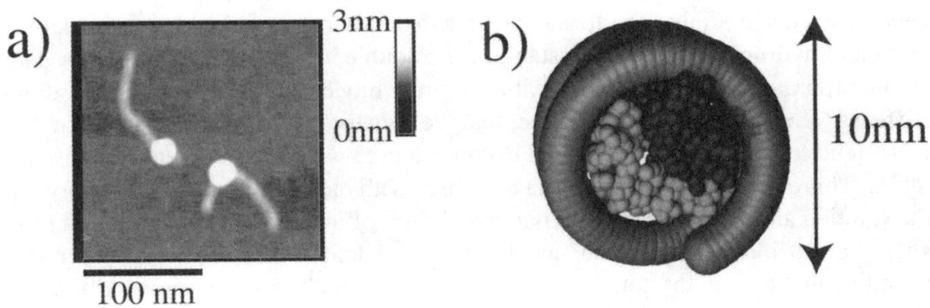

Fig. 3. a) AFM pictures of two complexes consisting of a positively charged histone protein and a partially wrapped DNA strand of length *130 nm*. Adapted after [18]. b) View of the three-dimensional structure of the histone protein; the approximate path of the DNA consists of two full turns (approximately 146 base pairs) and is indicated by a tube (adapted after [19]).

Fig. 3.a shows atomic-force-microscopy pictures of two reconstituted complexes consisting of histone proteins and DNA strands of length *130 nm* [18]. The structures were obtained at a salt concentration of $c_s = 50\ mM$. The precise path of the DNA on the proteins can not be resolved, but from the length of the unwrapped DNA portion it can be deduced that roughly one whole turn of DNA is wrapped around the proteins. Fig. 3.b shows the structure of the complex as obtained from X-ray diffraction on crystallized DNA-histone complexes. Indeed, in the crystalline state (which is not necessarily exactly equal to the solubilized state at room temperatures) the DNA is wrapped twice around the proteins. A huge body of experimental evidence [20]–[34] suggests that the complex is only stable for intermediate salt concentrations between *20 mM* and *500 mM* (the so-called physiological salt concentration in the body is roughly *150 mM* which corresponds to the salt concentration in the sea). For salt concentrations outside this finite range the complex falls apart. Since salt modulates the electrostatic interactions, it follows that an electrostatic mechanism is the cause for this interesting behavior.

Designing a simple model that can explain the salt and charge dependent behavior of the nucleosomal core particle requires a number of approximations concerning the structure of the DNA, the histone octamer and their interactions. Our strategy was to formulate the simplest possible model which captures the characteristic features responsible for the salt-dependent nucleosomal core structures [35, 36, 37]. In the following we report calculations of the ground state of a single semiflexible polyelectrolyte of finite length which is in contact with a single oppositely charged sphere in the presence of salt [38]. Polymer form fluctuations away from the ground state are neglected, which is an acceptable approximation for rather stiff or highly charged polymers. We specifically consider parameters appropriate for the DNA-histone system, approximating the histone as a uniformly charged, impenetrable and solid sphere and the DNA as a uniformly charged semiflexible polymer of length L [41, 42]. We deal with the so-called nucleosomal core particles consisting of DNA with 146 base pairs (bp) of length *0.34 nm* each, leading to a total DNA length of $L = 49.6$ *nm*. The stiffness of the DNA contains charge independent and charge dependent contributions. The former are due to the energy associated with the deformation of hydrogen bonds which stabilize the double helical geometry of DNA and are incorporated by using a semiflexible polymer model with a mechanical bending stiffness. The latter stem from the fact that the negatively charged monomers of the DNA tend to maximize their mutual distance and thus prefer an extended configuration. This electrostatic contribution decreases with increasing salt concentration of the solution and vanishes at the hypothetical limit of infinite salt concentration [43, 44]. The salt independent mechanical persistence length of DNA is therefore the infinite salt limit of the total persistence length, which has experimentally been determined as $l_0 \approx 30$ *nm* [45–47]. A discussion of different methods to determine the persistence length of DNA is found in [42]. The electrostatic contribution to the persistence length we take into account by explicitly including the electrostatic self energy of DNA conformations, we therefore accurately include the scale-dependence of the electrostatic contribution to the persistence length [48] which is particularly important in our case since the scale of bending (the histone diameter) becomes of the order or even smaller than the screening length.

The histone octamer is approximated as a rigid sphere of radius $R_{hist} = 4$ *nm*. This is of course only a very rough approximation of the real structure, which is not a perfect sphere and also possesses a corrugated surface with specific binding sites for the DNA. Also, any conformational changes of the histone octamer, which do occur for extremely low or high salt concentration, are neglected. The DNA is modeled as a polymer with radius of *1 nm*. In the actual calculation we fix the minimal distance between the sphere center and the DNA monomer centers to be $R = 5\ nm$, which is the sum of histone and DNA radii.

The electrostatic interactions between charges on the DNA with each other and the sphere are described by Debye-Hückel (DH) potentials that neglect non-linear effects (such as counterion condensation or counterion release, for which one would need to use the full non-linear Poisson-Boltzmann theory [12, 49, 50]. The main reason for this approximation is that the calculation of the optimal DNA configuration within the PB approach is at present numerically not feasible [51]. For large salt concentrations, the DH approximation becomes valid, as has been shown by calculating the electrostatic contribution to the bending rigidity of a charged cylinder [49, 50].

The free energy for a given DNA configuration of contour length L, parameterized by the space curve $\boldsymbol{r}(s)$ and in units of k_BT, reads

$$F = \frac{l_0}{2}\int_0^L \ddot{\boldsymbol{r}}^2(s)ds - \frac{l_B Z\tau}{1+\kappa R}\int_0^L \frac{e^{-\kappa(|\boldsymbol{r}(s)|-R)}}{|\boldsymbol{r}(s)|}ds + l_B\tau^2\int_0^L ds\int_s^L \frac{e^{-\kappa|\boldsymbol{r}(s)-\boldsymbol{r}(s')|}}{|\boldsymbol{r}(s)-\boldsymbol{r}(s')|}ds' \tag{23}$$

where we implicitly assume that the DNA molecule is inextensible, i.e., $|\dot{\boldsymbol{r}}(s)| = 1$. The first term describes the mechanical bending energy, proportional to the bare persistence length l_0, where the curvature $\ddot{\boldsymbol{r}}(s)$ is given by the second derivative of $\boldsymbol{r}(s)$ with respect to the internal coordinate s. The second term describes the electrostatic attraction between the sphere and the DNA [52]. The charge of the sphere in units of the elementary charge e is denoted by Z and the linear charge density of the DNA (in units of e) is denoted by τ. A key ingredient of the Debye-Hückel-theory is the screening of electrostatic interactions, which is quantified by the Debye-Hückel screening length κ^{-1}. It measures the distance beyond which the interaction between two charges is exponentially damped. For monovalent salt one finds $\kappa^2 = 8\pi l_B c_s$, where c_s is the salt concentration. At 0.1 molar concentration in water, i.e. at physiological conditions, one has $\kappa^{-1} \approx 1\ nm$. The third term describes the electrostatic repulsion between charges on the DNA. We therefore have two terms that tend to straighten the DNA, namely the mechanic bending energy and the electrostatic repulsion between DNA monomers. The former is salt independent, whereas the latter looses importance with increasing salt concentration. These repulsions are balanced by the sphere-DNA electrostatic attraction, which favors bending of the DNA in order to wrap it around the sphere, but also becomes weaker for increasing salt concentration. It transpires that salt will determine the DNA structure in a rather subtle way, as will be demonstrated by our numerical results.

In analyzing the model defined by Eq. (23) we take advantage of the short length of the DNA strand, $L = 50\ nm$, compared to the effective persistence length, which is at least of the order of the bare persistence length, $l_0 = 30\ nm$. We therefore neglect fluctuations and undulation forces [35] between DNA and sphere and apply a ground state analysis to the model. This consists of finding the configuration of the

DNA that minimizes Eq. (23) with the constraints $|\dot{r}(s)| = 1$ (no stretching) and $|r(s)| \geq R$ (impenetrability of sphere). We also use Debye-Hückel-potentials with the full DNA line charge $\tau = 2/0.34\ nm$, corresponding to the maximal degree of dissociation of DNA (and neglecting counterion-condensation effects [12] which have been investigated in [37].

In Fig. 4. we show a series of DNA configurations obtained for a fixed sphere charge $Z = 25$ and for inverse screening lengths ranging from $\kappa = 0\ nm^{-1}$ to $6\ nm^{-1}$. In the pure Coulomb case with no added salt the sphere is located in the middle of the DNA, and the two arms are slightly bent towards the sphere. As the salt concentration is increased from $\kappa = 0\ nm^{-1}$ to $\kappa = 0.1\ nm^{-1}$, the deflection of the DNA arms increases continuously. Interestingly, the binding of the DNA onto the sphere becomes stronger as one weakens the electrostatic interactions. Upon further increase of salt concentration, the two-fold rotational and the mirror symmetries are broken, see the configuration at $\kappa = 0.30\ nm^{-1}$. One of the DNA arms is totally wrapped around the sphere. As κ is increased further, the extended arm is more and more pulled onto the sphere until at $\kappa = 1\ nm^{-1}$ the DNA is fully adsorbed onto the sphere and the two-fold rotational symmetry is restored. At $\kappa = 6\ nm^{-1}$ a strongly discontinuous transition occurs in which the DNA completely dewraps from the sphere. The dewrapped state at high salt concentration is markedly different from the state at zero salt. There is only one short region of nonzero bending of the DNA connecting two basically straight arms. This sequence of complexation structures demonstrates one of our main results, namely that the wrapped DNA conformation is only stable for intermediate salt concentrations, explaining a large set of experimental results for nucleosomal core particles [34].

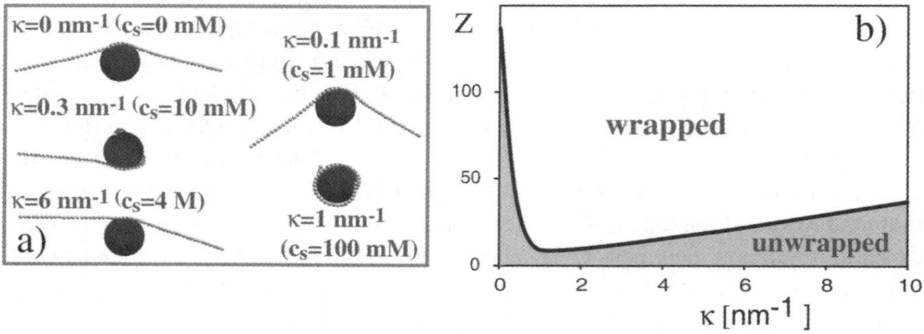

Fig. 4. a) DNA configurations as obtained by numerical minimization of the free-energy expression Eq. (23) for fixed sphere charge $Z = 25$ and various salt concentrations. As can be seen, the DNA is only wrapped for intermediate values of the salt concentration. b) Global phase digram, featuring the wrapping transition as a function of salt concentration and sphere charge.

We summarize our results for a DNA length of $L = 50\ nm$ in the phase diagram presented in Fig. 4.b. In the absence of salt, $\kappa = 0$, the wrapping transition occurs at

$Z = 133$ (in agreement with previous theoretical predictions [53]). In agreement with experiments [34], complexation is most pronounced at intermediate salt concentrations. For low salt concentration, the strong DNA-DNA repulsion prevents complexation, for high salt screening weakens the DNA-sphere attraction sufficiently so that the mechanical bending resistance induces dewrapping. The minimal sphere charge to wrap the DNA, $Z \approx 10$, is obtained for $\kappa^{-1} \approx 1\ nm$ ($cs \approx 0.1\ M$ for monovalent salt), corresponding to physiological conditions. Since the total charge on the DNA is about 300, the complex is strongly overcharged for all $Z < 300$, i.e., in the whole wrapped region shown in the phase diagram. The high-salt prediction for the wrapping transition can be obtained analytically by locally balancing the various terms in the free energy, Eq. (23), namely the bending energy per unit length, $F_{bend} \cong l_0 / 2R^2$, and the electrostatic attraction per unit length in the limit $\kappa R > 1$, $F_{att} \cong l_B Z\tau/\kappa R^2$, leading to $Z_{wrap} \cong l_0\kappa / 2l_B\tau$, in agreement with the numerical results [35].

These results highlight a peculiarity of electrostatic complexation phenomena, and is mirrored by an at first sight perplexing approximation used in our model calculation: We do use the Debye-Hückel approximation for the interaction between charges on the sphere and on the DNA, which amounts to taking into account positional fluctuations of the salt ions within a Gaussian approximation [54]. However, we do not take into account positional fluctuations of the DNA itself, but concentrate on the ground state instead. The reason for the different treatment of salt ions and DNA monomers is that the total amount of charge per statistically independent unit is $q = 1$ for monovalent ions but roughly $q \cong 180$ for one persistent segment of DNA of length $L \cong l_0$. Therefore fluctuations are rather unimportant for the DNA structure (except for very large salt concentrations where a desorption transition does occur which can be treated using similar methods as used for the desorption of polyelectrolytes on planar substrates in the preceding section [35]) but are of extreme importance for the counterion clouds.

DNA adsorption on charged cylinders

When semiflexible charged polymers are mixed with much stiffer oppositely-charged polymers, a complex forms where the more flexible polyelectrolyte (PE) wraps around the stiff polymer [55–57, 16]. Experimentally, such complexes itself fold up into toroidal or stem-like structures [58, 59]. In this section we try to understand the morphology of the underlying molecular complex, namely the conformation of the wrapping polymer: does it form a helix or does it adsorb in (one or more) parallel straight strands onto the cylinder? Using linear (Debye-Hückel) theory, supplemented by nonlinear (counterion release) arguments [56], we find transitions between both morphologies as the salt concentration (or other parameters) are varied.

Most of the current interest in such complexes comes from their potential applications in gene therapy: The main problem here is to introduce genetic material into patients' cell nuclei, a process called DNA transfection. The classical viral strategies are highly effective in transfecting DNA but may provoke immune reactions of the body, switch back to their lethal origin or lead to a stable transformation of target cells (advantageous in some cases though in general undesirable) [60, 61]. Nonviral transfection strategies avoid these difficulties at the price of much reduced effectiveness [61]. Still, they hold promising potential for further development and refinement. The polyfection scheme consists of complexing DNA with physiologically tolerated polycations, such as polypeptides or synthetic polycations [61, 62] and shows relatively high efficiency, especially with confluent (non-dividing) cells [63]. As a major advantage, the properties of these self-assembling *polyplexes* can be controlled rather reliably by for example varying the mass or charge density of the polycations [64], by using block-copolymers with a cationic block and an uncharged block which forms some type of protection layer against coagulation or degradation [65], and finally by linking target-specific ligands to the polymer chains [66].

The microscopic structure of polyplexes is not very well understood. Electron micrographs of DNA-polylysine complexes exhibit highly condensed torus or stemlike structures [58], very similar to what is seen with DNA condensed by multivalent counterions [67]. More recent AFM studies demonstrated that condensation involves five to sixfold overcharging of the DNA by peptide charges at elevated salt concentration [59]. The underlying molecular structure of polycation-DNA complexes (toroidal or stemlike), which involve multiply packed DNA loops, is not resolved in these experiments. X-ray diffraction measurements, on the other hand, showed that polylysine wraps helically around the DNA molecule (and at low salt concentrations neutralizes the DNA charges), while polyarginine, a cationic polypeptide with a different backbone flexibility, shows a different wrapping mode [68]. Similar complexation is obtained by mixing DNA with a rather bulky cationic dendrimeric polymer: AFM pictures demonstrate that in this case the DNA wraps around the dendro-polymer [69]. This essay, with potential gene-therapeutical applications, holds the advantage that the physical properties of the complex and the effects of various parameters can be studied in great detail and with comparative ease. From all the above listed experiments, it is clear that the salt concentration of the surrounding medium, the charge of the complexing polycations, and their flexibility can induce different morphologies of the polyplex. The possibility and mechanism of DNA-overcharging by adsorbed polycations is interesting from a fundamental point of view [56, 53, 16, 35], although it has been shown that optimal transfection yield is obtained with neutral complexes [64, 66] (despite the naive expectation that cationic complexes would interact more favorably with the typically negatively charged cell and endosomal walls).

In Fig. 5.a a DNA-dendropolymer complex is shown which consists of a DNA strand of length $L_P \approx 850\ nm$ wrapped around a cylindrical dendropolymer of length $L_C \approx 50\ nm$ and radius $R_C \approx 3\ nm$. The DNA is complexed as a single strand with the dendropolymer, from the wrapped DNA length (which can be determined by measuring the length of the non-wrapped DNA sections) it follows that the DNA almost fully covers the dendropolymer surface. Since no loops are seen that emerge from the complex, it is suggested that the DNA wraps helically around the core. In fact, more DNA wraps than is needed to neutralize the dendropolymer core, the precise amount of wrapped DNA turns out to be salt dependent [70].

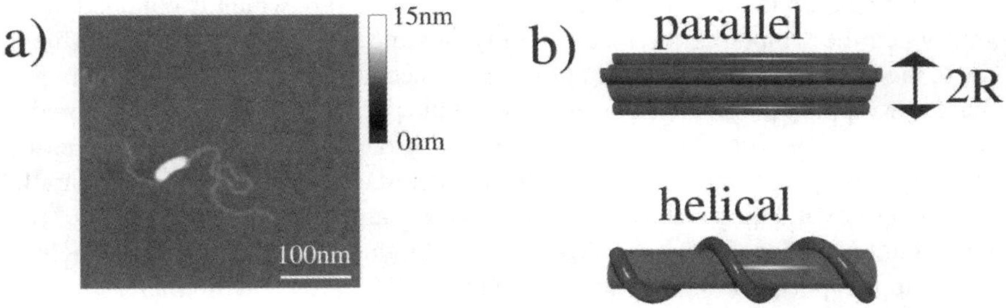

Fig. 5. a) AFM pictures of a DNA strand of lengh $L_P = 850\ nm$ which is partially wrapped around a $L_C = 50\ nm$ long cationic dendropolymer (reproduced from Ref. [69]). b) The two posibble wrapping morphologies that are theoretically studied and compared.

In this section we analyze the complexation between a charged rigid cylinder and an oppositely charged semiflexible polyelectrolyte. The parameters in our model are the linear charge densities of the uniformly charged cylinder and the PE, τ_C and τ_P, the cylinder radius R_C and the bare PE persistence length l_0. On the linear level the interaction between all charges is given by the bulk Debye-Hückel (DH) potential $v_{DH}(r) = l_B exp(-\kappa\ r)/r$ where r is the distance between charges, l_B is the Bjerrum length, $l_B = e^2/4\pi\varepsilon\varepsilon_0 k_B T$, and κ^{-1} is the screening length. The DH approximation is valid for weakly charged PEs, elevated salt concentrations and, as is explained in Ref. [57], for complexes close to electroneutrality. Effects due to dielectric boundaries, additional non-electrostatic interactions, inhomogeneous charge distributions on the cylinder and on the PE, polymer confinement (which are all neglected) and counterion release are of only secondary importance for the resulting phase diagrams, since we always compare different morphologies of roughly the same amount of adsorbed PE. It is the free-energy difference between different morphologies that we are most interested in, not their absolute values [57]. The thermodynamic ensemble we consider is the one where PE is present in excess, i.e., we minimize the free energy per cylinder unit length, neglecting end effects which will only be important if the screening length becomes larger than the cylinder length. This is the ensemble that is indeed relevant to describe the experimental situation in Fig. 5.a. We compare two morphologies, namely a helical arrangement of the PE, where a single helix, characterized by the length ratio of the wrapped polymer section and the cylinder length, wraps around the cylinder (see Fig. 5.b), and the straight morphology, where n parallel strands of PE adsorb on the cylinder. We minimize both configurational energies with respect to the relative amount of wrapped polyelectrolyte (neglecting configurational fluctuations around the ground state which are unimportant for rather stiff and highly charged PEs) and compare the two resulting free energies to determine the stable phase. This comparison, which is done numerically in the general case, shows that both morphologies compete closely with each other.

As can be seen in the phase diagram Fig. 6.a, page 206, which is obtained in the limit when the wrapping polymer is totally flexible and has no bending stiffness, $l_0 = 0$, the helical phase is favored at low salt concentrations (to the left) and highly charged wrapping polymers ($\tau_P/\tau_C \gg 1$), while the parallel morphology is favored at high salt concentrations. As the charge density of the wrapping polymer increases, as one moves up in the phase diagram, the number of adsorbed strands in the parallel phase goes down. Fig. 6.b shows for the specific charge density ratio $\tau_P/\tau_C = 0.5$ that the amount of wrapped polymer, characterized by the ratio of the contour length of the wrapped polymer and cylinder length, L_P/L_C, grows with increasing salt conentration (the desorption transition, which is expected to occur at high salt concentrations in the absence of additional non-electrostatic attractive forces is not shown but follows the same rules as outlined in the previous sections). For line charge ratio $\tau_P/\tau_C = 0.5$ the complex would be neutral for a wrapping ratio $L_P/L_C = 2$. As a matter of fact, more polymer wraps around the core cylinder than is needed to actually neutralize the complex (in agreement with the experimental results [69]). As the salt concentration increases the overcharging is even further enhanced, also in agreement with experiments [70].

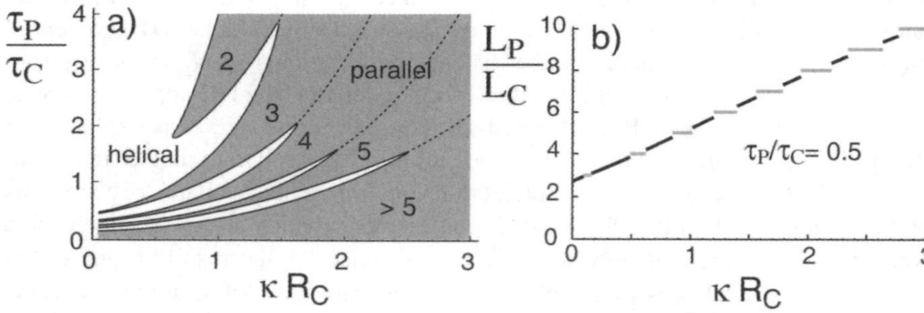

Fig. 6. a) Phase diagram of the polyelectrolyte-cylinder complex as a function of the linear-charge density ratio τ_P/τ_C and the inverse rescaled screening length κR_C for vanishing persistence length of the wrapping polymer, $l_0 = 0$. The helical phase (white) dominates at low salt concentration (small κR_C) while the straight configuration (yellow; number of adsorbed PE strands is indicated and direct transitions between states with different numbers of adsorbed strands are denoted by broken lines) is realized at large κR_C. b) Relative amount of wrapped PE, L_P/L_C, for zero persistence length $l_0 = 0$ for a line charge density ratio of $\tau_P/\tau_C = 0.5$, corresponding to a horizontal cut through the phase diagram Fig. 6.a. In the helical phase the wrapping parameter (and thus the overcharging of the cylinder) continuously increases as κR_C grows, while the straight configuration is characterized by integer values.

The overcharging in the low-salt limit is easily understood analytically. For simplicity we consider the parallel morphology with n adsorbed polymers with line charge density τ_P at a cylinder of line charge density τ_C. In the limit of low salt, $\kappa R_C \to 0$, the electrostatic potential of a charged cylinder, which is reproduced in Eq. (9) on the Debye-Hückel level, shows a logarithmic behavior. The potential at the cylinder surface is given by $-2l_B\tau_C ln\kappa R_C$. The total attractive free energy of n adsorbed polyelectrolytes is thus in the low-salt limit per unit length given by

$$F_{att} \cong 2nl_B\tau_C\tau_p\ln(\kappa R_C) \tag{24}$$

Between the n adsorbed polymers there are $n(n-1)/2$ repulsive pair interactions, all of the same logarithmic type. Clearly, the distances between the various pairs are all different, but since the repulsion is logarithmic these differences give negligible additive contributions to the resulting total repulsive free energy, which can be written as

$$F_{rep} \cong -n(n-1)l_B\tau_P^2\ln(\kappa R_C) \tag{25}$$

The number of adsorbed polymers results from minimization of the sum of the repulsive and attractive contribution, $\partial(F_{att} + F_{rep})/\partial n = 0$, and is given by

$$n = L_P/L_C = 1/2 + \tau_C/\tau_p \tag{26}$$

It equals the ratio of the wrapped polymer length and the cylinder length, L_P/L_C, which is the quantitiy that is plotted in Fig. 6.b. The number of adsorbed strands n is an integer quantity. However, the analogous calculation for the helical phase in the low-salt limit gives the same result as Eq. (26). Both phases turn out to be degenerate for integer values of the wrapping ratio L_P/L_C. The result Eq. (26) is in agreement with the numerical data displayed in Fig. 6.b, and predicts for the case $\tau_P/\tau_C = 0.5$ the wrapping length ratio $L_P/L_C = 5/2$ in the zero salt limit $\kappa R_C \to 0$ (note that the asymptotic approach of this limit is logarithmically slow, see [57]). The effective charge of the complex is in the same low-salt limit from Eq. (26) predicted to be

$$\tau_{eff} = n\tau_P - \tau_C = \tau_p/2 \tag{27}$$

This result shows that in the low-salt limit the complex will have the same charge sign as the wrapping polymer, the usual wording for this is that the complex is overcharged. The effective charge density of the complex amounts to half the one of the wrapping polymer. Therefore, if the negatively charged DNA wraps around a cationic dendropolymer, the complex will have a net negative charge, if however a flexible cationic polypeptide wraps around the negatively charged DNA, the resulting complex will be positively charged. These qualitative trends are in agreement with experiments, and they show how to tune the charge of a polyelectrolyte complex by changing the ratio of the bending rigidities of the cationic and anionic polymers involved in forming the complex.

Acknowledgements

I thank I. Gössl, J.P. Rabe, K. Takeyasu, K. Yoshikawa for discussions and for sending data and pictures.

References

1 Nature Genetics Supplement (1999) 21:1
2 Decher G (1997) Science 277:1232
3 Sukhorukov GB, Donath E, Davis SA, Lichtenfeld H, Caruso F, Popov VI, Möhwald H (1998) Polym Adv Technol 9:759
4 Caruso F, Caruso RA, Möhwald H (1998) Science 282:1111
5 Fang Y, Yang J (1997) J Phys Chem 101:441
6 Maier B, Rädler JO (1999) Phys Rev Lett 82:1911
7 de Meijere K, Brezesinski G, Möhwald H (1997) Macromol 30:2337
8 von Berlepsch H, Burger C, Dautzenberg H (1998) Phys Rev E 58:7549
9 Netz RR, Joanny JF (1999) Macromol 32:9013
10 Netz RR (1999) Phys Rev E 60:3174
11 Netz RR (2000) Eur Phys J E 3:131; Netz RR (2001) Eur Phys J E 5:189
12 Manning GS (1996) Ber Bunsenges Phys Chem 100: 909; Manning GS, Ray J (1998) J Biomolecular Structure and Dynamics 16:461
13 Fleck C, Netz RR, von Grünberg HH (2002) Biophys J 82:76
14 Odijk T (1984) Macromol 17:501
15 Wiegel FW (1977) J Phys A 10:299
16 Nguyen TT, Grosberg AY, Shklovskii BI (2000) J Chem Phys 113:1110
17 Dan N (1997) Biophys J 73:1842
18 Sakaue T, Yoshikawa K, Yoshimura SH, Takeyasu K (2001) Phys Rev Lett 87:078105
19 Alberts B, Bray D, Lewis J, Raff M, Roberts K, Watson JD (1994) Mol Biol Cell. Garland Publishing, New York
20 Gordon VC, Knobler CM, Olins DE, Schumaker VN (1978) Proc Natl Acad Sci U S 75:660
21 Russev G, Vassilev L, Tsanev R (1980) Molec Biol Rep 6:45

22 Burch JBE, Martinson HG (1980) Nucleic Acids Res 8:4969
23 Ausio J, Seger D, Eisenberg H (1984) J Mol Biol 176:77
24 Harrington RE (1981) Biopol 20:719
25 Harrington RE (1982) Biochem 21:1177
26 Uberbacher EC, Ramakrishnan V, Olins DE, Bunick GJ (1983) Biochem 22:4916
27 Dieterich AE, Axel R, Cantor CR (1979) J Mol Biol 129:587
28 Libertini LJ, Small EW (1980) Nucleic Acids Res 8:3517
29 Libertini LJ, Small EW (1982) Biochem 21:3327
30 Libertini LJ, Small EW (1987) Nucleic Acids Res 15:6655
31 Oohara I, Wada A (1987) J Mol Biol 196:399
32 Brown DW, Libertini LJ, Small EW (1991) Biochem 30:5293
33 Khrapunov SN, Dragan AI, Sivolob AV, Zagariya AM (1997) Biochim Biophys Acta 1351:213
34 Yager TD, McMurray CT, van Holde KE (1989) Biochem 28:2271
35 Netz RR, Joanny JF (1999) Macromol 32:9026
36 Kunze KK, Netz RR (2000) Phys Rev Lett 85:4389
37 Kunze KK, Netz RR (2002) Phys Rev E 66:011918
38 Similar models have been formulated for longer DNA where more than one sphere complexes with a single DNA, see [39, 40]
39 Nguyen TT, Shklovskii BI (2001) J Chem Phys 114:5905
40 Schiessel H, Bruinsma RF, Gelbart WM (2001) J Chem Phys 115:7245
41 Grosberg AY, Khokhlov AR (1994) Statistical Physics of Macromolecules. AIP Press, Woodbury New York
42 Hagerman PJ (1988) Ann Rev Biophys Biophys Chem 17:265
43 Odijk T (1977) J Polymer Sci 15:477
44 Skolnick J, Fixman M (1977) Macromol 10:944
45 Borochov N, Eisenber H, Kam Z (1981) Biopol 20:231
46 Manning GS (1981) Biopol 20:1751
47 Sobel ES, Harpst JA (1991) Biopol 31:1559
48 Barrat JL, Joanny JF (1996) Adv Chem Phys 94:1
49 le Bret M (1982) J Chem Phys 76:6243
50 Fixman M (1982) J Chem Phys 76:6346
51 Even for the relatively simple problem of the adsorption of a single DNA, modelled by a rigid charged cylinder, on an oppositely charged plane, the accurate numerical solution of the PB equation is nontrivial, see [13]
52 Verwey EJW, Overbeek JThG (1948) Theory of the Stability of Lyophobic Colloids. Elsevier, Amsterdam New York
53 Mateescu EM, Jeppesen C, Pincus P (1999) Europhys Lett 46:493
54 Netz RR, Orland H (1999) Europhys Lett 45: 726
55 Odijk T (1980) Macromol 13:1542
56 Park SY, Bruinsma RF, Gelbart WM (1999) Europhys Lett 46:454
57 Kunze KK, Netz RR (2002) Europhys Lett 58:299
58 Laemmli UK (1975) Proc Natl Acad Sci US 72:4288
59 Golan R, Pietrasanta LI, Hsieh W, Hansma G (1999) Biochem 38:14069
60 Felgner PL et al (1987) Proc Natl Acad Sci US 84:7413
61 Kabanov AV, Kabanov VA (1995) Bioconjugate Chem 6:7
62 Pouton CW et al (1998) J Control Release 53:289

63 Zauner W, Brunner S, Buschle M, Ogris M, Wagner E (1999) Biochim Biophys Acta 1428:57
64 Erbacher P, Roche AC, Monsigny M, Midoux P (1997) Biochim Biophys Acta 1324:27
65 Dash PR, Toncheva V, Schacht E, Seymour LW (1997) J Controlled Release 48: 269.
66 Wagner E et al (1990) Proc Natl Acad Sci USA 87:3410
67 Bloomfield VA (1996) Curr Opin Struct Biol 6:334
68 Suwalsky M, Traub W, Shmul U (1969) J Mol Biol 42:363
69 Gössl I, Shu L, Schlüter AD, Rabe JP (2002) J Am Chem Soc 124:6860
70 Gössl I, Shu L, Schlüter AD, Rabe JP, to be published

Future Radiobiology at Accelerators: From Heavy Ion Therapy to Basic Research

Gerhard Kraft

Biophysik GSI Darmstadt, Planckstr. 1, 64291 Darmstadt, Germany
g.kraft@gsi.de

Introduction

Some years ago, in early molecular biology, it was believed to be a very stringent rule that information always proceeds in one direction: from DNA to RNA. After retrovirus were isolated, the opposite way, the exception from this stringent rule was established. In science it is well known that application is always a spin-off from basic science and there is no way back from the applications to basic science. This was also true to some extend for the development of particle radiotherapy as a practical application of radiobiology, physics and accelerator technology. Without the basic research in these fields, particle radiotherapy never would have happened. But as in the case of the DNA pathways the science transfer starts now in the opposite direction and the application of radiotherapy will intensify the basic research in cellular repair and signal transduction as well in the physical aspects of track formation and nuclear fragmentation.

In the following sections the fundamentals of radiotherapy in the different disciplines are briefly reported and the clinical realisation is outlined together with the possible development of particle therapy in Europe. Finally the radiobiological research presently carried out for particle therapy and its possible future development will be discussed.

Radiobiological fundamentals of heavy ion therapy

In 1946 R. Wilson measured ionisation profiles of protons from the newly constructed cyclotron at Berkeley and found that the energy deposition, i.e. the dose, was increasing with penetration depth up to a maximum at the end of the particle range, the Bragg maximum named after W. Bragg who had reported in 1903 the same behavior for the dose profiles of a particles emitted from radioactive sources. Wilson recognized the potential of the "inverse dose profile" for radiotherapy in comparison to the exponential decay with penetration depth of other electromagnetic radiation like photons in electron bremsstrahlung and published his results in a medical journal [1]. But it took nearly one decade before the first patient was treated with proton beams at Berkeley by J. Lawrence and C.A. Tobias [2]. Tobias later on recognized the importance of ion beams heavier than protons like neon where the relative biological effectiveness RBE varies with particle energy i.e. with beam penetration. For the light ions the maximum RBE values were at the end of the range that gave the higher doses of the Bragg maximum an additional biological impact for tumor cell inactivation.

At this time the RBE values for light ions were known from the classical experiments of Barendsen, Skarsgard and Todd [3, 4] but for therapy the RBE had to be evaluated systematically for a larger set of ions from carbon to argon [5].

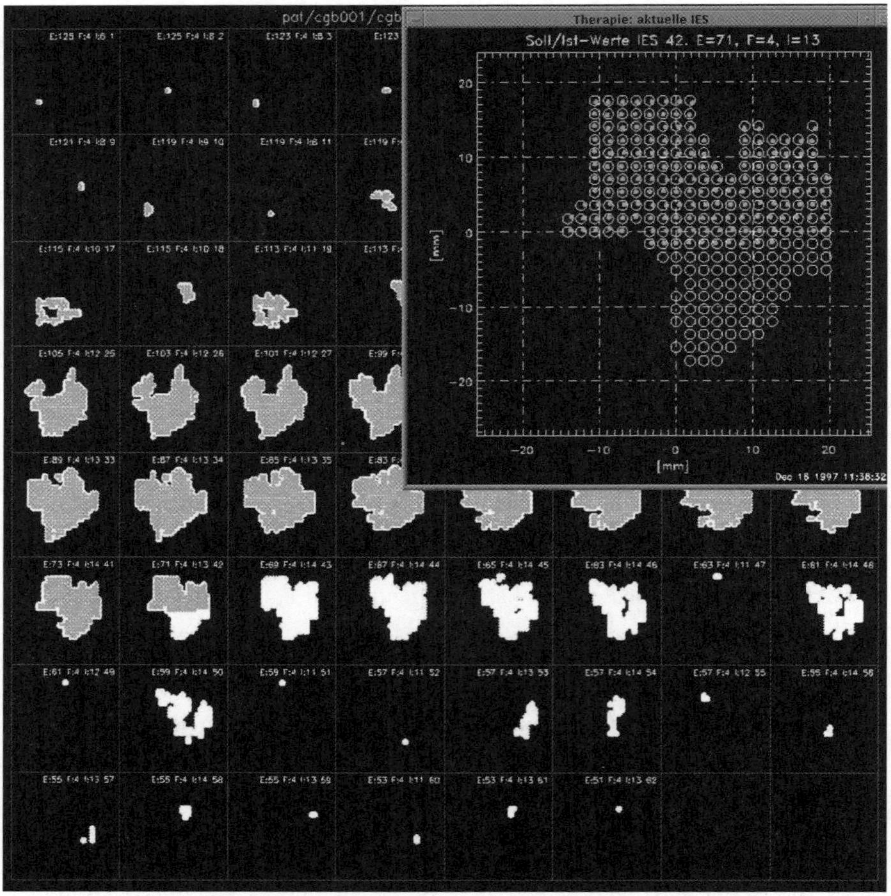

Fig. 1. Compilation of the different range slices of one treatment volume. In each panel one isoenergy-slice is shown. In the magnified panel the circles represent the calculated center positions of the beam that are filled with the measured centerpositions. The diameter of the beam is larger than the distance between three positions, yielding a large overlap and a great homogeneity.

In these measurements it became evident that the RBE depends not only on the dose with the tendency for high doses to come close to one, but RBE depends also on the energy or LET and the atomic number and probably more importantly on the repair capacity of the cells or tissue. Cells, radioresistant to X-rays as expressed in a large shoulder in the survival curves did also show larger RBE values when exposed to heavy ions. However, the difference of RBE between different cells was measured mostly with immortalized cell lines where the X-ray dose response curves differ only a little. Consequently the energy dependence of RBE as measured in vitro was believed to be universal and taken for the design of field modulating equipment like the range shifters that allowed to spread out the Bragg peak in depth. Using this approach for the range shifters, static dose distributions could be produced that were superior to the X-ray fields at that time used in the clinical practice.

In the following years it became more and more evident that a better conformity of the treatment field to the target volume is expected to produce better clinical results. For photons the intensity modulated radiation therapy (IMRT) produced very conform fields by choosing many ports with beam collimation and intensity modulation as requested by the projection of the target volume to the source and finally distributing the not wanted entrance dose over the many entrance channels [6]. In particle therapy the intensity modulated voxel- or rasterscanning was introduced into the practice of patient treatment at PSI, Villigen for protons [7] and at GSI, Darmstadt for carbon ions [8]. This new technique produces a perfect congruence between the treated and the planned volume by dividing the treatment volume into small volume elements (voxels), that are treated separately with a individually pre-calculated number of particles in order to achieve the homogenous effect wanted over the complete treatment volume (see Fig. 1., page 212).

This was an extreme challenge for the understanding and calculation of RBE: The dose in each of the 10 000 to 30 000 voxels in a treatment field of a patient is made up of particles of different energies and atomic numbers. Consequently individual RBE values have to be assigned to each voxel. These RBE values vary not only in depth but in all three dimensions of the treatment field. For such a complex distribution a deeper understanding of all dependencies of RBE on the various parameters was necessary to calculate point- or voxelwise the RBE.

Model calculation of the relative biological effectiveness

The basic assumptions of RBE models changed in parallel to the development of a deeper scientific understanding of the RBE. First RBE was assumed to be an universal factor, depending only on the particle' LET. Later on the repair dependences as the main biological variable was established, too.

In the beginning there were only physical and statistical considerations like in target theory or in microdosimetry yielding a strict physical explanation of the coefficients α, β of the linear quadratic approach of cell survival by the formula [9, 10, 11]:

$$S = S_0 \exp - \{\alpha D + \beta D^2\}$$

Newer models tried to incorporate the biology in two different ways: In the first type the attempt is made to define the pathway of the lesion over repair and non-repair channels and to calculate survival curves for both, sparsely and densely ionizing radiation from basic assumptions on the probabilities of these pathways [12, 13].

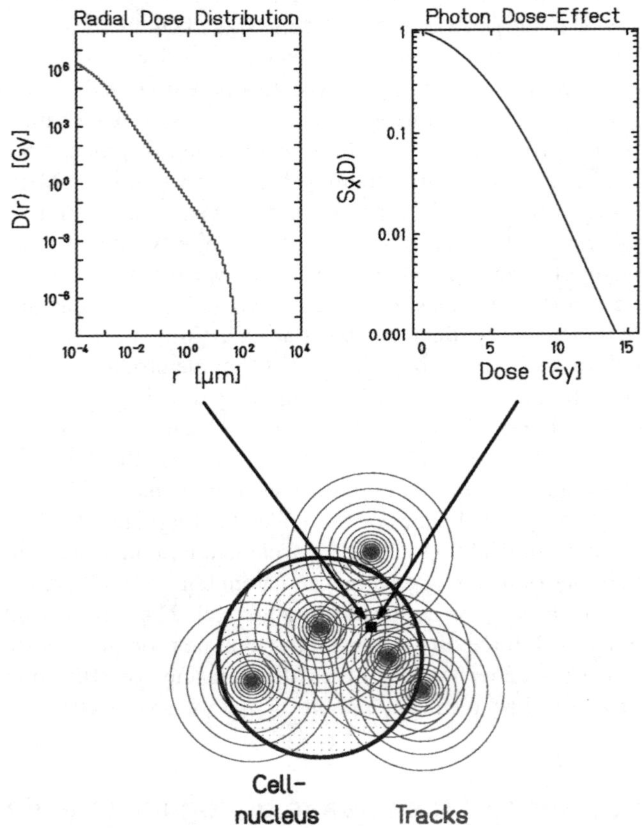

Fig. 2. Principle of the local effect model: The radial dose distribution around a track (top left) is divided into zones of nearly equal dose. The probability to produce a lethal lesion in each compartment in the cell nucleus is calculated according to the X-ray dose effect curve (top, right)

In contrast, in the so-called track structure models the RBE of densely ionizing radiation is calculated using the dose effect curve for sparsely ionizing radiation as a repair containing template for the calculation without trying to understand X-ray dose effect curves. The basic principle of the track structure models is to calculate the dose deposited in the critical target of the biological object, according to the radial dose distribution and to compare this dose in its efficiency to the corresponding X-ray dose effect curve. In its most sophisticated version, the Local Effect Model LEM [14] this comparison is performed for small subunits of the cell nucleus as critical targets (see Fig. 2.). This means that for the cell nucleus a non-linear X-ray dose response curve is folded point by point with a non-linear dose distribution inside individual particle tracks to calculate the inactivation probabilities. Using Monte Carlo calculations the fluence dependence of the inactivation effect can be calculated even for fields of non identical particles as they occur in tumor therapy.

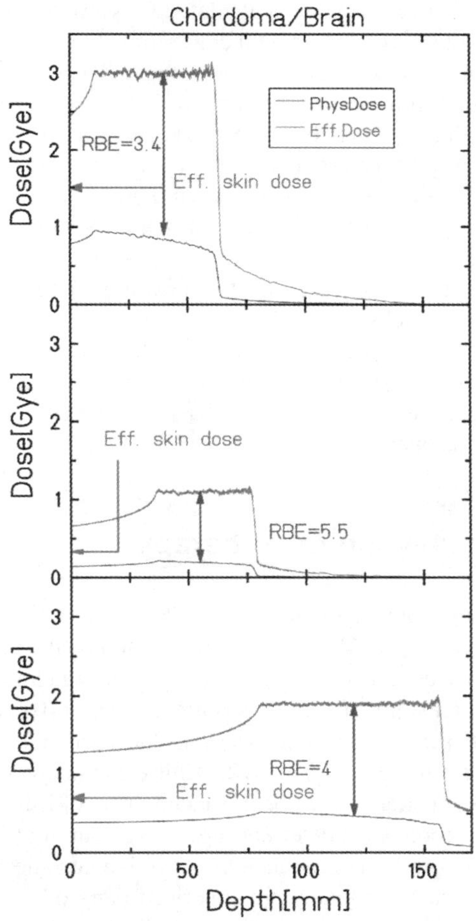

Fig. 3. Comparison of the distribution of the physical dose with the biological effective dose for three chordoma patients treated at GSI. The RBE values differ between 3–5 because of the location and size of the tumors and the different dose levels.

The results of cell and animal experiments, as well as the success of the carbon therapy at GSI confirm the LEM model. In Fig. 3., page 215 the physical dose and the biological effective dose given to chordoma patients are shown. According to the depth, the width and the dose level the RBE changes for these tumors although they are of the same histology. The RBE values in the tumor center vary between three and five, indicating the large importance of RBE. The correctness of the LEM calculation is indicated by the fact that no patient did develop necrosis in the treatment field because of over-dosage nor a recurrent tumor because of under-dosage.

This type of model calculation can also be extended to other endpoints than cell inactivation. When the X-ray effect curve of late effects in the normal tissue is known then their incidence can alsobe calculated for particles. This has been tested successfully for the tolerance of the spinal cord of rats [15]. The principles of LEM have been also applied successfully to non biological material like photographic emulsions or thermolumminescent detectors [16, 17]. In this non biological detectors no efficiency values greater than one exists because of the pure linear X-ray dose dependence. But the decrease in efficiency for the higher LET values towards the end of the tracks or for very heavy ion is reproduced very well.

In all of these examples LEM worked very well and explained the variation of the efficiency for biological and non biological material, simply as the effect of a non linear dose response curve when exposed to an extreme inhomogeneous dose field as produced by particle tracks.

Although LEM is able to predict RBE values with an accuracy needed for therapy one has to be aware that LEM is only a reduction of the particle RBE problem to the non linear dose effect curves for X-rays. The basic question of DNA repair after ionizing radiation remains open.

Status of the heavy ion radiotherapy

The majority of ion beams patients have been treated with protons. There only the ballistic properties of ions, the inverse depth dose profile, represents the major advantage. But protons do not have a significantly elevated clinical RBE and radiobiological considerations are of minor importance. The effects of high RBE for tumor killing are only present for the heavier ions like carbon. Presently tumor therapy with carbon ions is in operation at NIRS Chiba, Japan and GSI, Darmstadt. At NIRS more than 1000 patients have been treated, at GSI 150 [18]. In both cases tumor control is very good and side effects are lower than in other therapy modalities when a better conformity is produced by an active dose delivery systems. The clinical results of the carbon therapy are superior to any other modality [19]. For instance at GSI a three year tumor control rate of 100% has been achieved for chordoma tumors, a rare but very radioresistant tumor with an expected control rate of approx. 40% with conventional therapy. Also for other tumors, a superior response has been found for carbon therapy [19].

This extreme positive results of the carbon therapy have triggered a strong interest of the European and Japanese radiotherapy community in heavy ion radiotherapy. At Heidelberg the radiological clinic that is operating the GSI pilot project managed to allocate the money for a dedicated carbon-proton therapy unit to be build and operational for patient treatment in 2006. In Italy, the TERA project has recently received half of the totally required money to construct a proton-carbon therapy unit at Pavia in Northern Italy. The same happens for the Austron project at the Wiener Neustadt, where the money for the construction of a proton-carbon therapy unit has been allocated by the state and federal governments. It is expected that these facilities will be operational one year after the Heidelberg facility. Other activities in Lyon, France and Stockholm, Sweden did not yet reach this critical level of an approved funding but are supported by governmental funds to prepare heavy ion therapy [18].

In addition to these more research triggered activities, the European industry developed some interest in particle therapy. IBA, a Belgian company originating from the Louvain radiotherapy is selling proton facilities as well as ACCEL, a German company near Cologne. ACCEL recently made a feasibility study for proton-ion therapy complex for Siemens and the Rhön-Klinikum, one of the largest German private hospital operators, for a modular system that could be build in a larger scale and adapted to the special needs of the local clinics. In Japan after the positive results from Chiba, the Hyogo facility went into operation with a proton-carbon machine and a new facility is discussed at Gumna University (Fig. 4.).

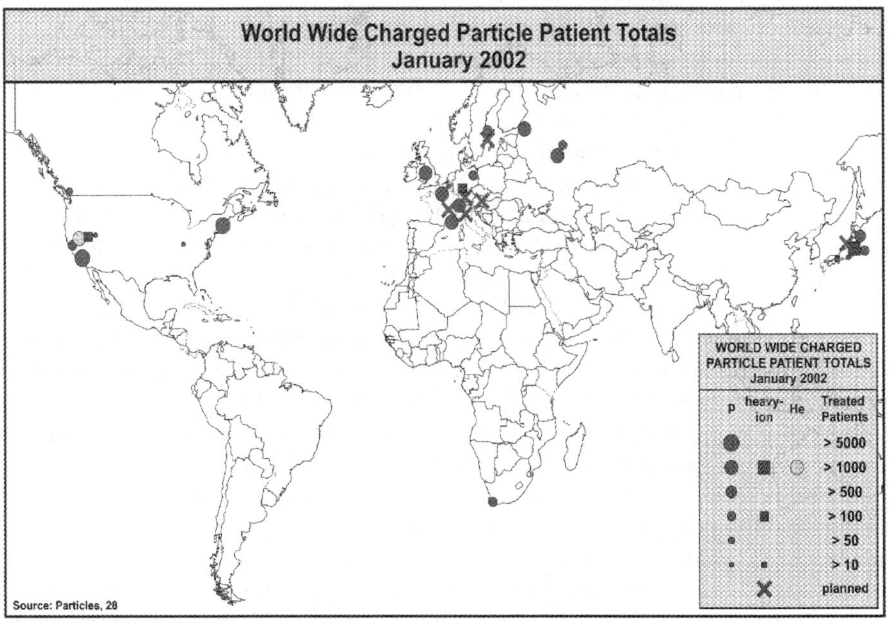

Fig. 4. Worldwide patient treatment including plans for heavy ion centers.

Although only three heavy ion therapies are operating world wide, the number of carbon facilities seem to be increasing rapidly. This evolution will also trigger research for a deeper understanding of the basic radiobiological processes.

Radiobiology triggered by the therapy

Track structure models fit most properly the needs of conforming ion therapy where local RBE values are needed for mixed fields. But track structure models, especially the LEM, are based on the silent assumption that the local ionisation density and the consequent local damage is not washed out but stays local even when viewed in a sub micrometer scale. This is not only in some contradiction to the early ideas of Neary that an interaction inside the cell nucleus over a range of 1 μm should take place and the consequent assumption of a 1 μm^3 critical target size in microdosimetry [10]. It is also in some contrast to the recently found bystander effect i.e. the fact that cells that are not hit at all by a particle can be inactivated when cells nearby are hit [20].

On the other hand in recent experiments using immunofluorescent methods the very local activity of proteins involved in DNA repair confirmed the high localisation of the DNA damage in the cell nucleus (Fig. 5.). For instance, immediately after particle irradiation very local foci of p21 concentration can be detected. In separate experiments it was demonstrated that the location of these foci correlate precisely with the location of the heavy ion traversals [21]. It could be also shown that different repair proteins are found at the same locations of track traversal that occur much later [22]. From these experiment it is quite evident that the non homogeneous dose distribution in a particle track is directly translated into hotspots of DNA damage. Local DNA damage is not distributed over a larger part of the DNA matrix. From other experiments a wash out effect of the dose on a much smaller scale in the very center of the track seems to be necessary to reproduce the observed DNA damage. A deviation of the strict localisation was found in an application of LEM type calculations to a large set of data describing the induction of double and single stand breaks in SV40 DNA by a representative set of ions between He and U and energies between 3 and 1000 MeV/u [23]. The calculation did only yield a good agreement with the measurement when the extremely high dose in the center of the track was homogenised over a distance of approx. 100 nm. But this could be due to the specific situation of the exposure of the SV40 in solution but it could also indicate a general feature of a small washout effect in ion tracks. However, it cannot be expanded to produce the long range effects, the bystander effect, described before.

However, it is possible to assume that in the very early state of electron transport and radical formation a small diffusion effect of the primary energy deposition takes place. In the center of the track then the dose contours are stabile after these early time shorter than 10 sec after the particle hit. Later on, in the course of the DNA repair process diffusion of inactivation potential in form of cytokines may occur on a very much larger time scale than the primary effects. However, such proposals are pure speculations without any experimental evidence. The time course of the inter and intra cellular signalling process and the sequence of the different repair enzymes are not known in such a detail after particle damage. The same is true for the primary process of electron emission and transport in condensed matter.

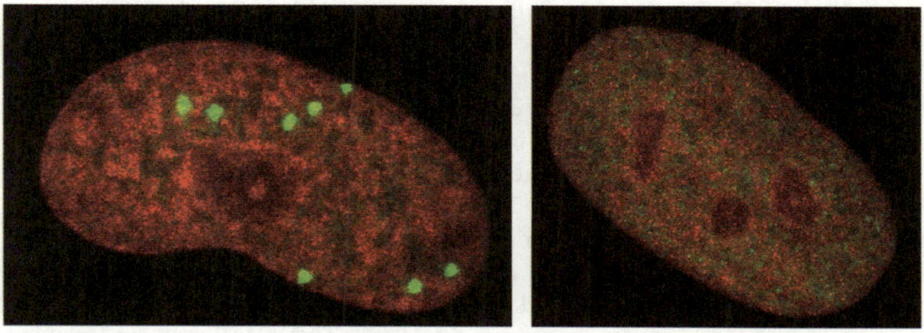

Fig. 5. Activity of p21 a protein involved in the DNA repair left after exposure, right after hits of heavy ions. The DNA is stained in red, the 21-protein in yellow. The local activity of the damaged spots after particle exposure are evident [21].

Fig. 6. Chromosomal translocations in a human-hamster hybrid cell. The human chromosome (in yellow) had least two breaks and exchanged parts of the chromatin with a hamster chromosome.

Although our present knowledge is sufficient to calculated a reasonable and obviously correct RBE for tumor therapy, the underlying general process of DNA damage caused by ionizing radiation and its repair is not understood. Particle beams, especially micro-beams offer the great potential of an intensified research in this field. With micro-beam facilities single ions can be directed with an accuracy of 1 μm at best on single cells and cellular compartments like the nucleus. Using ions of different atomic numbers and to some extend of different energies the intercellular communication, the production and the diffusion of cytokines could be studied in detail. This would be one of the great challenges in consequence of the increased application of ion beams to humans in therapy.

Furthermore the physical processes involved in radiation action and in therapy have to be studied in greater detail. The nuclear aspect is the beam fragmentation and there the neutron production. First measurements are under way and confirmed the large probability of the created neutron to escape the near interaction in the patient body. Therefore it is expected that this will be more a measurement for completeness rather than for new innovative directions. This is different for the other physical aspect, the electron emission in solids. Electron emission is the fundamental process for track formation and important not only for the biological response but also for all processes of track formation like material modification and material processing. First experiments concerning electron emission at low energies indicated interesting differences between gas and solids, that will be now studied in greater detail [24].

Another and very serious problem is the genetic action of densely ionizing radiation. The most frequent exposure to ionizing radiation is caused by medical diagnosis and therapy which use mostly sparsely ionizing radiation like X-rays and other photons. But the densely ionizing radiation like neutrons, alpha particles and heavy ions become also an important component of radiation exposure. With the increasing use of heavy particles in therapy there is also normal tissue that will be exposed to a low level of carbon ions. Neutrons represent the main exposure during the operation of nuclear power plants and the long living α-emitters are the major risk component in the waste disposal and storage over many thousand years. Heavy ions are a major component of the cosmic radiation and consequently of concern for all manned space flights outside the magnetic shielding of the earth.

In all these cases the dose is low and cell inactivation is of minor importance for the risk analysis. There, the risk is determined by the possibility of mutations of the DNA code to produce cancer in the exposed person or to produce long term genetic changes in the following generations. For these questions only a very limited answer can be given for sparsely ionizing radiation. For heavy particle there is no sound data base at all. From cell experiments we know that particles produce a significantly different genetic response from X-rays. First there is a strong interference with the cell cycle for the early effects like the genetic alterations indicated by chromosomal aberrations. Cells that are hit by particles are mostly retarded in the expression of the chromosomal damage. Scoring these cells in the experiment at the early time that is used in X-ray experiments most of the particle induced damage will be missed and the particle efficiency will be completely underestimated [27]. However, when in a cell culture experiment the chromosomal damage is scored over a long period and also the late occurring aberration events are monitored it seems to be very premature to apply these data to risk estimation for humans. First experiments indicated that there is already a significant difference between partially immortalized cell lines and primary cells. Immortalized cells show a delayed but large amount of chromosome damage after particle exposure. When primary cells like fibroblast are hit by particles only a very minor fraction of cells like 5% proceed to the next mitosis at all. The majority of more than 90% directly differentiate and undergo an premature aging process yielding post mitotic phases. Thus the majority of cells is prevented from cycling and will not contribute to the radiation risk.

As primary cells are closer to the situation of a total body exposure one has to question all the results from tissue culture of immortalized cells. But also primary cells are a simple model, that will not give a firm answer.

A correct answer to the question of the genetic changes in man after ion exposure can be only given from the observation of their incidence in humans. The growing number of cancer patients treated with ion beams represents a collective of people where we know the location, the dose and the fractionation of the exposure very well. However, the source of information from these patients is limited to the analysis of blood lymphocytes that can be easily and frequently obtained. First explorative tests showed that an increased aberration rate can be observed in these patients. But the comparison between different patients or at different time points in the same patient showed that the aberration rate depends on many more parameters than just particle fluence and LET. There seems to be a strong influence of biological parameters that have to be identified before their influence can be understood in molecular studies of the mechanisms involved. Consequently, the research on the genetic changes after heavy charged particle exposure has to be done on two different levels. First the possible influence of global parameters on the physical and biological level have to be performed and secondly detailed studies to understand their mechanisms on a molecular level. The systematic study of these parameters will be very important for a safe prediction of the genetic consequences of exposure to densely ionizing radiation. The combination of in vitro data with the in vivo data of the patient analysis is a very unique situation and will help to verify the models of the exposure risk. In addition these type of experiments could be also helpful to understand the radiation sensitivity of the patients of ion beam therapy and will help to optimize their treatment.

Summary

Radiobiological research combined with new delivery technologies of heavy ion beams have produced carbon therapy as an ultimate tool for external radiotherapy of deep seated, radioresistent tumors that are not operable. Local tumor control rates superior to any other irradiation techniques have been achieved and many new projects for heavy ion therapy all over Europe and in Asia are underway. As a consequence of this success, the radiobiological research will be intensified to understand cell inactivation process in greater detail on a molecular level. This deeper understanding includes also a better knowledge of the early physical processes like the electron emission in condensed phase that finally yields the extreme dose variation inside particle tracks.

In the case of the understanding of the genetic action of particles we have not yet reached this level of knowledge. The genetic effects play an important role of the long term effects induced in people that are exposed to low levels of densely ionizing radiation. These are in practice the exposure to neutrons, α-particles, ions and cosmic radiation that are of concern to a more and more growing fraction of our population.

For the risk analysis of these people, research on two levels is needed. First a general "screening" of the parameters that dominate the probability of cancer induction as well as that of genetic mutations has to be performed. Secondly mechanistic studies to understand the biological action are necessary.

Both these levels are of great importance but as we have now the new tools of molecular biology, the mechanistic research is frequently rated too high. In this situation it might be useful to remember that in the evolution of the human culture the experience always proceeded the understanding of mechanisms:

Greek philosophers like Sokrates had great symposia in the original meaning of the word symposium as drinking a lot of well made wine about 2300 years [28] before the mechanistic aspect, the biochemistry of wine making, was explained by Louis Pasteur.

Acknowledgment

I would like to thank the organizers of the Giessen workshop, Prof. J. Kiefer and Dr. E. Gelfort for their invitation and their encouragement to fight for a renewal of radiobiology in Germany.

In addition I would like to thank my colleagues at GSI for material and helpful discussions and Angela M. Phalen-Weiß and Marcus Winter for the preparation and editing of the manuscript.

References

1 Wilson RR (1946), Radiological Uso of Fast Protons, Radiology 47:487–491

2 Tobias CA, Alpen EA, Blakely EA, Castro, Chatterjee A, Chen GTY, Curtis SB, Howard J, Lyman JT, Ngo FQH (1979) Radiobiological Basis for Heavy-Ion Therapy. In: Volkmer B, Heller H (eds) Treatment of Radioresistant Cancers, Abe M, Sakamoto K, Philips TJ, eds., Elsevier, 159

3 Barendsen GW, Walter HM, Fowler JP, Bewley DD (1963) Effects of Different Ionizing Radiation on Human Cells in Tissue Cultures, Radiat Res 15:106–119
4 Todd PW(1965) Heavy-Ion Irradiation of Human and Chinese-Hamster Cells, Radiat. Res 61:288–296
5 Blakely EA, Tobias CA, Ngo FQH, Curtis SB (1980) Physical and Cellular Radiobiological Properties of Heavy Ions in Relation to Cancer Therapy Application. In: Tobias CA (ed) Biological and Medical Research with Accelerated Heavy Ions at the Bevalac. Pirncello MD, LBL 11220:73–88
6 Webb S (1993) The Physics of Three Dimensional Radiation Therapy, Inst. of Physics Publ, Bristol, Philadelphia
7 Pedroni E, Blattmann H, Böhringer T, Coray A, Lin S, Scheib S, Schneider U (1991) Voxel Scanning for Proton Therapy. In: Itano A, Kanai T (eds) Proceedings of the NIRS, Int. Workshop on Heavy Charged Particle Therapy and Related Subjects, July 4–5, Anagawa, Japan NIRS-M-81:94–109
8 Haberer Th, Becher W, Schardt D, Kraft G (1993) Magnetic Scanning System for Heavy Ion Therapy. Nucl Instr Meth in Phys Res A 330:296–305
9 Lea DE (1956) Actions of Radiation on Living Cells, Cambridge University Press
10 Kellerer AM, Rossi HH (1978) A Generalized Formalism of the Dual Radiation Action. Rad Res 75:471–488
11 Chadwick KH, Leenhouts HP (1981) The Molecular Theory of Radiation Biology. Springer Verlag, Berlin
12 Tobias CA, Blakely EA, Ngo FQH, Yang TCH (1980) The Repair-Misrepair Model of Cell Survival, Radiat. Biology and Cancer Res. Meyn RA, Withers HR, eds., Raven Press, New York, 195–230
13 Kappos A, Pohlit W (1972) A Cybernetic Model for Radiation Reactions in Living Cells. I. Sparsely Ionizing Radiations: Stationary Cells. Int J Radiat Biol 22:51–65
14 Scholz M, Kraft G (1994) Calculation of Heavy Ion Inactivation Probabilities Based on Track Structure, X-ray sensitivity and target size. Radiat Prot Dosimetry 52:29–33
15 Karger CP, Debus J, Peschke P, Münter MW, Heiland S, Hartmann GH (2002) Dose-Response Curves for Late Functional Changes in the Normal Rat Brain after Single Carbon-Ion Doses Evaluated by Magnetic Resonance Imaging: Influence of Follow-up Time and Calculation of Relative Biological Effectiveness. Radiat Res:545–555
16 Spielberger B, Scholz M, Krämer M, Kraft G (2001) Experimental Investigations of the Response of Films to Heavy Ion Irradiation. Phys Med Biol 46:2889–2897
17 O.B. Geiss, M. Krämer and G. Kraft (1998), Efficiency of thermoluminescent detectors to heavy-charged particles. Nucl Instr Meth B142:592–8
18 Sisterson J (2002) Particles Newsletter, 30:20
19 Schulz-Ertner D, Nikoghosyan A, Thilmann C, Haberer T, Jäkel O, Karger C, Scholz M, Kraft G, Wannenmacher M, Debus J (2003) Carbon Ion Radiotherapy for Chordomas and low grade chondrosarcomas of the skull base: Results in 67 Patients. subm. to Strahlenther Onkol
20 Österreicher J, Prise KM, Michael BD, Vogt J, Butz T, Tanner JM (2003) Radiation-Induced Bystander Effects, Mechanisms. Biological Implications and Current Investigations at the Leipzig LIPSION Facility, Strahlenther Onkol 2:69–77
21 Scholz M, Jakob B, Taucher-Scholz G (2001) Direct Evidence for the Spatial Correlation between Individual Particle Traversals and Localized CDKN1A (p21) Response Induced by High-LET Radiation. Rad Res 156:558–563

22 Jakob B, Scholz M, Taucher-Scholz G (2000) Immediate Localized CDKN1A (p21) Radiation Response after Damage Produced by Heavy-Ion Tracks. Rad Res 154:398–405
23 Brons S, Taucher-Scholz G, Scholz M, Kraft G (2003) A track structure model for simulation of strand breaks in plasmid DNA after heavy ion irradiation. Radiat Envirn Biophys in press (DOI: 10.1007/500411-003-0184-9)
24 Zäpfel T, Hagmann S, Rothard H, Ullrich J, Kraft G, Schmidt-Böcking H, Groeneveld KO (2002) Experimental study of electron ejection by heavy ion irradiation of solids: Observation of forward and backward emitted electron jets. Meth.in Phys B 193:651–655
25 G. Kraft (2000), Tumor therapy with heavy charged particles, Progr. Part. Nucl. Phys., 45, 473–544
26 M. Krämer, M. Scholz (2000), Treatment planning for heavy ion radiotherapy: calculation and optimization of the biologically effective dose, Phys. Med. Biol., 45, 3319–3330
27 M. Scholz, S. Ritter, G. Kraft (1998), Analysis of chromosome damage based on the time-course of aberrations, Int. J. Radiat. Biol., 74, 325–331
28 Platon, Symposium (380 BC) Athens

Targeting Radiation at the Subcellular, Cellular and Tissue Levels: Future Strategies

Kevin M. Prise, M. Folkard, Boris Vojnovic and B.D. Michael

Gray Cancer Institute, PO Box 100, Mount Vernon Hospital, Northwood, HA6 2JR, UK
prise@gci.ac.uk

Abstract.

In the last 10 years there has been a sea-change in the knowledge of how radiation interacts with biological systems. Much of this has questioned the DNA centric approach which has been a common feature of not just radiation biology but molecular and cell biology studies in general. With the realization that global gene expression changes in cell populations are not fully predictive of cell function the new emphasis is on single cell processes and proteomics and cell to cell communication. The recent ability to target radiation very precisely to individual cells or subcellular locations is providing a unique opportunity to dissect processes, not only to give a better understanding of radiation effects but also to use radiation as a probe for stress-induced processes.

Several groups worldwide have commissioned or are developing microbeams to allow targeted studies with charged particles, X-rays and electrons. Our own experience with both a charged particle microbeam and a focused soft X-ray microprobe has allowed us to develop novel approaches to studying low dose effects in cell and tissue models. Studies performed with these approaches allow us to outline future strategies for how these technologies will impact on radiation research. Key to this is a need to encourage other areas of research to interface with radiation biology and to learn from the multidisciplinary approach taken to understanding radiation action.

Introduction

Since the elucidation of DNA as the genetic material of a cell in 1953 by Crick and Watson a wealth of studies have elucidated the central role of this molecule in the propagation of individual cells and the precise mechanisms which have evolved for its maintenance and repair. A central tenet of the study of cell effects has been that direct damage to DNA was caused by genotoxic agents and this led to serious biological consequences if the integrity of the genome was not maintained. Ionizing radiation interacts with biological systems via stochastic processes dependent on the track structure characteristics of the type of radiation. This essentially separates the mode of action of radiation from most other genotoxic agents and provides unique challenges in determining its effects and understanding mechanisms of its action. Although much is known regarding mechanisms of radiation action at the subcellular, cellular and tissue levels, much of this has been obtained by studies using relatively high doses or dose-rates which are not relevant to many of the situations where radiation exposure occurs in the human population. A key change has been to develop approaches to specifically allow low doses and dose-rates to be studied. Also the ability to locally direct radiation to cells and targets of interest is necessary to determine key mechanisms and damage signalling process in biological systems when they respond to radiation exposure. The development of radiation microbeams has been a major advance in achieving these goals and the types of studies performed with these to date will allow us to outline future strategies for how these technologies will impact on radiation research.

Development of microbeam facilities world-wide

The first development of a microbeam for cell irradiation was in 1912 by Tchahotine (reviewed by [1]). He described a technique for focusing, using quartz lenses and objectives, a 280nm UV beam down to around a few microns in diameter. These studies have evolved over the years to the sophisticated laser based UV microbeams in use to day. For ionizing radiations, Seed and colleagues in 1960 developed an approach for the focusing of an X-ray beam for biological studies [2]. Zirkle and others concentrated on the use of particle irradiation approaches (reviewed in [1]). These utilized the limited range of charged particles as a means of titrating effect throughout a cell population (macrobeams) or the ability to precisely irradiate regions of an individual cell using, for example, polonium-tipped micro needles [3]. Similar approaches have been used with partially penetrating low energy electrons [4]. In recent times Braby [5] developed collimation approaches which allowed the development of systems which could both target cells individually with counted particles and allow these to be delivered to within a few microns of the required target. The work of Braby has fuelled an explosion in interest in the development and application of microbeams for radiation biology studies. This has coincided with tremendous strides in computing, imaging, electronics, radiation delivery and detection. In parallel, the ability to monitor cellular processes on an individual cell basis, coupled to sensitive microscopy techniques has allowed tremendous strides to be made. Currently at least 18 systems are being developed worldwide (see Table 1.) with 5–6 of these currently in use for cell irradiations. These include not just charged particle microbeams, but focused X-ray beams and more recently, electron microbeams.

Essentially, a microbeam is a means of localizing a precise radiation dose to a specific target. It should be noted however, that several other useful approaches have been used for targeting radiation to individual cells are possible. One such approach is the Biostack technique where cells are plated on a material which can record the traversal of a charged particle. The locations of the attached cells are recorded and then later related to the location of a population of cells present at the time of irradiation when exposed to low fluences of conventionally delivered charged particles. Mostly, track etch plastic is used [6–8], which can be etched after irradiation. Although this approach does not allow individual cells or locations to be actively targeted, it does allow low fluence studies to be performed with good reproducibility. To date these have mainly measured cell killing after exposure with charged particles and allowed the effect of single particles to be unfolded from the Poisson distribution.

Table 1. Current microbeams in use or under development world-wide.

Charged Particle		Soft X-rays	Electrons
Gray Cancer Institute, London, UK	Texas A&M, USA	Gray Cancer Institute, London, UK	Texas A&M University, USA
Columbia University, New York, USA	MIT, Boston, USA	Lawrence Berkley National Laboratory, USA	Pacific Northwest Laboratories, Richland, USA
JAERI, Takasaki, Japan	LNL-INFN, Padua, Italy		
PTB Braunschweig, Germany	Univ. of Leipzig, Germany	Photon Factory, Tsukuba Japan	
GSI, Darmstadt, Germany	Univ. of Munich, Germany		
NIRS, Chiba, Japan	Univ. of Hefei, China		
CENBG, Bordeaux, France			

For subcellular targeting, radioisotopes attached to specific biomolecules have long been used. For example, the classical studies of Warters and colleagues comparing incorporation of I^{125}deoxyuridine incorporated into DNA versus I^{125} Concanavalin A bound to the cell membrane [9].

Several approaches have utilized grids or masks to restrict dose delivery to areas or regions of cells or multicellular systems. For example, Nelms *et al.* [10] used a mask system for localizing soft X-rays to a striped pattern which could be applied to cellular irradiations. These consisted of a striped gold mask with 1.35 μm exposed and 1.85 μm shielded regions. When human cells were exposed to high doses of 1.34keV X-rays delivered through the mask and then fixed and stained for BUdR uptake, and the repair protein mre11, a co-localized striped pattern of staining within the cell nucleus was observed. This was an important observation of the localization of repair activities to sites of localized damage and is an important tool which can also be utilized by the modern generation of microbeams to follow DNA damage and repair processes at the individual cell level.

Also, a grid made from a gold mesh has been used to restrict dose delivery to fixed percentages of cells present in a culture dish, Using this approach, Lorimore *et al.* [11], compared the effect of irradiating 50 or 100% of mouse haemopoietic stem cells with α–particles from a ^{238}Pu source. They observed an important link between radiation-induced genomic instability and bystander responses using this approach.

Localization and delivery of UV-radiation to individual cells is relatively straightforward using lens [12], lasers [13] or even approaches with membrane filters with defined pore sizes [14]. These approaches have been utilized with a high degree of success to probe the sequence of repair recruitment to the sites of UV-damage [15] and to probe chromosomal organization within cells [16].

Experimental approaches with microbeams

For radiation biology, several types of approaches are possible with microbeams which can be useful tools in understanding basic mechanisms of exposure as summarized in Fig. 1. These four basic approaches are discussed in more detail as follows:

1 The modern generation of charged particle microbeams can deliver a single particle with 100% efficiency, allowing the effects of low doses to be carefully studied. This is an important advance to previous studies where the best that could be obtained was to deliver only an average of one particle traversal to a cell population. Under these conditions, 37% received no particles, 37% a single particle and 26% more than one particle. Although it was assumed that the effect of a single particle could be unfolded from the Poisson distribution, the discovery of bystander responses has shown that this approach is fundamentally flawed. Particle detection systems currently available either use post-cell detection via gas proportional counters (see for example, [17]) or pre-cell detection via scintillation plastic and photo-multiplier tubes (see for example [18]).

2 The radiation can be delivered to precise locations, using either a collimation system or a focusing system, via electrostatic lens (charged particles) or diffraction optical devices (X-rays). Typically for the current generation of particle microbeam systems using collimation, resolution at the cell is around 1–2 μm [19]. Newer electrostatic focusing devices may improve on this, but a fundamental limitation is the scattering inherent to all particles when they interact with matter. For soft X-rays however, resolutions below 1μm are easily achievable and scattering is not a problem [20].

3 Individual cells can be targeted within populations using the ability to localization the dose, allowing cell to cell communication processes to be studied relative to intracellular signalling processes. This is easily achievable using the current generation of microbeams, even without the use of sophisticated cell finding and

revisiting systems.

4 In solid tissues, 3-D targeting approaches are possible using the restricted ranges available with charged particles. For example localized areas of a tissue section can be targeted and the depth of energy deposition into the tissue carefully controlled.

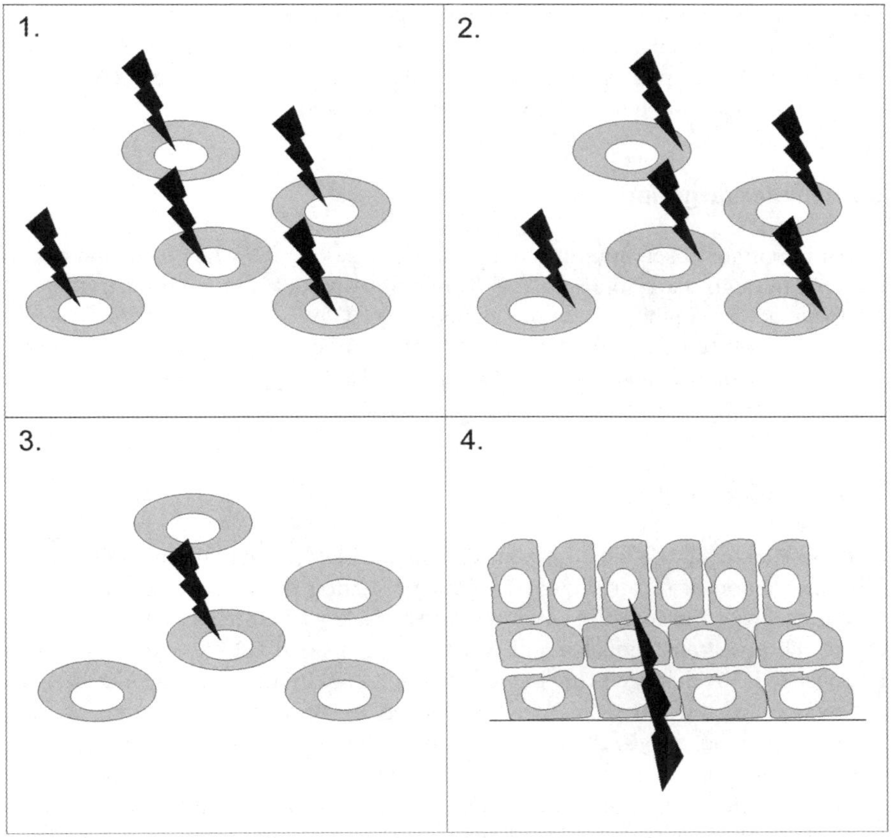

Fig. 1. Schematic of the main types of experiment which can be performed with the current generation of microbeams; (1) single particle effects, (2) subcellular targeting, (3) single cell studies and (4) tissue studies.

These features give microbeams a tremendous flexibility for the performance of radiation studies in a range of biological models and are a major advance. Coupled with new molecular and cell biology techniques, these approaches are being rapidly utilized by the expanding numbers of microbeam sources. In general they are in 4 main areas; low dose studies, signalling mechanisms, DNA damage studies and studies of non-targeted responses.

Low dose studies

From the approaches that can be used with microbeams, a range of effects have been reported in various, predominantly cellular, irradiation studies. Most studies have concentrated on the effect of low doses of relevance to radiation protection by following the effects of low numbers of charged particles to assess radiation risk [21]. Typically, for helium ions these still equate to relatively high doses (0.2–0.5 Gy) depending on cell thickness at the time of exposure. However some studies have utilized low energy protons and have measured the effects of a single 3.2 MeV proton delivering a dose to the cell nucleus of 20 mGy [22].

Subcellular targeting

An important observation has been the response of cells when these have been targeted through the cytoplasm [23]. Targeting α-particles through the cytoplasm led to an increase in the yield of mutations at the $CD59^+$ locus in an "exotic" human hamster hybrid cell line. These mutations were similar in their molecular characteristics to spontaneous mutations and they appeared to be a product of reactive oxygen species (ROS).

DNA damage studies

In parallel to studies which have used localized UV irradiation to monitor localized association of repair proteins, ionizing radiation microbeams are now being used in this way. For ionizing radiation exposure a key advance has been the discovery of the post-translational modification of the histone H2AX, which occurs in response to DNA damage being induced within a cell. In particular, H2AX becomes phosphorylated to form γH2AX and this occurs in chromatin close to the site of a dsb leading to the visualization of a focal spot when fluorescent antibody staining is used. The number of foci produced is reported to equate to the number of dsb induced. The phosphorylation of H2AX appears to be a key initiator of the recruitment of repair proteins to the site of DNA damage. Using a charged particle microbeam it is relatively easy to deliver particles to precise regions of the nucleus. Fig. 2. shows an example of some primary human fibroblasts where the nucleus was irradiated with a single 3 MeV helium-3 ion at three separate locations in an equilateral triangular pattern with sides of 5μm. The cells were fixed immediately after irradiation and fixed and stained with antibodies against γH2AX. A single focus is observed at each of the targeted locations. This clearly shows the power of the targeting ability of the microbeam and the specific role that γH2AX plays as a marker for localized DNA damage.

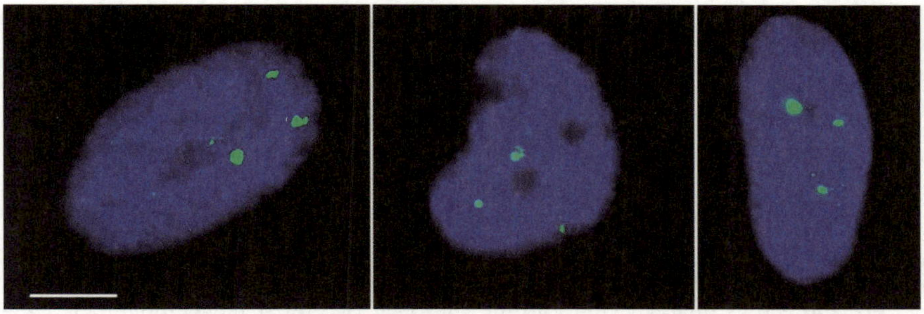

Fig. 2. Image of γH2AX staining in human fibroblasts irradiated with a single helium ion each at three separate locations using the Gray Cancer Institute Charged Particle Microbeam. The bar represents 5 μm.

Studies of non-targeted effects

Non-targeted responses have currently become a major area of application for microbeams. These are responses which do not follow the classical paradigm of direct DNA damage leading to a proportionate response. Archetypical of these is the bystander response where cells, which have not been directly targeted with radiation, respond to their neighbors being exposed. The first direct studies of bystander effect detected using a charged particle microbeam reported that these could be induced when only a single cell within a population was exposed to a single helium ion [24, 25]. More recently, these phenomena have been observed in multicellular tissue models when these are locally irradiated or an explant outgrowth technique is used [26, 27].

Future studies and developments

Development of microbeams continue apace in parallel to improvements in both the technology underpinning them and the advances in cell and molecular biology research. Central to this are improvements in particle detection and delivery systems, coupled to advances in imaging and processing. These will allow other types of imaging to be used for both localization of targets to be irradiated and post-irradiation revisiting and analyses. Most of the current systems utilize fluorescent dyes and imaging, which require care to reduce photobleaching and the damaging effects of the illuminating light. For other radiations, several groups are developing approaches to allow the targeting of different energies of X-rays and electrons. Also particle studies are being extended to heavier ions of relevance to, for example, space research. Together, these developments will allow a more complete biophysical analysis of targeted microbeam approaches to be obtained. For endpoints of low frequency, ultra fast stages and scanning systems are being developed to increase throughput of cells and novel approaches now allow non-attached cells to be analyzed [28].

The ability to localize the action of a genotoxic agent such as ionizing radiation within cells or tissue has several implications for performing mechanistic studies in cellular and multicellular models. It allows the role of subcellular organelles to be carefully studied within specific processes. For example separating nuclear versus cytoplasmic based signalling processes is relatively easy and in the future the ability to target regions where, for example, mitochondria or other cell organelles are present will be feasible.

For molecular analysis, the current approach of global measurement of gene expression in up to 10^6 cells can only give an average picture of the response of a system to irradiation. New *in situ* approaches for gene expression studies allow the response of individual cells to be carefully followed after targeted irradiation. Also with the sequencing of the human genome it is clear that the number of genes present is considerably less than that predicted for the function of a cell and that much of cellular response is regulated at the protein level. Therefore the real challenge will be in the proteomics arena. Already, the ability to follow protein expression in individual cells is possible for limited numbers of proteins [29].

At the level of intact tissue models and organisms microbeams have the potential to make significant contributions. Increasingly, microbeams will be used in cell surgery techniques particularly related to understanding developmental systems [30]. For example studies of developmental processes in the nematode C. elegans can be easily performed by targeting individual cells within the organism at different stages of development and monitoring changes in neural, digestive and reproductive pathways. In mammalian tissue models, the role of radiation-damage to endothelial cells versus stem or other tissue cells will be possible to answer important questions related to radiation mechanisms of relevance to therapy [31].

In summary, microbeams provide an important tool not just for radiation biology, but for other areas of biology which should guarantee them a central part in future research.

Acknowledgements

The authors are grateful to Cancer Research UK, the European Community, the Department of Health and the US Department of Energy for supporting the development of their microbeam program.

References

1 Zirkle RE (1957) Partial-cell irradiation. In: Lawrence JH, Tobias CA (eds) Advances in Biology and Medical Physics, Academic Press: New York 103–146

2 Smith CL (1964), Microbeam and partial cell irradiation. Int Rev Cytol 16:133–153
3 Munro TR (1970) The site of the target region for radiation-induced mitotic delay in cultured mammalian cells. Radiat Res 44: 48–757
4 Cole A, Humphrey, RM, Dewey WC (1963) Low-voltage electron beam irradiation of normal and 5-bromouridine deoxyriboside-treated L-P59 mouse fibroblast cells in vitro. Nature 199:780–782
5 Braby LA (1992) Microbeam studies of the sensitivity of structures within living cells. Scanning Microsc 6(1):164–174
6 Pugliese M, Durante M, Grossi GF, Monforti F, Orlando D, Ottolenghi A, Scampoli P, Gianella G (1997) Inactivation of individual mammalian cells by single α-particles. Int J Radiat Biol 72:397–407
7 Soyland C, Hassfjell SP (2000) Survival of human lung epithelial cells following *in vitro* alpha-particle irradiation with absolute determination of the number of alpha-particle traversals of individual cells. Int J Radiat Biol 76:1315–1322
8 Soyland C, Hassfjell SP, Steen HB (2000) A new alpha-particle irradiator with absolute dosimetric determination. Radiat Res 153(1):9–15
9 Warters RL, Hofer KG, Harris CR, Smith JM (1977) Radionuclide toxicity in cultured mammalian cells: elucidation of the primary site of radiation damage. Current Topics in Radiat Res Quarterly.12:389–407
10 Nelms BE, Maser RS, Mackay JF, Lagally MG, Petrini JHJ (1998) In situ visualisation of DNA double-strand break repair in human fibroblasts. Science 280:590–592
11 Lorimore SA, Kadhim MA, Pocock DA, Papworth D, Stevens DL, Goodhead DT, Wright EG, (1998) Chromosomal instability in the descendants of unirradiated surviving cells after alpha-particle irradiation. Proc Natl Acad Sci US 95:5730–5733
12 Rustad, RC (1968) A simple UV-microbeam for partial cell irradiation. Experientia 24(9):974–5
13 Cremer C, Cremer T, Fukuda M, Nakanishi K (1980) ion of laser-UV microirradiation-induced DNA photolesions by immunofluorescent staining. Hum Genet 54(1):107–10
14 Katsumi S, Kobayashi N, Imoto K, Nakagawa A, Yamashina Y, Muramatsu T, Shirai T, Miyagawa S, Sugiura S, Hanaoka F, Matsunaga T, Nikaido O, Mori T (2001) In situ visualization of ultraviolet-light-induced DNA damage repair in locally irradiated human fibroblasts. J Invest Dermatol 117(5):1156–61
15 Volker M, Mone MJ, Karmakar P, van Hoffen A, Schul W, Vermeulen W, Hoeijmakers JH, van Driel R, van Zeeland AA, Mullenders LH (2001) Sequential assembly of the nucleotide excision repair factors in vivo. Mol Cell 8(1):213–24
16 Cremer T, Cremer C, Baumann H, Luedtke EK, Sperling K, Teuber V, Zorn C (1982) Rabl's model of the interphase chromosome arrangement tested in Chinese hamster cells by premature chromosome condensation and laser-UV-microbeam experiments. Hum Genet 60(1): 46–56
17 Randers-Pehrson G, Geard CR, Johnson G, Elliston CD, Brenner DJ (2001) The Columbia University single-ion microbeam. Radiat Res 156(2):210–4
18 Folkard M, Vojnovic B, Hollis KJ, Bowey AG, Watts SJ, Schettino G, Prise KM, Michael BD (1997) A charged particle microbeam: II A single-particle micro-collimation and detection system. Int J Radiat Biol 72:387–395
19 Peng S, Folkard M, Gilchrist S, Locke RJ, Yu Z, and Micheal BD (2001) Measurements of the targeting accuracy of the Gray Laboratory charged-particle microbeam. Nuclear Instruments and methods in Physics Research B(179):145–150

20 Folkard M, Schettino G, Vojnovic B, Gilchrist S, Michette AG, Pfauntsch SJ, Prise KM, Michael BD (2001) A focused ultrasoft x-ray microbeam for targeting cells individually with submicrometer accuracy. Radiat Res 156(6):796–804
21 Miller RC, Randers-Pehrson G, Geard CR, Hall EJ, Brenner DJ (1999) The oncogenic transforming potential of the passage of single alpha particles through mammalian cell nuclei. Proc Natl Acad Sci US 96(1):19–22
22 Prise KM, Folkard M, Malcolmson AM, Pullar CH, Schettino G, Bowey AG, Michael BD (2000) Single ion actions: the induction of micronuclei in V79 cells exposed to individual protons. Adv Space Res 25(10):2095–101
23 Wu LJ, Randers-Pehrson G, Xu A, Waldren CA, Geard CR, Yu Z, Hei TK (1999) Targeted cytoplasmic irradiation with alpha particles induces mutations in mammalian cells. Proc Natl Acad Sci US 96(9):4959–4964
24 Prise KM, Belyakov OV, Folkard M, Michael BD (1998) Studies of bystander effects in human fibroblasts using a charged particle microbeam. Int J Radiat Biol 74:793–798
25 Belyakov OV, Malcolmson AM, Folkard M, Prise KM, Michael BD (2001) Direct evidence for a bystander effect of ionizing radiation in primary human fibroblasts. Br J Cancer 84(5):674–9
26 Belyakov OV, Folkard M, Mothersill C, Prise KM, Michael BD (2002) Bystander-induced apoptosis and premature differentiation in primary urothelial explants after charged particle microbeam irradiation. Radiat Prot Dosimetry 99(1):249–251
27 Belyakov OV, Folkard M, Mothersill C, Prise KM, Michael BD. (2003) A proliferation-dependent bystander effect in primary porcine and human urothelial explants in response to targeted irradiation. Br J Cancer in press
28 Kadhim MA, Marsden SJ, Goodhead DT, Malcolmson AM, Folkard M, Prise KM, Michael BD (2001) Long-term genomic instability in human lymphocytes induced by single-particle irradiation. Radiat Res 155(1 Pt 1):122–6
29 Zhang HT, Kacharmina JE, Miyashiro K, Greene MI., Eberwine J (2001) Protein quantification from complex protein mixtures using a proteomics methodology with single-cell resolution. Proc Natl Acad Sci US 98(10):5497–502
30 Kobayashi Y, Watanabe H, Taguchi M, Kiguchi K (1997) Microbeam system for local irradiation of biological systems and effect of collimated beams on insect egg. In: Goodhead DT, O'Neill P, Menzel HG (eds) Microdosimetry: An interdisciplinary approach, The Royal Society of Chemistry: Cambridge, UK. p. 343–346
31 Paris F, Fuks Z, Kang A, Capodieci P, Juan G, Ehleiter D, Haimovitz-Friedman A, Cordon-Cardo C, Kolesnick R (2001) Endothelial apoptosis as the primary lesion initiating intestinal radiation damage in mice. Science 293(5528):293–7

The PTB – a Competent and Reliable Partner for Research and Development Projects in the Field of Ionizing Radiation

Jürgen Böhm

Physikalisch-Technische Bundesanstalt, 38116 Braunschweig, Germany
juergen.boehm@ptb.de

Abstract.

It is purpose of this paper to introduce a German federal research establishement and to demonstrate its abilities in a number of radiation research areas. This is intended to serve as an illustration of the variety of applications and also give an insight into the many different fields of fascinating science where radiation is used or applied.

Introduction

The mission of Division 6 *Ionizing Radiation* of the Physikalisch-Technische Bundesanstalt (PTB) in Germany is to ensure progress and reliability of the metrology of ionizing radiation for the benefit of society, economy and science. The division regards itself as a provider of services, the division's "clientele" in Germany being characterized as follows:
In medicine: Each year, approximately 300 000 patients undergo a radiotherapy. On an average, every inhabitant is about twice yearly exposed to X-rays for diagnostic purposes.
In radiation protection: Approximately 300 000 persons who are exposed to radiation at their workplace are monitored with personal dosemeters. 150 000 persons are exposed to increased natural radiation.
In research: Universities and research institutes use the division's calibration and measurement capabilities and its metrological competence for joint projects.

Solid and continuous "grass-roots" work is a prerequisite for metrological competence in the field of ionizing radiation. Cornerstones for successful work are:
Staff motivation. Emphasis is laid on community spirit, implementation of the PTB's management guidelines, regular laboratory/project discussions and advanced training.
Good instrumentation. This includes in particular the provision and development of

reference fields for ionizing radiation and of the associated metrological infrastructure. Ion accelerators, electron accelerators, X-ray devices and radionuclide sources, which must be maintained or renewed and calibrated at regular intervals, are used to generate the reference fields. 25 to 40 years can be regarded as standard values for the service life of accelerators, whereas a service life between 10 and 15 years can be expected for X-ray equipment. All computer-controlled measuring set-ups must normally be renewed after approximately 10 years. It thus is obvious that a considerable part of the division's resources (about 38 percent) is to be used for the development and provision of instrumentation, including the reference fields.
Quality management. The division has adopted a quality management system in accordance with DIN ISO 17025.

Fields of activity

The tasks of Division 6 Ionizing Radiation can be assigned to four interrelated fields of activity:

Fundamentals of metrology

This area of tasks covers the realization, maintenance and dissemination of the SI units of activity, dose and fluence. Primary and transfer standards meeting both, today's requirements and future requirements are developed and made available.

Metrology for economy

The main emphasis of this area of tasks lies on the promotion of industry by research and development activities, calibrations as well as advice and information in the field of ionizing radiation.

Metrology for society

The provision of measuring techniques and measuring methods in the field of ionizing radiation is of particular social relevance, as it helps the citizen to preserve and/or recover his/her health; the provision of measuring techniques and measuring methods, however, also serves to protect man and his environment against the hazards of ionizing radiation.

International affairs

In this area of tasks, the division contributes to the international uniformity of measurements in the field of ionizing radiation and furthers cooperations with other national metrology institutes.

Competence centre for metrology of ionizing radiation

The PTB has particular competence in a number of fields summarized below (MI standing for *major investment*). A few fields are described in some more detail. For further information, references and contact persons, see http://www.ptb.de.

Neutron metrology and dosimetry

Monoenergetic neutron reference radiation fields with neutron energies from 1,2 keV to 20 MeV are generated with the aid of two ion accelerators (MI: 12 million €); a neutron irradiation facility with radionuclide sources (MI: 1 million €) is available. With these neutron reference radiation fields generated at the PTB and the fields operated together with other institutions (GKSS/Germany, NAC/South Africa, UCL/Belgium), neutrons in the very wide energy range from meV to 150 MeV are available for calibration, research and development. Various spectrometers for measurements of the spectral neutron fluence have been developed.

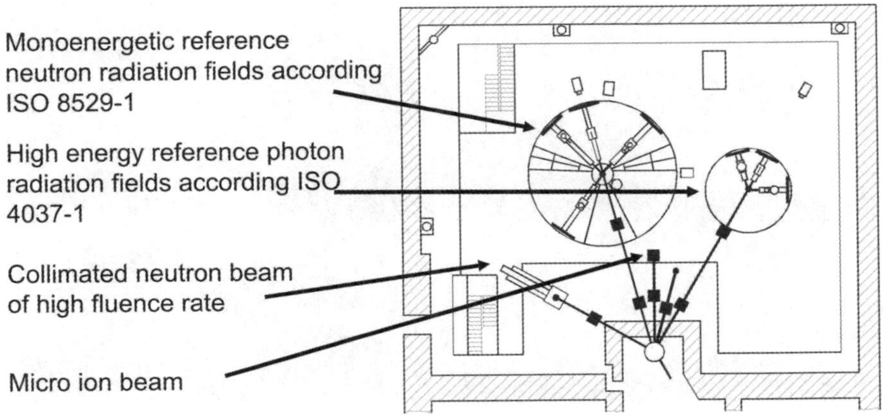

Fig. 1. Ground plan of the low scatter experimental hall at the PTB accelerator facility for fast neutron research. Different reference radiation fields are produced at various locations. The black squares indicate the positions of the quadrupole magnets.

Figure 1 shows the ground plan of the low back scatter experimental hall (area 24 m x 30 m and height of 14 m) with the locations in which the reference fields are produced. A 3,75 MeV Van de Graaf generator for low energy ion beams and a variable energy cyclotron for high energy ion beams are connected to the hall via a sophisticated beam transport system.

In 2001, a key comparison of neutron fluence measurements was carried out in the neutron reference radiation fields at the PTB. The PTB is the only official body to carry out neutron calibrations in Germany. This is why the task of measuring the radiation field in the environment of the first CASTOR transport and storage container stored in Gorleben and of a POLLUX mockup model in the Asse Salt Mine was also entrusted to the PTB.

Beta dosimetry in the field of radiation protection

The extrapolation chamber developed at the PTB (main component of a primary measuring device) and the beta secondary standard are internationally used and have been transferred to industry for fabrication under licence.

Fig. 2. Calibration set-up to generate filtered X-radiation spectra of different spectral widths and air kerma rates. The filters are mounted in the big wheels and are selected remotely by computer.

Photon personal dosimetry

At the 400 kV X-ray irradiation unit (MI: 1 million €) which can be used 24 h a day and has a complex quality management system, photon reference radiation fields are generated in accordance with ISO 4037-1. The equipment is mainly used for type tests. Figure 2 shows a calibration set-up of the PTB to generate filtered X-radiation spectra of different spectral widths and air kerma rates.

The PTB is the only National Metrology Institute in Germany with is bound by law to perform type tests. Type tests for area, personal and diagnostics dosemeters are governed by the Verification Ordinance, type tests for X-ray sources, X-ray facilities and stray radiation emitters by the X-ray Ordinance, and type tests for devices and facilities containing radioactive substances by the Radiation Protection Ordinance.

High-energy photon reference radiation fields with energies between 4 MeV and 7 MeV are generated in the experimental hall shown in Figure 1, page 237. Due to the great importance of this high-energy photon radiation in nuclear power plants, legal control of radiation protection dosemeters was recently extended to cover these radiation energies.

At the PTB, the measurand $H_p(10)$ is also realized and disseminated on the basis of spectral fluence measurements with an HPGe spectrometer. The personal dose equivalent $H_p(10)$ transfer chamber recently developed at the PTB for the dissemination of $H_p(10)$ is unrivalled.

Dosimetry in the environment

Unique measuring and calibration capabilities are offered by the underground laboratory UDO (925 m deep, MI: 1 million €) located in a salt mine (only one hundredth of the dose rate on the earth's surface), by an outdoor reference measuring set-up on the PTB terrain, integrated into the federal IMIS measurement network, and by special measuring arrangements for flight dosimetry. In 2002, an international comparison of early dose warning systems was carried out at the PTB.

The importance of dosimetry in the environment has increased during the past few years as protection against increased natural radiation has been provided for in the EU by EURATOM Directive 96/23 and therefore had to be implemented in the national legislation on radiation protection. In this implementation PTB played an important role, in particular as far as radiation protection due to increased exposure to natural radiation from ground-level to flight altitudes (all components, including radon) is concerned.

Fig. 3. View into the radon reference chamber (outer dimensions 4,8 m x 2,19 m x 2,5 m) through its air lock 2,44 m in height.

Radon concentration measurements

Systematic investigation of the radon and radon progeny activity concentration in air and special calibration of radon measuring instruments are carried out in a man-sized radon reference chamber (Figure 3), the temperature and relative humidity of which can be varied. An aerosol generator developed at the PTB is used to produce reference aerosols with adjustable aerosol size distribution and concentration. In the radon reference chamber, at equilibrium factors between 0,1 and 1 and with an unattached fraction between 0,01 and 0,9, radon activity concentrations in air of 1 kBq m^{-3} to 100 kBq m^{-3} can be produced with a relative standard uncertainty of 5 percent.

- Recently, an additional calibration facility for measuring the thoron (Rn-220) activity concentration in air has been developed. Exposures to thoron gas generally do not reach values as high as those for radon gas but cannot be disregarded in a number of working and living areas.

Pioneer in the field of the measurand absorbed dose to water in radiotherapy

About 15 years ago, PTB was the first National Metrology Institute which in the field of radiotherapy changed from a dosimetry based on the measurand exposure to a dosimetry based directly on the quantity absorbed dose to water. Meanwhile, teletherapy is based almost globally on the absorbed dose to water for all types of radiation. Today, PTB plays a similar pioneer's role in brachytherapy for low-energy photon radiation. For measurements in teletherapy, the PTB operates a clinical 20 MeV linear accelerator (MI: 1 million €). Replacement of the obsolete accelerator is planned (MI: 14 million €). The reference radiation field for teletherapy is generated by a ^{60}Co irradiation device (MI: 0,5 million €).

Fig. 4. Disassembled multi-electrode extrapolation chamber. Left: multi-electrode in the centre of the upper plate. Right: electrode translation system with ring of electrode connectors.

Beta dosimetry in radiotherapy

At present, the strongly increasing number of patients in the new field of intravascular brachytherapy is a great challenge to the PTB as far as exact calibrations in terms of absorbed dose to water are concerned. A beta secondary standard recently developed at the PTB for detector calibrations and developments in the greatly expanding field of endovascular brachytherapy was left to industry for fabrication. A novel multi-electrode extrapolation chamber (Figure 4) was tested and will soon be used as a primary standard measuring device. It meets in particular the requirements for a small measurement uncertainty. In contrast to conventional extrapolation chambers, the central part of this chamber is a segmented collecting electrode which was manufactured at the Clean Room Centre of the PTB by electron beam lithography on a wafer. The chamber allows three-dimensional dose distributions to be measured with high spatial resolution and without having to fall back on an additional relative dosimetry system, thus avoiding additional uncertainties.

Diagnostic radiology

The PTB was the first national metrology institute which offered air kerma calibrations for diagnostic radiation qualities in general and, at a later date, air kerma calibrations for mammographic radiation qualities. 30 percent of today's calibrations is performed for foreign customers. Calibrations in terms of the measurand "air kerma" are carried out on the quality-assured X-ray equipment designed for 24 h operation. The development of the new measurand "practical peak voltage" by the PTB in 1998 was a prerequisite for the elaboration of an IEC standard which has meanwhile been published and covers requirements for non-invasive measuring instruments for the high voltage of X-ray equipment.

Radioactivity standards

For the production of standard solutions for the realization and dissemination of the unit of activity, a sophisticated type B radionuclide laboratory (MI: 2,5 million €) is available in which radioactive substances with up to 10^5 times the maximum permissible level specified in the Radiation Protection Ordinance can be handled. For dissemination of the activity, a laboratory of the DKD (Deutscher Kalibrierdienst) has been affiliated to the PTB. The PTB is engaged only in the fabrication and provision of special activity standards which cannot be made available by this DKD laboratory.

Nuclear and atomic data of radionuclides provide the basis for any activity measurement. Experimental determinations and evaluations of nuclear and atomic data with smallest uncertainties are therefore important areas of work for the PTB. The tight resources for evaluations are today optimized by cooperation in projects with international partners.

Radioactivity in environmental samples

For the German "Integrated Measuring and Information System" (IMIS) for environmental radioactivity – a system set up as a consequence of the metrological situation after Chernobyl – the PTB is the body giving reference values for comparison measurements among laboratories engaged in the field of environmental radioactivity.

A sophisticated radiochemical laboratory (MI: 2 million €) is available for the processing of environmental samples. Its facilities allow these samples to be completely decomposed and plutonium, strontium, thorium and uranium isotopes to be radiochemically separated. To avoid contaminations from outside, the supply air for the air-conditioning is cleaned using high-quality dust filters. This allows smallest activities to be reliably analyzed. The detection limits achievable for ^{238}Pu and $^{(239+240)}$Pu lie around 50 pBq/m^3. The measurements themselves are carried out with semiconductor alpha spectrometers and semiconductor photon spectrometers at the PTB itself. In addition, two semiconductor photon spectrometers are available at the underground laboratory UDO. Since 1963, the PTB has determined the activity concentrations of ^{7}Be and ^{137}Cs and of different plutonium isotopes in the ground-level air with the aid of air dust samplers.

Fundamental knowledge of the track structure of ionizing radiation

The knowledge of the track structure of ionizing radiation occurring as a result of interaction processes may open up new prospects for radiation metrology and is closely linked with radiobiology. A generally accepted model which can predict radiobiological effects from the physics of radiation interaction does not yet exist. The "Optical Particle Avalanche Chamber" OPAC developed at the PTB comprises a nuclear track imaging system designed to study in detail the ionization topology of charged particle tracks in a low pressure gas (triethylamine). The micro ion beam measuring arrangement established in the last few years as well as the Monte Carlo simulation calculations of ionization cluster formation in nanometric volumes are promising means for future innovations.

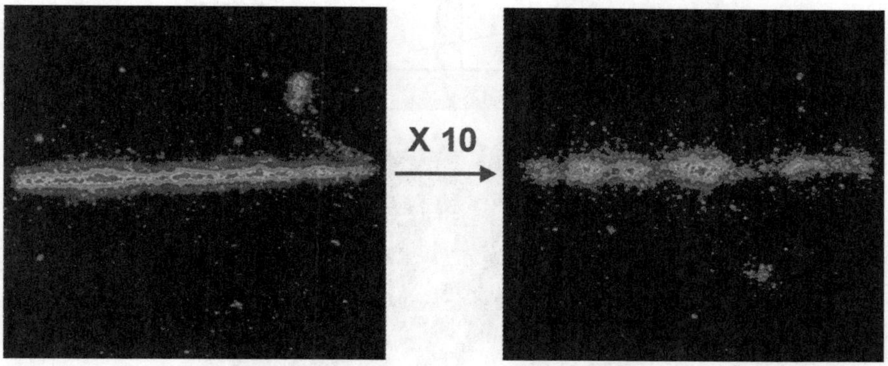

Fig. 5. Images of C-12 ion tracks produced by ions with energies of 200 MeV/u at different triethylamine pressures. Left: 40 hPa, right: 3,5 hPa. The flight direction of the ions is from the left to the right side.

OPAC chamber. The method is based on a time projection chamber with a parallel drift field, parallel-plate charge and light amplification steps and optical readout. The spatial resolution of the chamber is limited by electron diffusion during the drift in the 10 cm wide sensitive volume of the chamber. Spatial ionization distributions of particle tracks in gas, of equivalent length in tissue of 4 mm and 40 nm (RMS), respectively, are obtained. Figure 5 shows tracks of 200 MeV/u C-12 ions for two different pressures of the chamber. The 10 times lower gas pressure used for taking the particle track picture of the right side has the effect of a tenfold magnification of the track and furnishes additional details.

Micro ion beam for high-energy protons and ^{4}He ions. This special particle beam is adjustable to a beam diameter as low as 1 mm; the schematic cross-section of the facility and the experimental set-up are shown in Figure 6. In the extreme case, a single ion can be applied. The detection probability for ions is between 95 and 100 percent depending on the energy loss in the detector used. The micro ion beam is especially suited to investigate ionizing radiation damages to living cells at low doses.

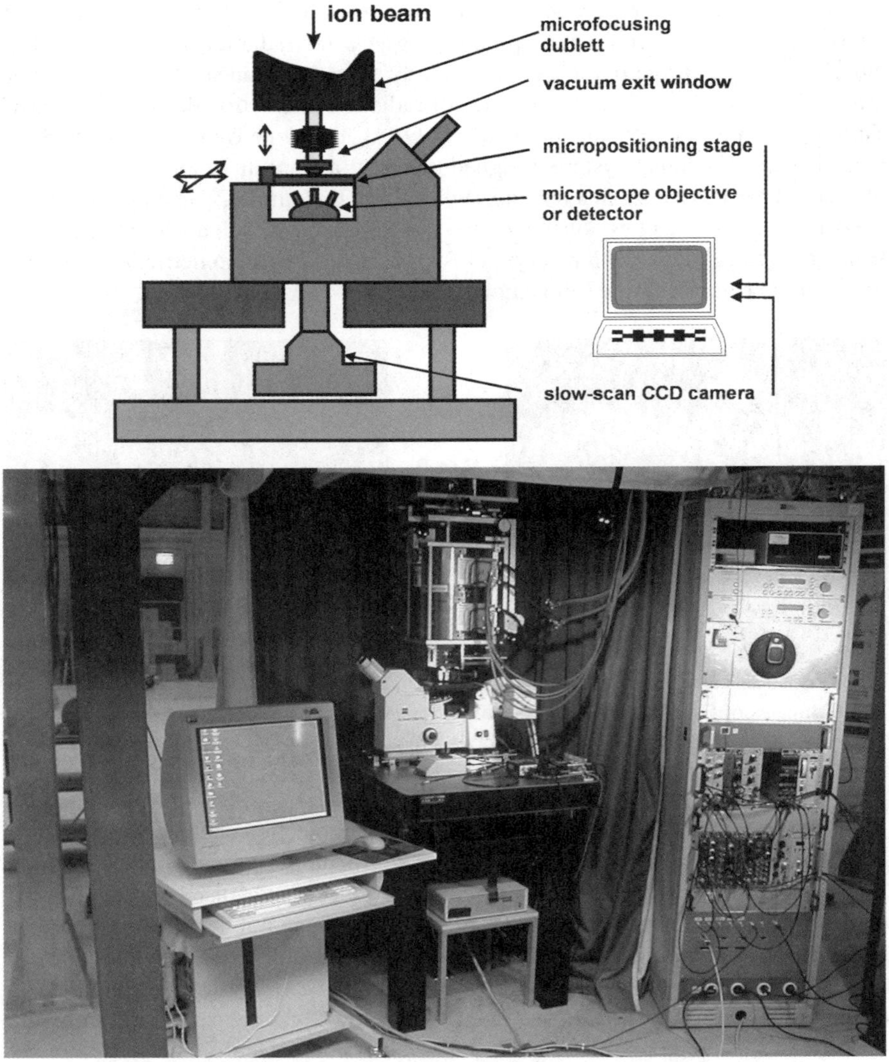

Fig. 6. Micro beam facility. (a) Schematic cross-section of the target place at the end of the beam transport system. (b) Experimental set-up. In the middle the vacuum exit window with the microscope mounted on an optical table can be seen.

Nanodosimetry. The DNA molecule with its diameter of 2 nm is considered as the most important radiation-sensitive target in a biological system and thus radiation interactions in volumes of a few nanometer (nanodosimetry) should be investigated. Attempts are made at realizing Monte Carlo simulation calculations of ionization cluster formation in nanometric volumes, in connection with the determination of scattering cross sections required for that purpose. These are expected to help to develop new metrological concepts for ionizing radiation.

Basic data. The link between nanodosimetry and traditional dosimetry (ionization dosimetry) is established at the PTB by the comparison of ionization cluster production in nanometric volumes and charge carrier production in an infinitely extended medium. For this purpose, not only the Monte Carlo simulation of the formation of ionization clusters but also the measurement of the so-called W-value (mean energy to produce pairs of charge carriers) of electrons and ions in gases are used. The determination of improved basic data is important for radiation transport calculations to be reliable.

Conclusions

The PTB has outstanding competence in numerous fields of ionizing radiation. This competence was built up during many years and is maintained by continuous "grass-roots" work. Staff motivation, good instrumentation – supplemented if necessary by major investments – and quality management are essential for the success of its work. The PTB's competence and facilities should be utilized as much as possible in future research and development projects. A number of PTB's competence areas are listed in this paper.

The PTB, and thus also its Division 6 *Ionizing Radiation*, was recently subjected to an evaluation initiated by the Ministry of Economics and Labour. The competence was fully confirmed in the Final Evaluation Report dated December 16, 2002.

General Issues

Life Sciences and Radiation Research – Examples from the Past

Jürgen Kiefer

Strahlenzentrum der Justus-Liebig-University, Giessen, Germany
juergen.kiefer@strz.uni-giessen.de

Abstract.

The evolution of modern life sciences is closely linked with radiation research. This paper aims at demonstrating this by having a closer look at three examples: molecular genetics, repair processes and the cell cycle.

Introduction

The title of the introductory talk given at the workshop and which is the basis of this paper was "These were thy days my friend, we thought they never end...", paraphrasing a song quite popular in the England of 1970. It describes the situation of radiobiology at that time and in a way prophesies the present state. Reasons for this are outlined elsewhere in this volume but it appears appropriate to look also back into the merits of radiation research – not to be unduly proud but as guidance and a stimulus for further activities.

The history of radiation biology has still to be written but there are a few very good summaries: To celebrate the 59th anniversary of the "Radiation Research Society" a number of articles on "Historical and current highlights in radiation biology" were commissioned. In the present context the most relevant is that by Bedford and Dewey [1], a more general view is taken by Eric Hall [2]. The present paper does not want to compete with these works but attempts to draw the connecting lines between radiation research and modern life sciences by chossing three examples: the contribution to molecular genetics, the role of repair processes and the cell cycle.

Some aspects of the present paper were already discussed at the celebration of the 100th anniversary of Timofeeff-Ressovski's birthday [3].

From radiation targets to genes and molecular biology

After the discovery of X-rays by Wilhelm Conrad Röntgen in 1895 the great potential of this "new type of radiation" for medical applications was immediately realised, not only in diagnostics but also in therapy. Already in 1896 the first radiation treatment of a tumour was performed. When radioactive compounds became available, particularly radium, all kinds of ailments were hoped to be curable by radiation. Detrimental influences on human health were also noted but did not arouse so much interest at the time. They were, however, studied experimentally and the first "dose-effect curves" were established. They were of a comparatively simple shape and presented a good case for theoretical interpretation. The mechanism, however, how radiation could damage biological systems, were completely unclear. The total amount of energy was much to small to lead to gross chemical alterations. The solution was found by combining the ideas of the recently established quantum theory with biological phenomena, creating a new discipline "quantum biology" [5]. One of the leading pioneers in this field was Friedrich Dessauer who developed what was later termed "target theory" [6, 7]. This was not only the birth of "biophysics" but also of quantitative radiobiology and paved the way to today's' molecular biology.

The underlying idea was simple: The number of photons to kill a cell was much too small to cause massive destruction but the absorption of a single quantum in one essential molecule was large enough to change it irreversibly. Dessauer and his coworkers concluded that the radiation-induced inactivation of cells must be due to alterations of essential targets of molecular dimensions whose nature remained, however, obscure.

A historical breakthrough marked H. J. Muller's discovery that mutations are caused by X-rays [8]. It demonstrated that genes can be changed by radiation. The resulting dose-effect curves could also be analysed by applying Dessauer's approach suggesting that genes are some kind of large molecules.

These ideas were further developed by others and culminated in a series of three voluminous papers by Timofeeff-Ressovsky, Delbrück and Zimmer [9]. In this publication one finds the first clear quantitative definition of the gene in modern terms: This was taken up by another eminent scientist, Erwin Schroedinger, and prompted him to write a wonderful little book, the essay "What is life?" durig his exile in Dublin [10]. James Watson describes in his account of the "Double Helix" [11] that Francis Crick was so impressed by Schroedinger's ideas that he left physics and entered biology – the results are history. There is an uninterrupted line from Roentgen to the structure of DNA – not to speak about the fact that this structure was unravelled by X-rays (but this is pure physics and hence not part of our story).

There is another contribution of radiobiology to molecular genetics which is only rarely acknowledged and hardly mentioned in biology textbooks: In 1930 the American Gates measured the wavelength dependence of the inactivation of bacteria by UV [12] and found a striking similarity with the absorption spectrum of nucleic acids. A few years later similar action spectra were found for mutation induction

[13, 14] clearly ruling out proteins which were at that time generally assumed to be the constituents of genes. The chemical proof was only given in 1943 by Avery *et al.* [15] – more than 10 after Gates! The technique of action spectroscopy has more recently succeeded also in the solution of an enigma of present days molecular biology – the clarification of the "scrapie" agent, a close relative of the substance causing BSE. Tikvah Alper showed already in 1963 that it does not consist of nucleic acid [16] and contributed thus significantly to the currently accepted "prion" hypothesis [17]. Stanley Prusiner mentions in his Nobel lecture [18] the great impact these studies had on his thinking and the further development of his work.

Figure 1 depicts in a diagrammatic manner the historical relation between radiation studies and the development and modern molecular genetics.

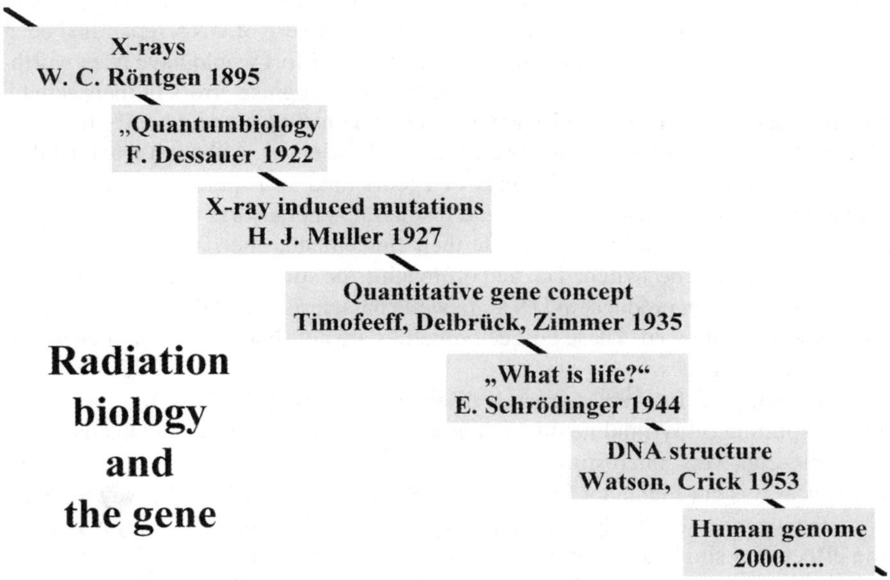

Fig. 1. The influence of radiation research on the development of molecular genetics.

More recently, there is another marriage of physics and biology: With the identification of essential DNA alterations the term "hit" received a new and more concrete meaning, namely its identification with radiation induced biologically significant molecular alterations and the mechanism of its formation. The actual process of lesion formation is, however, in most cases not yet clear. This is particularly true with different types of ionising radiation. It is now well known that dose (i.e. energy deposited per unit mass) does not solely determine the biological outcome. Densely ionising particles usually display a higher effectiveness per unit dose than more conventional radiations like X- or gamma-rays. Obviously the microscopic pattern of energy deposition plays a decisive role. Considerable progress has been made in the understanding of this both experimentally and by using sophisticated theoretical computational approaches. It started with realising that dose in its conventional definition looses its unambiguous meaning in very small mass elements which lead to the establishment of "microdosimetry" as a special subdiscipline [19]. It is built on the principles of target theory and extends the ideas of Dessauer, Timofeeff-Ressovsky [20] and Lea [21] who aimed at explaining quantitatively the formation of chromosomal aberrations by different types of radiation – still an important and largely unsolved task of theoretical radiobiology.

Repair

Bedford and Dewey in their extensive review [1] call DNA repair the "jewel in the crown" of radiobiology and continue: "If the discovery of DNA repair had been the only thing learned by irradiating cells, the whole effort would have been worthwhile". Indeed, the discovery that cells are able to recognise errors in their genetic material and to repair them with great accuracy constitutes undoubtedly the most important contribution of radiation research to life sciences. Although repair studies are now commonplace with all kinds of agents (and still quite often invoked to explain poorly understood experimental results) it is not always realised that radiation is still the best tool to investigate their fundamental mechanisms. It is easily quantifiable, it can be switched on and off making the study of time courses possible and many radiation products in DNA are well characterised and their disappearance may be easily followed. These unique properties are not shared by most of chemical genotoxic substances.

Historically the first repair process, discovered in 1949 [22], is the light-activated in-situ splitting of pyrimidine dimers known as "photoreactivation". This phenomenon presents very interesting biophysical problems but has little relevance for humans as it is not found not found in this species. A landmark on the way to repair were the ingeniously simple experiments by Elkind and Sutton [23]. They compared the effect of a single acute and fractionated exposures with several hours interval and found that the cells which survived the first dose behaved after the second dose as if they had not seen any radiation before. The sublethal damage which they suffered before had obviously disappeared, the cells had recovered. Consequently the phenomenon was called "recovery from sublethal damage". Further studies demonstrated that this process depended on active cellular metabolism. It was suggestive of some kind repair but the nature was completely unknown. Later its relationship to the repair of DNA double strand breaks was shown, first in yeast [24].

The real year of birth of repair studies was, however, 1964 when excision repair was discovered. Two papers in the same issue of the Proceedings of the US National Academy of Sciences reported the "disappearance of thymine dimers from DNA" [25, 26] and postulated an error-correcting mechanism. This finding was possible because the radiation product – pyrimidine dimers – was well characterised and its biological role already known. In the course of years much was learned about the ubiquity of excision repair and it is still the best understood repair mechanism (for review see e.g. [27]). It was also the first one where a relationship between a genetically determined repair capacity and human health was demonstrated [28]. This area is presently under very active investigation (see below) and an important branch of cancer research where it forms an essential part of the new discipline "Molecular Epidemiology".

In the beginning and still partly today repair studies concentrated mainly on the action of ultraviolet radiation. They did not only lead to better understanding of basic mechanisms but also to the discovery of new repair pathways. The genes involved were first determined in microorganisms, particularly in yeast. Later investigations in rodent and human cells made clear that repair pathways are astonishingly conserved during biological evolution demonstrating their great importance for life at all levels. It is now well known that cellular DNA is not as stable as assumed by Erwin Schroedinger who characterised this molecule (without knowing its chemical nature) as "aperiodic crystal" [10]. This statement is still true, the structure of DNA gives it a high degree of stability but it is also subject to attacks from quite a number of metabolic intermediates as well as from environmental toxicants and prone to errors during replication. Repair ensures the conservation of genetic information. It is, however, not completely error-free and mutations may occur. They are the driving force of evolution but there must be a well-tuned balance between variability and stability. For a mutant to be viable the structure of the genetic apparatus has to be maintained. Bulk damage to DNA is not a mutation but kills the cell. Repair leads in the first place to a reconstitution of viable structures and secures survival. There are no viable mutants without preceding repair which leads to the paradoxical conclusion that repair is a necessary requirement for mutagenesis. The maintenance of the stability of the genetic machinery can often only be achieved at the expense of correctness. There is a number of DNA polymerases (termed ξ to κ) which allow the continuation of replication even in the presence of structural DNA alterations which normally block its further progression [29, 30]. This "translesion synthesis" obviously plays an important role as its misfunction in XP-variants predisposes the affected individuals to UV-induced skin cancer [31].

Stimulated by the original discovery of repair processes an era of very active research started which is still going on and an end is not in sight. A recent literature search yielded more than 35 000 citations from journals, not counting the books published on the subject! 13 400, i.e. nearly one halve, involved in some way radiation. How any further proof for the importance of radiation research could be provided remains a mystery.

Excision repair, or to be more correct, nucleotide excision repair (NER), is still the best understood process (and despite its seemingly simple mechanism incredibly complicated in its dependence on a multitude of genetic factors) but in the meantime much has also been learned about other pathways which in most cases were found with the help of radiation, although mainly ultraviolet. To name the more important: base excision repair (BER), transcription coupled repair (TER), mismatch repair (MMR) and the already mentioned translesion synthesis (TLS) which is a subgroup of "replicational bypass". A comprehensive summary is not intended here (see e.g. [32], the examples given should serve only as illumination of the impact of radiation research on current up-to-date life sciences.

Investigations on repair of damage by ionising radiation appeared later. One reason is certainly the fact that it was difficult to identify the critical lesion, contrary to UV where the relevance of cyclobutane pyrimidine dimers was firmly established. The interest concentrated particularly on strand breaks. The reconstitution of radiation induced single strand breaks (SSB) was first demonstrated in bacteria by McGrawth and Williams in 1966 [33]. It was soon clear that SSB were relatively unimportant what is, of course, not the case for the far more deleterious double strand breaks (DSB). In the beginning they were thought to be always lethal. We know today that they can be repaired, mainly by endjoing or homology driven recombinational repair. They are discussed in some detail at many places in this volume so that further deliberations are unnecessary here. It must be said, however, that the developments of recent years increased the knowledge not only in terms of a better foundation of radiation risk assessment but also of fundamental mechanisms of carcinogenesis in a general way, not only linked to radiation. The importance of "physiological" DSB in the body's immune response and the role of joining processes in the generation of antibodies constitutes another interesting example.

Radiation and the cell cycle

Cell division is indispensable for all multicellular organisms. Its balanced regulation secures the function of organs, failure to respond properly to the signals of the body leads to cancer. Faithful replication of the genetic information is the key issue in this context. When and how this is achieved was a great enigma in the 1950ies, a look at older biology textbooks illustrates the problem in hindsight. In 1953 two radiobiologists, Alma Howard (a biologist) and Stephen Pelc (a physicist), studied the problem in more detail by incorporating ^{32}P into DNA (certainly a poor precursor but ^{3}H-thymidine was not yet known at that time) and came to the (in those days) surprising conclusion that DNA replication occurs during a limited period in interphase which they called "S-phase", the preceding "gap" was termed G_1, the subsequent one G_2 [34] This all is now common place, the inventors are largely forgotten and rarely quoted. In 2001 the Noble Prize for Medicine or Physiology was given to L. Hartwell, P. Nurse and Timothy Hunt for their discovery of "key regulators of the cell cycle". Radiation played also here a decisive role. To quote from the press release by the Nobel Foundation [35]: *"Hartwell studied the sensitivity of yeast cells to irradiation. On the basis of his findings he introduced the concept checkpoint, which means that the cell cycle is arrested when DNA is damaged. The purpose of this is to allow time for DNA repair before the cell continues to the next phase of the cycle. Later Hartwell extended the checkpoint concept to include also controls ensuring a correct order between the cell cycle phases."* Most of the relevant investigations were – by the way – conducted with yeast, interesting for those who think that only mammalian cells count. The concept of checkpoints goes back to a paper by Hartwell and Weinert where it was shown for the first time that radiation induced division delay is under genetic control [36].

Checkpoints and the genes involved in their regulation is at present a "hot" topic of research (see e.g. [37], [38], [39]). Nearly 3000 papers were published on this issue during the last years, with more than a quarter of them involving radiation. It needs no special sense of prophecy to predict that these numbers will grow substantially as it is generally felt that a better understanding of the fundamental mechanism will bring us closer to the solution of the cancer problem. There is already an impressive number of human malignancies – not only cancers – where the relation to defects in repair and/or checkpoint disturbances has been shown. Table 1. lists some examples.

Table 1. Examples of human malignancies related to repair or checkpoint deficiencies (partly based on [33])

Malignancy or syndrome	Symptoms	Affected mechanism
Xeroderma pigmentosum (XP)	UV induced skin cancer	NER
Cockaynes Syndrome	Growth retardation, neurological failures	TCR
Ataxia telangiectasia	Lymphomas	CP/DNA repair
Nijmegen Breakage Syndrome	Lymphomas	DSB repair
Familial Breast Cancer	Breast (ovarian) cancer	HR
Familial Colorectal Cancer	Colorectal cancer	MMR
XP variants	UV induced skin cancer	TLS
Blooms Syndrome	Leukaemia	HR
Familial malignant melanoma	Malignant melanoma	CP
Retinoblastoma	Eye cancer	CP

Abbreviations: NER: nucleotide excision repair, TCR: transcription coupled repair, HR: homologous recombination, MMR: mismatch repair, CP: checkpoint

Concluding remarks

Th few examples given up were intended to serve as an illustration how deeply results from radiation research influenced our knowledge of fundamental biological phenomena. Their relevance was not always obvious and sometimes they even went unnoticed. This is not a rare event in this discipline between physics and biology and unfortunately the tendency for this to happen is increasing. Specialisation and the growing flood of publications is certainly one reason but not a sufficient explanation. There seems to be still a border between physics and biology preventing in many cases real joint work and exchange of ideas. At least in Germany biologists traditionally dislike physics, and with biology being declared the science of the millennium they feel it is no longer necessary to bother with this discipline far away from real life and filled with non-digestible mathematical formulae. This description is, of course, a satirical exaggeration but may not be too far from reality although this will rarely be freely admitted.

Radiation biology has also recently added to our knowledge of biological phenomena. Genomic instability, adaptive response and the bystander effect are just a few examples which would not have been discovered without the help of radiation. But there is more to offer. Although DNA damage is still of utmost importance it is now being realised that the cell's response to genetic alterations is a very dynamic process with many players. The interaction with the supporting tissue and signal transfer between cells must in no way be neglected. Radiation biology offers tools for these studies which can hardly be surpassed by any other agent. It can be applied at and for very definite times and precisely directed at specific locations and , with the new microbeam techniques, even at subcellular dimensions. For taking full advantage of these promising developments it is indispensable for the radiation biologist to be fully aware of the current biological ideas but also for the "pure" biologist to be aware of the modern trends in radiation biology. There is clearly a need and demand for close cooperation. This will, however, only be possible if research institutions in radiation research do not cease to exist and the expertise (and the experts) in this field do not die out.

References

1 Bedford JS, Dewey WC (2002) Historical and current highlights in radiation biology: Has anything important been learned by irradiating cells? Radiat Res 158:251–291

2 Hall EJ (1996) Clinical and radiobiological research in "A history of the radiological sciences: radiation oncology". (Gagliardi, RA, Wilson JF, eds.) Radiology Centennial, Reston VA

3 Kiefer J (2001) Radiation Biology – Glory of the past. Chances and challenges of the future in:"Modern problems of Radiobiology, Radioecology and Evolution". Proc. of the Intern, Conf. dedicated to the Centenary of the Birth of N.W. Timofeeff-Ressovsky,

Dubna, 6–9 September 2000. (Korogodin VI; ed. Comp. by Korogodina VL, Dubrovina NI). – Dubna: JINR:.130–139
4 Röntgen WC (1895) Über eine neue Art von Strahlen. Sitzungsberichte physik.-med. Ges. Würzburg 137, engl.: Stanton A (1896) Nature 53: 274
5 Dessauer F (1954) Quantenbiologie, Springer-Verlag Berlin-Göttingen-Heidelberg
6 Dessauer F (1922) Über einige Wirkungen von Strahlen I. Z Phys 12:38–47
7 Blau M, Altenburger K (1922) Über einige Wirkungen von Strahlen II. Z Phys 12:315–329
8 Muller HJ (1927) Artificial transmutation of the gene. Science 66:84–87
9 Timofeeff-Ressovsky NW, Zimmer KG, DELBRÜCK M.(1935) Über die Natur der Genmutation und der Genstruktur: Nachrichten Ges. Wiss. Göttingen. Biologie. Neue Folge. 1:189–245
10 Schroedinger E (1942) What is life?, Cambridge University Press 1942
11 Watson JD (1968) The double helix, Weidenfeld and Nicolson London
12 Gates FL.(1930) A study of the bactericidak action of ultraviolet light. III. The absorption of ultraviolet light by bacteria. J Gen Physiol 14:31–42
13 Hollaender A, Emmons CW (1939) The action of ultraviiolet light on dermatophytes II. Mutations induced in cultures of dermatophytes by exposure of spores to monochromatic ultraviolet radiation. Amm J Bot 26:467–475
14 Knapp E, Schreiber H (1939) Mutations induced in Sphaerocarpus by ultraviolet light. Proc. 7th Int. Congress Genetics 175
15 Avery OT, MacLeod CM, McCarty M (1944) Studies on the chemical nature of the substance inducing transformation in pneumococcal types. J Exp Med 79:137–158
16 Alper T (1993) The scrapie enigma: insights from radiation experiments.Radiat Res 135:283–292
17 Safar J, Prusiner SB (1998) Molecular studies of prion diseases. Prog Brain Res.117:421–434
18 Prusiner SB (1998) Prions. Proc Natl Acad Sci USA 95:13363–13383
19 Rossi HH, Zaider M (1996) Microdosimetry and its applications. Springer-Verlag Heidelberg
20 Timofeeff-Ressovsky N., Zimmer KG. (1947) Das Trefferprinzip in der Biologie. S. Hirzel Verlag Leipzig
21 Lea, D. E. (1946) Actions of radiations on living cells. Cambridge University Press
22 Kelner A (1949) Effect of visible light on the recovery of Streptomyces griseus conidia from ultraviolet irradiation injury. Proc Natl Acad Sci USA 35:73–79
23 Elkind MM, Sutton H (1959) X-ray damage and recovery in mammalian cells in culture. Nature 184:1293–1295
24 Frankenberg D, Frankenberg-Schwager M, Harbich R (1984) Split-dose recovery is due to the repair of DNA double-strand breaks. Int J Radiat. Biol 46:541–53
25 Setlow RB, Carrier WL (1964) The disappearance of thymine dimers from DNA: an error-correcting mechanism. Proc Natl Acad Sci USA 51:226–231
26 Boyce RP, Howard-Flanders P (1964) Release of ultraviolet induced thymine dimers from DNA in E. coli K-12. Proc Natl Acad Sci USA 51:293–300
27 Friedberg EC, Walker GC, Siede W (1995) DNA repair and mutagenesis. ASM Press Washington DC
28 Cleaver JE (1969) Xeroderma pigmentosum: A human disease in which an initial stage of DNA repair is defective. Proc Natl Acad Sci USA 63:428–435

29 Goodman MF, Tippin B (2000) Sloppier copier DNA polymerases involved in genomic repair. Curr Opinion Genet Dev 10:162–168
30 Kunkel TA, Bebenek K (2000) DNA replication fidelity. Ann Rev Biochem 69:497–529
31 Masutani C, Kusumoto R, Yamada A, Dohmae N, Yokoi M, Yuasa M, Araki M, Iwai S, Takio K, Hanaoka F (1999) The XPV (xeroderma pigmentosum variant) gene encodes human DNA polymerase eta. Nature 399:700–704
32 Hoeijmakers JHJ (2001) Genome maintenance mechanisms for preventing cancer. Nature 411:366–374
33 McGrawth RA, Williams RW (1966) Reconstruction in vivo of irrradiated Escherichia coli deoxyribonucleic acid; the rejoining of broken pieces. Nature 212:534–535
34 Howard A, Pelc SR (1953) Synthesis of deoxyribonucleic acid in normal and irradiated cells and its relation to chromosome breakage. Heredity (Suppl.) 6:261–273)
35 Nobel Foundation Press Release (2001): www.nobel.se/medicine/laureates/2001/press.html
36 Hartwell LH, Weinert TA (1989) Checkpoints: Controls that ensure the order of cell cycle events. Science 246:629–534
37 Piwnica-Worms H (1999) Fools rush in. Nature 401:535–537
38 Carr AM (2000) Piecing together the p53 puzzle. Science 287:1765–1766
39 Lowndes NF, Murguia JR (2000) Sensing and responding to DNA damage, Curr Opin-Genet De 10 :17–25

Need and Importance of Further Research in the Field of Radiation

P. N. Srivastava, R. N. Sharan

Jawaharlal Nehru University, New Delhi 110067, India, North-Eastern Hill University, Shillong 793022, India
rnshara@nehu.ac.in, pns07@yahoo.com

Introduction

It would be very difficult to summarise and do justice in a small report the work done in radiation research and the support that it received from various governments and other financing agencies during the last century. It is, however, sad that its support during the later part of the twentieth century and in the present century, has been decreasing. In the beginning few decades of the twentieth century, the results of radiation studies were mainly presented and discussed at meetings of medical radiology. In the second half of the century it was realized that further progress in the understanding of biological effects of ionizing radiation could only be obtained from studying these effects at all levels of biological systems, from molecules, cells, organs to the whole organism in a multidisciplinary manner. Consequently, meetings and discussions amongst physicists, chemists, biologists and clinicians led to the development of radiation research as a field as we know it today.

Contributions of radiation research

Radiation research has contributed to a better understanding of action mechanisms not only of ionizing radiation but also of other toxic agents on cells and organisms. By such studies fundamental life processes have been discovered; DNA repair and the cycle of cell proliferation were observed and defined by radiobiologists. Through radiation research, many contributions have been made for the understanding of the mechanisms and regulation of these processes. Molecular and cellular biology in general have stimulated the understanding of these processes also with respect to radiation effects. The primary physical effects, radiochemical reaction and biological consequences have always been studied simultaneously and quite often in interdisciplinary groups. Our understanding of biology and its principles owes a great deal to the use of radiation as a tool. Recognising this, the Nobel Committee has bestowed a large number of Nobel Prizes to scientists for their work directly or indirectly involving radiation. A cursory look at the web site of the Nobel

Committee (www.nobel.se) makes the point obvious that a very large number of Nobel prizes in Physics went to the credit of 'radiation'. It is sad that the boom that set-in in the radiobiology research in 1950s, peaking around 1970s to 1980s, is gradually but significantly showing a declining trend across the globe in 1990s. India is no exception.

Due to technological advancements, radiation is increasingly becoming more useful in medical diagnosis and therapy than in the past opening new vistas of success in medicine. On the other hand, radiation is also finding desirable as well as undesirable uses in industries, in maintaining human safety (e.g., x-ray scanner at airports, etc.) and in warfare (nuclear devices and weapons of various kinds). It is also a futuristic substitute for conventional fuels (viz., nuclear powered industries, aircrafts, ships, submarines, etc.). A new dimension of radiation was revealed when Hiroshima and Nagasaki cities perished under the impact of atom bombs. Further, a series of accidents at different nuclear power plants in the world, from Three Mile Island in USA to Chernobyl, in the former USSR, exposed human populations to high doses of radiation. Human misery that followed in terms of immediate impact of radiation triggered destruction left lasting albeit negative impression on human minds about radiation. The late effects of radiation, some of which are still being evaluated, on the other hand, seemed not so drastic as was expected. In the heightened threat perception of the world in the present century wherein terrorist could use nuclear or 'dirty' devices to achieve their sinister goals has further accentuated the fear of radiation. Thus, human concerns on radiation safety of the population have increased significantly. Needless to say that historical experience with radiation, including its diagnostic and therapeutic potentials and tragedies, have been mixed. However, the potential usefulness of radiation in cancer diagnosis, radiotherapy and other medical applications remain high and is constantly growing with technological developments.

In future, as long as mankind lives in an environment with radiation exposure from natural and man-made sources, it is necessary to improve the understanding and estimation of radiation risk. Radiation research has a very high social impact for life throughout the world. The production of energy by nuclear power, the use of ionizing radiation for medical diagnosis as well as tumour therapy are very important factors for living standards in our countries world-wide. Under these circumstances it is very necessary that radiation research continues to obtain financial support from governmental as well as non-governmental agencies.

Needs of developing countries

Coming to the special problems of developing countries like India, unnecessary fear of radiation causes many problems. It is very well known that in many developed countries generation of power through nuclear energy is achieved from 35 to 75 per cent. They do not perhaps need more power plants. In countries like India it is of utmost necessity but it is opposed by the general public for fear of radiation risks. The assumption of a linear dose-effect curve without a threshold in the low-dose range has been made on a scientific basis by UNSCEAR for reasonable precaution in radioprotection. However, at least for carcinogenesis, the deleterious effect of very low dose of radiation has not been proven yet. But it has led to misunderstanding that very small radiation doses are also very deleterious to the human health, hence leading to opposition for power production through nuclear power plants. India has in the state of Kerala a large natural laboratory where the radioactivity measured in the 55 km long and 0.5 km wide belt is primarily due to the largest thorium deposit (and its decay products) ranging from 8–10 per cent, which is the largest anywhere in the world. The radiation levels in the region vary widely from 1 mGy to 35 mGy/year. The size and density of population in this region is also the largest anywhere in the high background areas of the world. Historically, human population has been living in these areas for over 1000 years. Computer-based cytogenetic studies along with dosimetry studies (chromosomal breakage and non-disjunction) which can be monitored as micronuclei in binucleated lymphocytes blocked at cytokinesis on over 10,000 newborns, based on an analysis of 850,000 cells at metaphase have shown no difference between babies born in normal as well as high background areas. Further continued work in this area is very necessary. Resolution of the effects as well risks of low dose radiation is extremely important. In our efforts to protect our people and the environment from the effects of low dose radiation, it should not be forgotten that the linear – no-threshold model is based on conservative assumptions.

Need for further studies in radiation

Let us take up another area of work radiation research is very necessary and would need continuous financial support. It is assumed that because of the political pressure and the international situation atomic bombs may not (?) be used in future. The unprecedented terrorist attack on the World Trade Centre on September 11, 2001 and the turn of events across the globe since then have heightened the perception of threat to the civilized world. The fear of a repeat of such events with increased intensity and possible use of nuclear weapons in such acts is not an impossible reality. Prompted by such threats, the scientific communities of the world have started to address the issue of planning appropriate strategies to minimize loss of life and property. One important workshop, "Interagency Workshop on the Radiobiology of Nuclear Terrorism" was held at Bethesda, USA in December 2001 [2] to address the nuclear related issues in the post-September 11 era. Incidentally, about one year earlier another workshop, "Modifying Normal Tissue Damage Postirradiation" also addressed issues relevant to radiation damage [6]. Both workshop deliberations have many overlaps, understandably due to common concerns, in the domain of the existing knowledge base of the consequences of human exposure to moderate and higher doses (1–10 Sv) of radiation. The Bethesda workshop, in addition, outlined possible strategies to be adopted over the next 5 years to improve our ability to diagnose, triage, prevent and treat radiation injuries [3]. While the academic contents and recommendations of these meetings are relevant to the nuclear disaster scenario, one has to keep in mind that in a nuclear tragedy situation, large segments of population are likely to be simultaneously exposed to a range of doses. Therefore, there is a need to look into the biological response to different qualities as well as to variable quantities of radiation. Further, the context of discussion in the two meetings mentioned above related to situations in developed countries where medical, paramedical and first-response teams are relatively better trained and the awareness level of the population is relatively high.

The circumstance in a large number of under-developed, third world countries, including India, is quite different. The response management and medical infrastructures are relatively poor and the awareness level of the population is low. Thus, it is important that other aspects, including the psychological state of the population in distress, are also discussed and evaluated so that the immediate impact of such a disaster could be kept to a minimum. With this in view, the Indian Society for Radiation Biology (ISRB) took the lead in organising a multinational, multiagency workshop, "Radiation Risk in the Age of Nuclear Terrorism", in collaboration with the Research Center, Juelich (Germany), Health Canada, Ottawa (Canada) and School of Life Sciences, Jawaharlal Nehru University (JNU), New Delhi (India).

Issues of discussion at the workshop:

The workshop was held on November 16, 2002 at the School of Life Sciences, JNU, New Delhi. The total of 40 participants included a multinational team of radiation biologists and allied scientists from Japan – the solitary country having first hand experience of management of the aftermath of nuclear bombings [1]. Prof. R. N. Sharan, President of the ISRB, in his opening remarks, highlighted the need and urgency of the workshop and discussed its scope in light of the increasing threat perception of nuclear or "dirty" devices falling in the hands of terrorists. Professor Emeritus P. N. Srivastava (JNU, New Delhi and founder President of the ISRB), in his inaugural address, recalled the chronological events of the past where large segments of population across the globe were exposed to heavy doses of radiation. He impressed upon the participants the need to come up with suitable recommendations, which may form part of the strategies of various governments to tackle radiation emergencies with dexterity. Prof. K. A. Dinshaw (Tata Memorial Center, Mumbai) delivered a lucid thematic talk entitled "'Managing Radiation Emergencies' – a physician's view" covering all aspects of medical and health management during and after a nuclear holocaust. The next four technical sessions covered, in depth, various aspects of the consequences of radiation exposure, new approaches to biological dosimetry, biological responses and radioprotection strategies. Dr. A. Trivedi (Health Canada, Ottawa) presented new strategies for radiation risk assessment in the dose range of 1–10 Sv – doses that may not be lethal but can potentially cause acute effects. Dr. K. P. Mishra (Bhabha Atomic Research Center, Mumbai) detailed the need for more extensive research to elucidate the role of free radicals in the induction of membrane and DNA damage, which may provide better understanding of apoptotic death and bystander effects. Prof. R. N. Sharan (NEHU, Shillong) presented evidence of nucleotide sequence determined vulnerability of segments of genomic DNA suggesting that genome instability or inherent radiosensitivity may be, at least in part, determined by the primary sequence of nucleotides. Professor Emeritus M. S. Sasaki (Kyoto University) presented a new model of biological dosimetry. The strategy takes into account the fact that dose distribution is not likely homogeneous in a person exposed to radiation in a nuclear disaster. Thus, the score of chromosome aberration recorded in lymphocytes is spread over a mixed Poisson distribution into dose component to get most probable dose distribution profile and a realistic biological dose assessment [4]. Dr. F. H. A. Schneeweiss (Institute of Medicine, Research Center, Juelich) offered an alternative to this approach in which early cellular response could be assessed by analyses of lymphocyte proteins by two-dimensional differential gel electrophoresis and mass-spectrophotometer coupled with COMET-FISH analysis of DNA damage. Prof. P. Uma Devi (J. N. Cancer Hospital, Bhopal) presented results of extensive research involving prenatal exposures of mice and cancer incidence. Prenatal exposure was shown to significantly increase genome instability. Prof. R. K. Kale (JNU, New Delhi) presented evidence of the xanthine oxidoreductase system producing free radicals in the post-irradiation period suggesting, thereby, that inhibition of the system may reduce radiation damage. Dr. B. S. Dwarkanath (Institute of Nuclear Medicine and Allied Sciences, Delhi) discussed the possible use of minor groove binding DNA ligands, such as Hoechst 33258 and 33342, for protection of radiation induced DNA damage. The Hoechst ligands were shown to scavenge free radicals as well as afford stabilization to DNA superstructure. Dr. A. Chatterjee (NEHU, Shillong) elaborated upon the use of the endogenous radioprotector, GSH, in reducing post-irradiation damage to proliferating cells.

Recommendations of the workshop:

The final session of the workshop was a round-table plenary discussion with panelists Prof. P. N. Srivastava (India), Prof. M. S. Sasaki (Japan), Dr. F. H. A. Schneeweiss (Germany), Dr. A. Trivedi (Canada) and Dr. Vijaylaxami (USA). The session was initiated by a short presentation on "Chemical, Biological, Radiological and Nuclear Research & Technology Initiative" by Dr. Trivedi. All panelists and the participants interacted extensively and freely on various aspects of the proceedings of the day and made recommendations to define a strategy to handle a nuclear disaster scenario. Briefly, the main points of the recommendations were [5]:

1 International Collaborations/Partnerships: Closely interactive joint efforts be initiated and strengthened for free exchange of information and collaborative re-

search among scientists and institutions engaged in radiobiological teachings and research across the globe.

2 Preparedness: The participants felt strongly that "preparedness" was an essential component of containing the damage of a nuclear disaster scenario. It was felt that special initiatives are needed for school children and public awareness. ISRB should play a pro-active role in the design of such programmes to prevent misinformation and unfounded alarm to children and the public.

3 First Response Team: First response teams should be created at various locations. The fire brigade, police and paramilitary personnel should be given training and orientation on the effects of radiations and handling radiation accident/disaster situations including decontamination procedures. Preparation of emergency situation 'manuals' for different teams was recommended.

4 Categorization of Hospitals: Hospitals should be categorized and equipped for handling various categories of patients and different degrees of emergencies. Triage Centres and Teams, Emergency Centres, Definitive Care Centres and Specialized Centres could be some of these categories of hospitals.

5 Trauma Control Hospitals: Specialized hospitals with psychotherapists and psychoanalysts should be developed for trauma control.

6 Biological Dosimeters with Field Applicability: Appropriate biological dosimeters, especially with field applicability, need to be developed. National and international initiatives on such research and development programmes and priority funding were recommended.

7 Radio-modulatory Drugs: New generation drugs for post-exposure treatment, such as endogenous or exogenous radioprotectors, free radical scavengers and nucleotide modifiers, should be developed and tested through international collaborations and priority funding.

8 Role of Bhabha Atomic Research Center (BARC): Active collaboration and exchange of views between scientists of BARC, Mumbai and Universities should be initiated and encouraged on a large scale.

9 Role of the Indian Society for Radiation Biology (ISRB): The ISRB should take initiatives to start appropriate contacts programs to achieve the above goals. Prof. Sharan, President, was entrusted with the task of preparing the report and forwarding the recommendations to appropriate agencies in Indian and other participating countries.

References

1 ISRB-Participants (2002): http://www.geocities.com/ISRB India/participantss main.html

2 Moulder JE (2002a) Radiobiology of nuclear terrorism: Report on an interagency workshop (Bethesda, MD, December 17–18, 2001) International Journal of Radiation Oncology Biology Physics 54:327–328

3 Moulder JE (2002b) Report on an interagency workshop on the Radiobiology of nuclear terrorism: molecular and cellular biology of moderate dose (1–10Sv) radiation and potential mechanisms of radiation protection (Bethesda, Maryland, December 17–18, 2001), Radiation Research 158:118–124

4 Sasaki MS (2003) Chromosomal biodiversity by unforlding a mixed Poisson distribution: a generalized model. International Journal of Radiation Biology 79:83–97

5 Sharan RN (2003) Preparedness to respond to possible acts of nuclear terrorism: Some strategies and recommendations, International Journal of Radiation Biology 79:217–219

6 Stone HB, McBride WH, Coleman C (2002) Modifying normal tissue damage postirradiation. Report of a workshop sponsored by the Radiation Research Program, National Cancer Institute, Bethesda, Maryland, Sept. 6–8th, Radiation Research 157:204–223

Reflections on the Acceptability of the Use of Ionizing Radiation to German Society

Jürgen Böhm

Physikalisch-Technische Bundesanstalt, 38116 Braunschweig, Germany
juergen.boehm@ptb.de

The situation is serious. At German universities, there is a dramatic decrease in the number of students registering for physics and chemistry. Also, a pronounced decline is observed for interdisciplinary courses of studies such as radiation biology, radiation physics and nuclear chemistry. Add to this that radiation research in Germany is being cut back – either by reductions in staff, materials, equipment and funding or by closing down chairs. The reasons students state for why they do not select ionizing radiation as a focus of their studies are above all that the job prospects are poor and that they do not see why they should deal with the application of ionizing radiation.

What are the origins of such negative visions? To what extent are these visions justified, and do prejudices play an essential role? What can be done to break down the prejudices? It is not the aim of this lecture to deal with this topic in all its facets; from this conference I expect contributions from different viewpoints. I here wish to refer to some points which in my opinion are important when reflecting on the acceptability of the application of ionizing radiation to German society.

First of all the ways in which many students take their decision are to be clarified. They certainly account for facts when assessing the advantages and disadvantages for their decision. But in the present case, the facts to be considered are not easily comprehensible, and in part they are contradictory. Also, students know from experience that facts alone do not offer adequate security for the planning of their future. So the voice from the subconscious plays an important part, and they try to reconcile "head and instinct".

The attitude towards the peaceful use of nuclear energy is certainly another important aspect. When about four decades ago we became aware of the global scarcity of fossil fuels, powerful campaigns for the peaceful use of nuclear energy were launched which ultimately led to nuclear energy accounting for 30,6 percent of the electrical energy produced in Germany (April 2001). But the Chernobyl nuclear power plant accident in 1986 continues to have psychic and social repercussions even today. This accident was caused by overheating of the nuclear fuel setting the graphite moderator in the reactor on fire during abnormal operating conditions of the reactor.

People are afraid of possible accidents in nuclear facilities and their short- and long-term effects on their health and the health of their children. Many problems in relation with the storage of nuclear waste produced in nuclear power plants are still unsolved. The production of bomb-grade plutonium in nuclear power plants producing electrical energy relocates the "peaceful" use. When people worry about the potential for a catastrophe, nuclear power is ranked as number one.

The present German government plans to opt out of nuclear energy. In view of the long residual operating times of the nuclear power plants, reliable technique and qualified staff are required not only until they are switched off but also for decades afterwards because of the decommissioning of the nuclear power plants with the ultimate goal of the removal or confinement of all radioactive materials and the unrestricted release of the sites. In the field of nuclear waste management, competence for nuclear technology will still be called for a long time.

From a student's viewpoint, this might look like administration of the estate rather than future-centred development. The vision of a new innovative generation of nuclear power plants which from their design and location are safe as regards faulty operation and terrorist attacks, and the radioactive waste from which can be stored *in situ*, and which cannot be used for the production of nuclear weapons is not opportune in view of investments – despite the missing CO_2 emissions. Electrical power can at present be produced at lower cost from oil or coal.

The beneficial character of ionizing radiation in medicine often is not recognized by the public and is explained in the following by one example, the radiation therapy (the important field of diagnostic radiology is not dealt with here). In Germany 300 000 persons develop cancer and every fifth dies of a tumour. For about 55 percent of all patients a local tumour is diagnosed, and about two thirds of these persons can be cured owing to the local treatment (surgery and radiation therapy). In the field of cardiology, in the last few years, promising results have been obtained in the treatment of patients suffering from restenoses using beta radiation. It is alarming that a medical doctor usually first has to convince the patient to be treated by means of radiation therapy that the ionizing radiation he uses is not as harmful as the patient commonly thinks (association with Hiroshima, Chernobyl, etc., see above).

I think that the acceptance of the use of ionizing radiation starts in the subconscious. Ionizing radiation is so frightening because we cannot understand it with our senses. Here, for example, visualization of the natural ambient radiation in a cloud chamber can do wonders. It is fascinating to see how a great number of particles of the natural cosmic radiation and the terrestrial ambient radiation permanently impinge on us – particles we do not realize at all (see Fig. 1., page 269).

Another point is that the effect of ionizing radiation can also be simply beautiful. What a wonderful spectacle are the polar lights (see Fig. 2., page 269). In the upper air layers of the atmosphere, ionizing radiation generates magical coloured images in the night sky, both in the northern and in the southern polar regions.

Small spinthariscopes whose outside resembles a microscope allow the phenomenon of radioactive decay to be observed with the naked eye by observing light flashes on a luminous mass such as, for example, zinc sulphide.

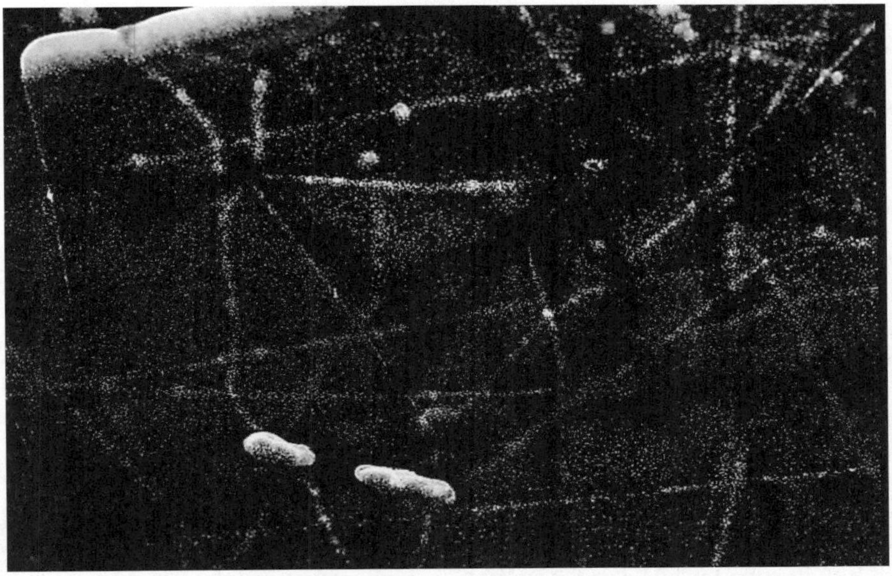

Fig. 1. Natural environmental radiation made visible by a diffusion cloud chamber. The tracks are produced by ionizing particles (photograph by Phywe Systeme GmbH, Göttingen).

Fig. 2. Structured polar light observed on the Andoya island (northern Norway).

Natural cosmic radiation and the terrestrial ambient radiation were already experienced by our ancestors – so can such a natural radiation level really be as harmful as some people try to make us believe? In principle, even a single particle of ionizing radiation could change an important molecule so that the cell could be damaged and the damage copied as the cell divides. There thus is a certain theoretical risk but in practice effects from small doses have not been observed. The long term biological effects of small radiation doses are so minute that even thorough epidemiological studies could not detect them. It is, for example, not known how many cases of cancer, if any, are caused by natural cosmic radiation and the terrestrial ambient radiation. On the contrary, some radiation may be beneficial to health. Our health will suffer, for example, if we do not get sunlight but massive doses of ultraviolet light would cause skin cancer.

If one wants to enhance the societal acceptance of the use of ionizing radiation in Germany, one should go in for the change of the negative image attached to ionizing radiation by the subconscious. From the events of Hiroshima and Chernobyl we have had to learn some things with grief – but not that ionizing radiation as such is bad. It is part of nature just as air and water and can be made visible to everybody in a cloud chamber. In the polar lights, its beauty blossoms in interaction with air. Spinthariscopes allow the statistic nature of radioactive decay to be experienced. In radiation therapy, ionizing radiation contributes to improvement and to preservation of life. It is worthwhile also for students looking into ionizing radiation even if at present there is no boom for the peaceful use of nuclear energy in Germany. Nevertheless, technologies associated with ionizing radiation contribute in many ways to the health, prosperity and security of the country. Investments in radiation research with its many interdisciplinary areas are necessary to be made in order not to lose contact with the international development and to maintain the competence acquired in Germany over many years.

Janus Face of Radiation Biology and Science for the Future

Ohtsura Niwa

Radiation Biology Center, Kyoto University, Kyoto 606-8501, Japan
oniwa@house.rbc.kyoto-u.ac.jp

Introduction

Radiation research is a multidisciplinary field of science including physics, chemistry, biology, medicine and risk science. Radiation biology, a branch of the field of radiation research investigates biological effect of radiation on living organism which starts from a physical event caused by a traversal of high speed electron in cells and tissues. Initial damage is induced by a physicochemical process in a stochastic manner and the amount of damage is determined by the dose of radiation. The classic radiation biology postulated that the biological effect of radiation is solely determined by the initial event. However, the development of modern radiation biology has excavated a variety of important biological processes and they, not the physical and chemical events, are the major determinant of radiation effects. In addition, it is becoming more and more apparent that biological processes in the irradiated cells are not designed to cope with radiation damage. They are in fact designed for more basic needs of the living organisms to cope with external and internal stresses. Therefore, study of radiation biology inevitably leads to the study of the basic design of life; thus complementing other disciplines of life science. Therefore, radiation biology is now contributing to the most fundamental understanding of how life perpetuates, although radiation biology started from the biophysics of radiation. Dual nature of radiation biology, strong biophysics and powerful tool to study basic biological processes, characterizes this discipline, like the dual face does for the god of Janus. In this article, some of the important issues of radiation biology are discussed to demonstrate that radiation research is science for the future which can contribute to the basic understanding of life.

Non-significance of ionizing radiation to life on earth

UV and ionizing radiations are detrimental to life due to the damage they inflict on DNA. Systems to repair DNA damage, and eliminate damaged molecule and cells are most important for the survival of the life. UV had been one of the most formidable DNA damaging agents especially when atomospheric ozone was non existent before oxygen generating cyanobacteria and blue-green algae evolved on earth around 3 billion years ago [21]. UV still is life threatening and therefore, organisms are well equipped to cope with UV damage. Two types of DNA repair evolved to repair UV damages; the photoreactivation and excision repair. The former specifically operates on UV dimer and 6-4 photoproduct, splits the damage and reverts them to the original conformation. The latter operates on a wider range of DNA damage, and functions by excising the region of the damage which is then filled by repair synthesis.

In contrast to UV, the level of ionizing radiations have never been high enough to affect the course of evolution of life, except in a rare occasion in which radioactive decay sometimes provides an energy source to bacteria in the depth of earth [9]. In any case, ionizing radiation has never been a threat to life throughout the history of earth. The same conclusion is reached when one consider the chance of radiation induced DNA damage in a genome. The classic work of radiation biology has established that radiosensitivity of organisms is inversely related to their genome size [14]. The genome size dependency of radiosensitivity can easily be understood since a chance to be hit by radiation is a linear function of the genome size. For example, 1 Gy of ionizing radiation gives an order of 10 double strand breaks and 1000 single strand damages per 10^9 bp DNA [31]. The genome size of 10^9 bp is common among higher animals with exception of amphibians some of which have DNA contents of one order of magnitude higher than mammals [12]. Also, the genome size of higher plants varies from 10^8 to 10^{11} bp. The annual dose in the natural environment is in the order of 1 to 10 mSv and this is translated to a traversal of single electron per nucleus of mammalian cells per year. Therefore, the level of DNA damage due to ionizing radiation in the natural environment is trivial to higher animals and plants, the latescomer on earth being evolved from organisms of smaller genome sizes to which the threat of natural radiation is almost non-existent. In addition, higher animals and plants have evolved in the last 500 million years, long after the basic design of DNA repair had been established in the preceding 2 billion years of evolution.

Enigma of well evolved repair system and cell cycle checkpoint

Even though ionizing radiation was not the threat, life is endowed with the capacity to evade efficiently the detrimental effects of radiation. Ionizing radiation is known to induce a variety of DNA damages, including double strand breaks, single strand breaks and base damages. Among these, double strand breaks are especially detrimental to cells [34].

Double strand break is repaired by two mechanisms: recombination repair and non-homologous endjoining repair. These two systems are found among and shared by prokaryotes and eukaryotes [6, 35]. In addition, radiation damage induces cell cycle checkpoint in eukaryotes which allows time for the cells to repair DNA damage. Almost all the molecular pathways of cell cycle checkpoint are shared among eukaryotes, from yeast to higher metazoan animals. In addition, metazoan species are endowed with two additional systems to eliminate irreparable cells by senescence mediated permanent cell cycle arrest and by apoptotic death [33]. So, in the face of non-existent threat of ionizing radiation, highly evolved repair and checkpoint systems pose an enigma of why evolution elaborated to create such costly mechanisms.

The other side of DNA repair: the real biological meanings

Recombination plays a major role in repairing radiation induced DNA double strand breaks in S and early G2 phase where two sister chromatids are aligned with the help of cohesin protein to serve as substrates of recombination. Many of recombination deficient mutants were isolated in E. coli and in yeast by selecting for their sensitivity to ionizing radiation [2, 8]. Mammalian cells exhibit split-dose recovery, also known as Elkind recovery, which is S phase dependent. Recent study using a Rad54 deficient chicken cell line has demonstrated that Elkind recovery is due to recombination repair of radiation damage [32]. Since radiation damage occurs rarely, if any, in natural environment, the biological significance of the highly evolved recombination repair have to be found in the more basic necessity of life.

Classically, recombination is thought to have evolved for the shuffling of genes to allow a wider genomic diversity. However, recent studies have indicated that the function of the recombination machinery is not for the shuffling of genes. DNA replication is most fundamental for life to perpetuate and is carried out by a specialized enzyme complex containing DNA polymerases. The template DNA is usually associated with numerous regulatory factors and systems for the structural maintenance of chromatin. Processivity of DNA polymerases varies, but during replication of template DNA, they frequently encounters obstacles such as tightly bound proteins, highly GC-rich regions and DNA damages, upon which polymerization stops and the replication fork collapses. Recent studies revealed that the basic function of recombination is to overcome the replication fork collapse by retrieving sequences from the sister strand [25]. This basic need explains why the recombination machinery is found among almost all forms of life, from bacteriophage to eukaryotes [5]. UV damage repair in contrast is not found in parasitic bacteriophages and viruses which make use of the host repair system for the economy of their genome size, with a rare exception of T4 phage which possesses a specialized glycosylase capable of excising pyrimidine dimmers in DNA. In addition to repair, the importance of recombination for more basic need is evident since recombination deficient mutants of any organism are quite sick, and sometime non-viable as is the case of higher eukaryotes [17]. So, although recombination was first identified by the study of radiation biology, it is a good example of the Janus nature of radiation biology; the damage repair system is more than for repair, but for essential need of life.

Since radiation gives to cells an insult which they have not encountered naturally, the cells have to utilize functions evolved for other purposes. Cell cycle checkpoints, senescence mediated permanent cell cycle arrest and apoptosis were evolved in metazoan animals when cells are irreparably damaged. Radiation induced DNA damage is sometime difficult to be handled by the system evolved to cope with natural DNA damage. Therefore, it is understandable that irradiated cells frequently undergo permanent cell cycle arrest and apoptosis. These examples again demonstrate that radiation biology is a science of dual nature; under the specific problem of radiation biology, there lies a more general mechanism of life. With this view in mind, we now move onto some of specific issues of radiation biology, radiation induction of cancer and low dose rate risk in the hope of finding the other side of the problem, namely the true biological significance of the phenomena.

Radiation carcinogenesis and the other side of the problem

Radiation has long been known to increase the frequency of cancer. It was previously assumed that radiation induced DNA damage is fixed to carcinogenic mutations directly. Numerous carcinogenesis models exist and the Armitage-Doll model was derived from the observation that the incidence of cancer among English men increases with age to 3rd to 7th power, depending on the type of cancer [1]. From this age-incidence relationship, the authors proposed a multistep carcinogenesis model in which accumulation of a certain number of age-related events is required for cancer development. More than thirty years later, molecular analysis of colon cancer excavated that the event and the step correspond to mutation of oncogenes and tumor suppressor genes [7]. Spontaneous mutation arises mainly from reactive oxygen species, a potent mutagen and inevitable byproduct of the bodily metabolism. So, mutation accumulates linearly with age, and cancer requiring "n" number of mutations would arises with age to $(n-1)^{th}$ power. The multihit nature of carcinogenesis is assumed to hold for any mutagen including radiation. Since the frequency of mutation is linearly related to the dose of radiation, one can assume multihit dose response of cancer incidence. In the case where radiation contributes only a limited number of mutations, say r, the rest of mutations have to be induced by the spontaneous process. In this case, cancer incidence of irradiated population would increase not by $(n-1)^{th}$ but by $(n-1-r)^{th}$ power of age.

Cancer among the atomic bomb survivors can be categorized into two distinct types, leukemia and solid tumors. The relative risk of leukemia is high among the survivors and fits a linear-quadratic dose response [27, 23]. This linear-quadratic dose response is somewhat reminiscent of the multihit dose response expected for the direct mechanism of radiation carcinogenesis. Leukemia exhibits short latencies of within 5 to 10 years after the bomb and this again is consistent with the direct mechanism.

In contrast to leukemia, risks of solid tumors are linearly related to the dose of radiation [27, 29]. Their latencies are generally long and exposed people tend to come down with solid cancer late in their life, similar to the non-exposed population. In fact, relative risk of solid tumors stays constant over attained age for those exposed in their adulthood [27]. Linear dose response and the constant relative risk of solid tumors among atomic bomb survivors indicate that contribution of radiation is not to induce carcinogenic mutation directly, but to increase the mutation frequency. Two mechanisms are known to give a higher frequency of mutation in cells and tissues after irradiation: chronic inflammation and genomic instability [19, 18, 3]. Both of the mechanisms lead to time dependent mutagenesis. Radiation induced mammary cancer in rats and radiation induced transformation of C3H10T1/2 cells in culture can best be explained by the time dependent induction of carcinogenic mutations [13, 15]. Whatever the mechanism would be, genomic instability and chronic inflammation can be viewed as the defense against biological insults such as infections and physicochemical insults to genome. So, behind the seemingly reasonable mechanism of direct radiation carcinogenesis, more intricate mechanisms of stress response are hidden.

Genomic instability, the last choice when all other defense mechanisms fail

DNA repair, cell cycle checkpoints, permanent arrest of cell cycling and apoptosis all evolved to protect life from DNA damage. One other ancient mechanism however exists to cope with DNA damage. Cell can either repair damage or die, but when cells choose survival in the presence of unreparable DNA damage, they have to do so with the cost of infidelity of replication and chromosome segregation. Thus, genomic instability is the inevitable consequence of the survival of damaged cells.

It is known in some organisms that they can change mutation frequency depending on the environment surrounding them. Stationary mutagenesis is one example in which mutation frequency is controlled by environmental condition. E. coli and yeast cells mutate more readily to acquire the ability to use a new energy source when they are kept in nutritionally poor condition [30]. Adaptive mutation is demonstrated in E. coli introduced into the mouse gut where they exhibit higher mutation frequencies until they stably colonize the flora [10]. In these instances, mutagenesis is not a passive process in which DNA damage is fixed to become mutation by some mistakes, but it is more of an active choice of the cells. Recent discoveries of a series of translesional DNA polymerases enriched our knowledge at the pathway of active mutagenesis [11].

Mutation by radiation induced genomic instability was first demonstrated in cells in culture [18] but it does also take place in vivo [20, 28]. However, the role of genomic instability in radiation carcinogenesis is yet to be demonstrated. In any case, radiation carcinogenesis, especially induction of carcinogenic mutations, is not a simple and passive process which is determined solely by the initial DNA damage. The process is deeply connected to important cellular and bodily regulations.

Low dose and dose rate radiation effect and the importance of tissue kinetics

Classical radiation biology postulated that the consequence of radiation on living organisms is determined stochastically. The proposal was made in which the number and the special distribution of DNA damage in a critical cellular mass determine the outcome of the biological effect [16]. Generalization of this hypothesis yielded the following equation of dose and the effect.

$$E = \alpha D + \beta D^2$$

In this equation, the biological effect is described as a combination of linear and quadratic terms with constant α and β values. The quadratic term assumes that damages interact each other when they are produced within a critical distance. Damage within the critical distance, but separated by a certain time period, do not to interact. Thus, the effect of low dose rate irradiation is the same as that of low dose and is described only by the linear term.

Since energy of radiation is quantumized, elementary dose, the lowest dose a cell can receive, has a limit; 1 mGy/nucleus/traversal of electron for low LET radiation such as gamma ray and 300 mGy/nucleus/traversal of a particle for high LET alpha particle. The elementary dose by an alpha particle is quite high and cannot be regarded as a low dose. In any case, dose rate effect should not operate on the linear term of the equation. However, we do know that the linear term of many of radiation effects, such as mutation induction and carcinogenesis by radiation is subject to dose rate dependent changes; a classic example is the dose rate dependent decrease of the linear dose response curve of radiation induced germ line mutation in mice [26].

In the tissue microenvironment, cellular proliferation/differentiation is conducted in a hierarchy fashion by stem cells, transit proliferating cells and terminally differentiated cells [22]. The terminally differentiated cells do not mutate since they are usually non-proliferating and are doomed to be excluded from the body after filling their function for a certain period of time. Transit cells mutate since they are proliferating rapidly. However, cells in the transit compartment accumulate mutations only until the cells exit from this compartment to become terminally differentiated cells. In addition, a hypothesis was recently proposed in which the stem cells keep immortal strands of DNA to avoid accumulation of mutations [4]. In this model, mutation in epithelial cells is possible only when mutated cells in the transit compartment take over the niche of stem cells which is super-sensitive to the killing effect of radiation. So, the low dose rate issue of radiation risk has yet to be elucidated not only from cellular mechanisms but also from the higher order mechanisms functioning at the tissue level.

More to come; radiation biology as science of future

Radiation biology has important applications to radiotherapy and the risk/safety problem. As discussed in this article, however, radiation biology is by itself science for future. Radiation biology owes its birth to radiation physics, but gradually it shifted to more of the basic life science. Research of radiation biology inevitably leads to deeper understanding of how life is coping with genomic stress. Janus face has been the nature of radiation biology in the past and it is for the future of this field.

Acknowledgement

The author sincerely thank Professor Jürgen Kiefer to give a chance to write this article.

References

1 Armitage P, Doll R (1954) The age distribution of cancer and multistage theory of carcinogenesis. Br J Cancer 8:1–12
2 Clark AJ., Margulies AD (1965) Isolation and characterization of recombination-deficient mutants of Escherichia cloi K12. Proc Natl Acad Sci USA 53:451–459
3 Bridges BA (2001) Radiation and germline mutation at repeat sequences: Are we in the middle of a paradigm shift? Radiat Res 156:631–641
4 Cairns J (2002) Somatic stem cells and the kinetics of mutagenesis. Proc Natl Acad Sci. USA 99:10567–10570
5 Cox MM (2001) Historical overview: Searching for replicarrinn help in all of the rec place. Proc Natl Acad Sci USA 98:8173–8180
6 Edelmann W, Kucherlapati R (1996) Role of recombination enzymes in mammalian cell survival. Proc Natl Acad Sci USA 93:6225–6227
7 Fearon ER, Vogelstein BA (1990) genetic model for colorectal tumorigenesis. Cell 61:759–767
8 Friedberg EC, Walker GC, Siede W (1995) in "DNA Repair and Mutagenesis" AMS Press, Washington
9 Field H (2003) Science Nuclewar-powered bugs. Science 299:1307
10 Giraud A, Matic I, Tenailon O, Clara A, Radman M, Fons M, Taddei F (2001) Cost and benefits of high mutation rates: adaptive evolution of bacteria in the mouse gut. Science 291:2606–2608
11 Goodman M F (2002) error-prone repair DNA polymerases in prokaryotes and eukaryotes. Annu Rev Biochem 71:17–50
12 Handbook of Biochemistry (1970) CRC Press, Sober H. H. et al. ed., Cleaveland
13 Kamiya K, Yasukawa-Barnes J, Mitchen JM, Gould MN, Clifton KH (1995) Evidence that carcinogenesis involves an imbalance between epigenetic high-frequency initiation and suppression of promotion. Proc Natl Acad Sci USA 92:1332–1336
14 Kaplan HS, Moses L (1964). Biological complexity and radiosensitivity. Science

145:21–25

15 Kennedy AR, Fox M, Murphy G, Little JB (1980) Relationship between x-ray exposure and malignant transformation in C3H 10T1/2 cells. Proc Natl Acad Sci USA:7262–7266

16 Kellerer AM and Rossi HH (1974) The theory of dual radiation action. Current Topics in Radiation Research 8:85–158

17 Lim DS, Hasty P (1996) A mutation in mouse rad51 results in an early embryonic death that is suppressed by a mutation in p53. Mol Cell Biol 16:7133–7143

18 Little JB (2000) Radiation carcinogenesis. Carcinogenesis 21:397–404

19 Neriishi K, Nakashima E, Delongchamp RR (2001) Persistent subclinical inflammation among A-bomb survivors. Int J Radiat Biol 77:475–482

20 Niwa O, Kominami R (2001) Untargeted mutation of the maternally derived mouse hypervariable minisatellite allele in F1 mice born to irradiated spermatozoa. Proc Natl Acad Sci USA 98:1705–1710

21 Ohno S (1997) The reason for as well as the consequence of the Cambrian explosion in animal evolution. J Mol Evol 44 suppl 1:S23–27

22 Loeffler M, Potten CS (1997) in "Stem Cells" Potten CS ed. Academic Press

23 Preston DL, Kusumi S, Tomonaga M, Izumi S, Ron E, Kuramoto A, Kamada N, Dohy H, Matsui T, Nonaka H, Thompson DE, Soda M, Mabuchi K (1994) Cancer incidence in atomic bomb survivors. Part III. Leukemia, lymphoma and multiple myeloma, 1950–1987. Radiat Res. 137:S68–97

24 Radman M (2001) Fidelity and infidelity. Nature 413:115

25 Radding C (2001) Links between recombination and replication: vital roles of recombination. Proc Natl Acad Sci USA 98:8172

26 Russell WL, Kelly EM (1982) Mutation frequencies in male mice and the estimation of genetic hazards of radiation in men. Proc Natl Acad Sci USA 79:542–544

27 Shimizu Y, Kato H, Schull WJ (1990) Studies of the mortality of A-bomb survivors. 9. Mortality, 1950–1985: Part 2. Cancer mortality based on the recently revised doses (DS86). Radiat Res, 121:120–141

28 Shiraishi K, Shimura T, Taga, M, Uematsu N, Gondo Y, Ohtaki M, Kominami R, Niwa O (2002) Persistent induction of somatic reversions of the pink-eyed unstable mutation in F1 mice born to fathers irradiated at the spermatozoa stage. Radiat Res 157:661–667

29 Thompson DE, Mabuchi K, Ron E, Soda M, Tokunaga M, Ochikubo S, Sugimoto S, Ikeda T, Terasaki M, Izumi S, Preston DL (1994) Cancer incidence in atomic bomb survivors. Part II: Solid tumors, 1958–1987. Radiat Res 137:S17–67

30 Torkelson J, Harris RS, Lombardo MJ, Nagendran J, Thulin C, Rosenberg SM (1997) Genome-wide hypermutation in a subpopulation of stationary-phase cells underlies recombination-dependent adaptive mutation. EMBO J 16:3303–3311

31 UNSCEAR Report (2000) In Annex F, "Sources and effects of ionizing radiation", United Nations Publication, New York

32 Utsumi H, Elkind MM (2001) Requirement for repair of DNA double strand breaks by homologous recombination in split-dose recovery. Radiat Res 155:680?688

33 Wahl GM, Carr A (2001) The evolution of diverse biological responses to DNA damage: insights from yeast and p53. Nature Cell Biol 3:277–286

34 Ward JF, Blakely WF, Joner EI (1985) Mammalian cells are not killed by DNA single-strand breaks caused by hydroxyl radicals from hydrogen peroxide. Radiat Res 103:383–392

35 Weller GR et al. (2002)Identification of a DNA nonhomologous end-joining complex in bacteria. Science 297:1686–1689

Postscript: Thoughts on the Future of Radiation Research and its Impact on Life Sciences

Jürgen Kiefer

Strahlenzentrum der Justus-Liebig-University, Giessen, Germany
juergen.kiefer@strz.uni-giessen.de

The problem

A large part of the workshop whose proceedings are summarised in this volume was devoted to discussions, both between presentations and in a dedicated session. As a free interchange of ideas was intended no recordings or notes were taken. It is thus difficult to give a reliable account and I shall not attempt it. The views presented are thus my personal deliberations although, of course, they are based on talks which I had with many colleagues.

It goes without saying that there was general agreement that science in general would loose if radiation research would die but the problem is "to convince the others". There is no doubt that radiation-related disciplines will always need the help of radiation research. Typical and important examples are radiation therapy (both for malignant and benign diseases), nuclear medicine and radiation protection. Many technical and environmental issues require the active involvement of radiation research and there is little discussion about these necessities. But even here one finds a lot of lip service and generally too little activity to secure at least the survival of existing institutions.

The main problem is, however, to demonstrate convincingly that progress in life sciences would be severely hampered in the absence of radiation research as a discipline in its own right. In the following it is attempted to draw a few lines. They relate to cellular and molecular biology. It is not intended to mean that this area is the most important but rather guided by personal interest and expertise and also influenced by the topics of the workshop. Neither exclusivity nor comprehensiveness is intended.

Life sciences and radiation. Why?

Repair

Even outside the radiation area it is appreciated that the knowledge about repair processes would be very meagre without radiation studies. But now the field has been taken over by biochemists and geneticists so radiation studies appear dispensable. It is certainly true that our understanding of the mechanisms of repair owes most to the intricate techniques of biochemistry but radiation is still a necessary tool. The main advantage is its exact delivery mode. It can be switched on and off at well defined time points. Only on this way it is possible to study time courses and the sequence of events. If DNA damage is induced by chemical agents it takes time for them to reach the target and they are not easily removed or diluted out. Dose rate studies are very difficult with them if not impossible. With the better knowledge of radiation products in DNA and powerful techniques to quantify them the disappearance of damage can be followed and the dependence on cellular parameters and added substances, e.g. repair inhibitors, can be easily followed.

The last years have seen the discovery of many repair-deficient mutants. They were nearly exclusively found with the help of radiation. We have just started to understand their impact on human health. This chapter is by no means closed. Genetic disposition to disease will increase in its importance in future and will never be understood without radiation studies.

Carcinogenesis

The fact that radiation induces cancer contributes essentially to the public fear about this agent and leads many scientists to the decision to avoid its use. On the other hand, it gives us a powerful handle to study the processes involved. As documented in the paper by Dahm-Daphi and also others in this volume there is a close connection between cancer and chromosomal aberrations, a signature of radiation impact. Currently discussed models of carcinogenesis were developed on the basis of radiation studies, they can only be tested with the help of radiation. Today many people believe that genomic instability plays a significant role in the long way from initiation to progression. How would we know about this phenomenon without radiation? The mechanism behind the scene is still obscure – it cannot be clarified without sophisticated radiation studies in cells, tissues and organisms.

Intra- and intercellular signalling

Radiation biology is often blamed to be too staid not taking into account the intricate web of cross talk between cellular compartments and in the tissue as a whole. The success of "target theory" contributed to this impression but we are nowadays well aware of the fact that the fate of a cell is not solely determined by hits in a target, although they are the starting points. Many contributions in this volume wit. ness a new way of thinking. In fact, modern radiation research has provided the tools to study the interaction of lesions in different parts of the cell. The "microbeam" technique offers opportunities to investigate the flow of signals between cells and the intercellular matrix. It may also have great potentials even in neurobiology which have not yet been explored at all. Another possible use is the study of topological aspects. In this way radiation biology can add enormously to the understanding of the cell as a web of communicating elements and eventually add to the knowledge of biological regulation and misregulation.

Others

The areas given above should just serve as examples, there are many more and the discussion would require a book of its own. The "brainstorming" at the meeting lead to a pictorial summary of topics which reproduced in Fig. 1. below. It was produced from the inputs of the participants by Friederike Eckardt-Schupp whom I thank for the permission to use it.

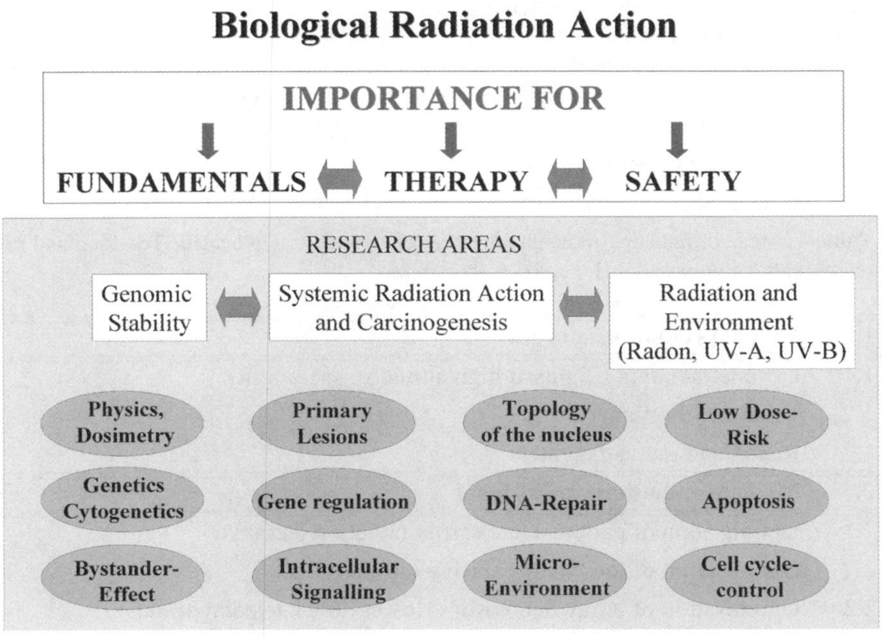

Fig. 1. An attempt to list areas of radiation research in relation to other branches of science. The lower "bullets" represents an unstructured collection of modern topics (courtesy of F. Eckardt-Schupp)

The discussions centered mainly around fundamental aspects but the good relations to radiation application are not to be underestimated. More than 50% of all cancer patients receive radiation as a part of their treatment, any further improvement will not only depend on new machinery but also on a better understanding of the effects on tumour and normal cells. The close cooperation between clinicians and radiobiologists has good tradition, it must not be given up, and medical faculties should be aware of the danger if they cut the links to the understanding of radiobiological fundamentals. This is not merely an academic issue but relates directly to the well-being of their patients. If a medical doctor does not know about genetically determined radiosensitivities he could easily end up with the wrong treatment.

Radiation protection is often seen only as a regulation for the working place. Although this is quite important it is often overlooked that this discipline has great merits far beyond the radiation area. The system served as a model for protection for many protection measures, both for the working and the general population. This expertise must be maintained for the sake of human health in an environment with steadily increasing contaminations from all kind of sources, not only from industry but also from traffic, communication (mobile phones) and spreading human activities.

It is thus clear that the significance of radiation research goes far beyond its importance for life sciences. H. Paretzke has prepared for the workshop a list of areas where research and the maintenance of expertise is necessary, it is given in a shortened version in table 1.

It is important to remember that radiation research has left its footprints not only in many branches of science but also generally in our society. These issues are by no means closed and they can only be properly addressed if we continue to teach young people at the best institutions.

Table 1. Areas of radiation research where special support is indicated (Teaching and research) (after a proposal of H. Paretzke, shortened)

1.	**Natural radiation sources**
1.1	Cosmic radiation (flights at high altitudes, spacecraft)
1.2	Uranium, thorium
1.3	Radon and radon daughters
2.	**Medical radiation exposures**
2.1	Optimisation of photon and electron radiation therapy
2.2	Optimisation of ion therapy (heavy ions, protons)
2.3	Optimisation of X-ray diagnostics (lower doses, telediagnostics)
2.4	Nuclear medicine

3.	**Internal dosimetry**
3.1	Bio- and pharmacokinetics of radionuclides
3.2	Development of voxel phantoms for both sexes and various ages
3.3	Improvement of computer codes for dose conversion factors
3.4	Improvement of risk estimates for inhaled radionuclides
3.5	Improvement of risk estimates for ingested radionuclides
3.6	Improvement of measurement techniques for incorporated radionuclides
4.	**External dosimetry**
4.1	Exposure by photons (radiation quality, dose rate)
4.2	Exposure by electrons (real time, contaminations, "hot spots")
4.3	Neutrons (e. g. personal dosimeters for mixed radiation fields)
4.4	New approaches for high sensitivity personal dosimetry
5.	**Retrospective dosimetry**
5.1	Improvement of luminescence methods for building and environmental materials
5.2	Improvement of ESR-methods for teeth (e.g. in situ measurements)
5.3	Improvement of biological dosimetry
6.	**Radioecology**
6.1	Better quantitative understanding of radionuclide distribution
6.2	Quantification of deposition and resuspension processes
6.3	Further improvement of current models of radionuclide distribution in the food chain
6.4	Identification and breeding of plants with reduced radionuclide up-take
6.5	Development of robust models for nuclear waste storage
7.	**Radiation effects**
7.1	Understanding of low dose carcinogenic action a on humans
7.2	Epidemiological and molecular studies on health effects other than cancer
7.3	Improvement of radiobiological models, including genetic instability, bystander effect etc., for low doses
7.4	Studies on individual radiosensitivity
7.5	Studies on molecular markers for past radiation exposures
7.6	Early recognition of tumour related processes in exposed individuals
7.7	Improvement of mathematical and biophysical methods in epidemiological studies
7.8	"Practical" radiation chemistry to assess waste storage problems
7.9	Radiation effects in ecosystems

8.	**Concepts of radiation protection**
8.1	Development of better health-related indices and of exposure limitations
8.2	Improvement in the derivation of operational quantities

What is to be done?

It does not suffice if we convince ourselves that we do interesting and important research – more has to be done and we have to take the message to the non-believers. A number of points come to mind which are listed below:

1 Be in the forefront of modern science!
This reminder is certainly unnecessary for most of us but this has not always been so. Radiation biology can only produce significant results if it contributes to the problems of current biology and proves its value in this way. This means, e.g., not to publish solely in radiation journals but to stand up to the competition in more general ones.

2 Do not hide!
We have to counteract the wide-spread attitude that radiation is bad. The contributions gathered in this volume give good examples about the value of research in our field. This is not paid by the nuclear industry, so we have no reason to feel guilty and hide our achievements under titles where the term "radiation" is suppressed.

3 Keep the links!
The close cooperation between radiation biology and radiotherapy or radiation protection should not only be kept in good shape but strengthened. We need them but they also need us.

4 Go public!
Radiation is everywhere and the questions about its action are by no means restricted to the use of nuclear energy. UV and skin cancer (see the paper by Greinert et al.) is just an example. The general public is increasing need of specialists who are able to give well founded advice. This field must not be left to "self-made experts". There is certainly no disagreement about this but the politicians and the public has to be told. The task is by no means easy, it takes a long time and many words to laymen that presumed dangers may not exist. On the other hands, if there are proven risks they must be named.

Radiation research has probably too late learned the lesson that publicly funded research must always justify its necessity as in earlier days money was comparatively easy to get. Present days competition is hard but our field is well enough equipped to stand it. It can survive, however, only if the torch is handed over to the next generation. More than in other branches we do not only have the obligation to support our students but we must attract them. This can only be achieved by attractive and exciting research in well-equipped laboratories. The governments and the funding agencies have to be convinced that providing them is a very good investment in the future – not just for our field but for society as a whole. If the workshop and these proceedings can help in a modest way to achieve these goals it has well served its purpose.

Printing: Mercedes-Druck, Berlin
Binding: Stein + Lehmann, Berlin